Technological Developments in Food Preservation, Processing, and Storage

Seydi Yıkmış
Tekirdağ Namık Kemal Üniversitesi, Turkey

A volume in the Advances in Environmental Engineering and Green Technologies (AEEGT) Book Series

IGI Global
DISSEMINATOR OF KNOWLEDGE

Published in the United States of America by
IGI Global
Engineering Science Reference (an imprint of IGI Global)
701 E. Chocolate Avenue
Hershey PA, USA 17033
Tel: 717-533-8845
Fax: 717-533-8661
E-mail: cust@igi-global.com
Web site: http://www.igi-global.com

Library of Congress Cataloging-in-Publication Data

Names: Yıkmış, Seydi, 1982- editor.
Title: Technological developments in food preservation, processing, and storage / edited by Seydi Yıkmış.
Description: Hershey, PA : Engineering Science Reference, an imprint of IGI Global, [2020] | Includes bibliographical references and index. | Summary: "This book explores the latest technological developments in food preservation, processing, and storage"-- Provided by publisher.
Identifiers: LCCN 2019035811 (print) | LCCN 2019035812 (ebook) | ISBN 9781799819240 (hardcover) | ISBN 9781799819257 (paperback) | ISBN 9781799819264 (ebook)
Subjects: LCSH: Food industry and trade--Technological innovations. | Food--Preservation--Technological innovations. | Food--Storage--Technological innovations.
Classification: LCC TP370 .T434 2020 (print) | LCC TP370 (ebook) | DDC 664--dc23
LC record available at https://lccn.loc.gov/2019035811
LC ebook record available at https://lccn.loc.gov/2019035812

This book is published in the IGI Global book series Advances in Environmental Engineering and Green Technologies (AEEGT) (ISSN: 2326-9162; eISSN: 2326-9170)

British Cataloguing in Publication Data
A Cataloguing in Publication record for this book is available from the British Library.

Advances in Environmental Engineering and Green Technologies (AEEGT) Book Series

ISSN:2326-9162
EISSN:2326-9170

Editor-in-Chief: Sang-Bing Tsai, University of Electronic Science and Technology of China Zhongshan Institute, China
Ming-Lang Tseng, Lunghwa University of Science and Technology, Taiwan
Yuchi Wang, University of Electronic Science and Technology of China Zhongshan Institute, China

MISSION

Growing awareness and an increased focus on environmental issues such as climate change, energy use, and loss of non-renewable resources have brought about a greater need for research that provides potential solutions to these problems. Research in environmental science and engineering continues to play a vital role in uncovering new opportunities for a "green" future.

The **Advances in Environmental Engineering and Green Technologies (AEEGT)** book series is a mouthpiece for research in all aspects of environmental science, earth science, and green initiatives. This series supports the ongoing research in this field through publishing books that discuss topics within environmental engineering or that deal with the interdisciplinary field of green technologies.

COVERAGE

- Renewable Energy
- Pollution Management
- Cleantech
- Contaminated Site Remediation
- Waste Management
- Electric Vehicles
- Air Quality
- Alternative Power Sources
- Biofilters and Biofiltration
- Industrial Waste Management and Minimization

IGI Global is currently accepting manuscripts for publication within this series. To submit a proposal for a volume in this series, please contact our Acquisition Editors at Acquisitions@igi-global.com or visit: http://www.igi-global.com/publish/.

The Advances in Environmental Engineering and Green Technologies (AEEGT) Book Series (ISSN 2326-9162) is published by IGI Global, 701 E. Chocolate Avenue, Hershey, PA 17033-1240, USA, www.igi-global.com. This series is composed of titles available for purchase individually; each title is edited to be contextually exclusive from any other title within the series. For pricing and ordering information please visit http://www.igi-global.com/book-series/advances-environmental-engineering-green-technologies/73679. Postmaster: Send all address changes to above address.

Titles in this Series

For a list of additional titles in this series, please visit:
https://www.igi-global.com/book-series/advances-environmental-engineering-green-technologies/73679

Handbook of Research on Resource Management for Pollution and Waste Treatment
Augustine Chioma Affam (University College of Technology Sarawak, Malaysia) and Ezerie Henry Ezechi (Universiti Teknologi Malaysia, Malaysia)
Engineering Science Reference • ©2020 • 795pp • H/C (ISBN: 9781799803690) • US $295.00

Handbook of Research on Globalized Agricultural Trade and New Challenges for Food Security
Vasilii Erokhin (Harbin Engineering University, China) and Tianming Gao (Harbin Engineering University, China)
Engineering Science Reference • ©2020 • 574pp • H/C (ISBN: 9781799810421) • US $285.00

Sustainable Waste Management Challenges in Developing Countries
Agamuthu Pariatamby (Sunway University, Malaysia) Fauziah Shahul Hamid (University of Malaya, Malaysia) and Mehran Sanam Bhatti (University of Malaya, Malaysia)
Engineering Science Reference • ©2020 • 586pp • H/C (ISBN: 9781799801986) • US $225.00

Nano-Phytoremediation Technologies for Groundwater Contaminates Emerging Research and Opportunities
Khushboo Chaudhary (Banasthali University, Tonk Newai, India) Suphiya Khan (Banasthali University, Tonk Newai, India) and Pankaj Kumar Saraswat (ICAR, Krishi Vigyan Kendra, Tamenglong District, Manipur, India)
Engineering Science Reference • ©2020 • 198pp • H/C (ISBN: 9781522590163) • US $165.00

Implementation and Evaluation of Green Materials in Technology Development Emerging Research and Opportunities

For an entire list of titles in this series, please visit:
https://www.igi-global.com/book-series/advances-environmental-engineering-green-technologies/73679

IGI Global
DISSEMINATOR OF KNOWLEDGE

701 East Chocolate Avenue, Hershey, PA 17033, USA
Tel: 717-533-8845 x100 • Fax: 717-533-8661
E-Mail: cust@igi-global.com • www.igi-global.com

Table of Contents

Detailed Table of Contents

energy storage systems with phase change materials are being used in cold storage systems. The new technologies are developed to reduce the energy consumption of cold storage systems. The quality of the perishable items stored in the cold storage may be retained by applying nano coating on the fruits. This chapter reviews cold storage systems, elements of cold storage, design of cold storage, application of phase change and nano particles in cold storage system.

The most commonly used meat preservation methods include cooling, freezing, drying, vacuum packing, and curing. Meat quality is impaired by a wide range of changes including physical, chemical, microbiological, and enzymatic reactions. Food manufacturers focus on processes that require fewer chemical additives to meet the increased demand of consumers and to obtain more natural, healthy, and nutritious meat products. Non-thermal food preservation methods are one of the new trends to minimise thermal effects on texture, nutritional value, and flavor losses of meats. The chapter focuses on two novel approaches; non-thermal (Pulsed Electric Field) and Atmospheric Pressure Cold Plasma (APCP) Technologies.

The packaging process is an important step in maintaining the quality characteristics of foods. Packaging foods protects products from external effects and provides product information to consumers. Due to the various changes occurring during the distribution and storage of the products, some significant quality characteristics can be lost. In recent years, novel packaging technologies have been developed to supply long shelf life, safety, and 'fresh-like' characteristics to the products. These novel technologies include nanotechnology, modified atmosphere packaging, active packaging, and intelligent/smart packaging. Since dairy products are generally vulnerable to biological, physical, and chemical changes, they lose their quality characteristics within a short term. Therefore, the use of these novel techniques in dairy products is greatly important. This chapter informs about general principles of the novel packaging techniques and their current applications in dairy technology.

Pasteurization is the most common processing method for microbial and enzyme inactivation to preserve foods. With this method, foods are exposed to high temperatures and there are disadvantages for many products: thermal treatments cause modifications of sensory attributes (for instance: flavour, colour, nutritional qualities). Now, another method can replace pasteurization: microbial inactivation by ultrasounds. It is a new alternative technology of food processing also called sonication, and it can be used coupled with pressure and/or heat. These techniques inactivate microorganisms in foods. They are effective and energy efficient to kill them, making the techniques promising for the food industry. In this chapter, the method of microbial inactivation by ultrasounds was explained, after that the applications in food industry for instance in milk, orange juice, wastewater, and whole liquid eggs were well-defined, and finally, the advantages, disadvantages, and the limitations of this method were examined.

Antibiotics have been responsible for the evolution of multidrug-resistant microbes. The side-effects of existing drugs and increased treatment costs have led to nutraceuticals gaining popularity. Nutraceuticals have therapeutic applications due to the ability of the probiotics to be viable in encapsulated pills and drinks. Due to their ability to exclude carcinogenic microorganisms by limiting the nutrients available and by competing for receptors nutraceuticals are useful against cancers. Nutraceuticals are useful against diabetes by controlling the genes involved in the insulin-signaling pathway. The future perspective for nutraceuticals includes an increase in production, reduction in manufacturing cost, and enhanced benefits.

This chapter presents the effect of ultrasound pretreatment on drying kinetics and product temperature. Drying provides a longer shelf life of agricultural products and low-cost transport and storage. The researchers performed pretreatments to the products to accelerate the drying process and to increase the quality for decades. The ultrasound applications considered as one of these are shown as the promising technology in the drying industry. The ultrasound application is used in drying processes in two different ways as ultrasound-assisted drying or as a pretreatment. In ultrasound-assisted drying, mechanical waves directly move to the cell wall of the product. On the other hand, the agricultural product to be dried is immersed in a liquid medium with different solution and ultrasound are applied at specified times by using as a pretreatment. They demonstrate that the ultrasound pretreatment method could be suitable for industrial production.

One of the major issues food technologists deal with is food preservation and safety. Growth of micro-organisms in food poses risk to its quality and safety. Moreover, resistance of food spoilage micro-organisms against various chemical food preservatives has led to an emergence of novel antimicrobial agents with improved action and low rates of microbial resistance. Development in nanotechnology has led to the production of nanoparticles that are not only safe but also effective to resolve the problem of microbial resistance. Nanoantimicrobials have shown improved bioactive performances and controlled toxicity to human beings. They are steadily gaining popularity and the trend will continue in coming years. The chapter gives a comprehensive view of nanoantimicrobials of organic and inorganic origin, various mechanisms adopted by these nanoparticles for the destruction of micro-organisms, factors affecting anti-microbial activities of these particles along with their applications in various fields of food technology.

Eating is a vital and essential part of life. The routine process is not straightforward but highly sophisticated in that it incorporates dynamic changes on the food structure that triggers different modalities of food sensation. Sensory studies are being expanded with contributions from physiology, psychology, dentistry, neuroscience, and food science. All these experts seek answers to catastrophic and complicated procedures created by the oral process. This chapter provides information about the past and present of sensory science in 5 main sections: Sensation and perception (to provide an insight to psychological and physiological approaches); psychophysical laws (to interpret the missing link between the stimulus and perception); sensory market

Preface

Food processing, preservation and safety have been a major concern for consumers in recent years. There has been a serious change in consumption habits. Heat treatment technology is a very effective method for inactivation of microorganisms, which has a very widespread use and allows many foods to be preserved in the food industry. The negative effects of heat treatment applications on the product increased the interest in new technologies. In recent years, in line with the preferences of the consumer has been processed to a minimum level, increasing demand for high-quality foods in terms of natural and nutritional physiology has increased interest in fresh or near-fresh food products. The negative effects of thermal preservation methods (such as pasteurization and sterilization) on nutrient losses have been questioned. Today, it is aimed to improve the sustainability of the food chain, to maintain the quality criteria of foodstuffs with new technologies developed in processes and to make them sustainable for a long time. In this respect, non-thermal technologies are used in the food industry (especially in fluid foods) as a method with a wide range of applications.

The content of this book includes basic issues and important applications for food industry (especially non-thermal applications). Each of the chapters in the book, prepared by scientists who are experts in their subjects is one of the features that make the book strong. In this context, the book is a pioneer among the English sources published in the field of food technology (especially new food preservation methods). At the same time, the book has been prepared to provide information on the new food processing and preservation methods to a certain level for undergraduate, graduate students and people working in food production by taking advantage of the latest information.

Technological Developments in Food Preservation, Processing, and Storage is organized into twelve chapters. The chapters include the recent research on novel and emerging non termal technologies for food preservation and survey the state of the art in this field. A brief description of each of the chapters follows:

Chapter 1 provides information on the principles of non-thermal protection in dairy products, recent developments and future expectations. Authors high hydrostatic pressure (HHP), pulsed electric field (PEF), ultrasound, cold plasma (CP), and pulsed light (PL) are the main non-thermal techniques that are used in the food industry. In this chapter, the authors aimed to give general principles of the non-thermal techniques, current applications in the dairy products and recent advances in the dairy industry.

Chapter 2 offers general information about cold storage systems. This book chapter reviews cold storage systems, elements of cold storage, design of cold storage, application of phase change and nano particles in cold storage system.

Chapter 3 evaluates non-thermal technology methods used in meat processing. Increasing technology and increasing human needs have also affected meat technology and processing. With the increase in the quality characteristics expected from the product, new searches have been started in meat processing methods. As a result of these new searches, the use of non-thermal processing methods is becoming increasingly common in many foods. This method minimizes food shelf life and reduces food quality deterioration.

Chapter 4 presents information about dairy packaging systems. The authors, this study aims to give information about general principles of the novel packaging techniques and their current applications in dairy technology.

Chapter 5 presents effects of ultraosun process on microorganisms. The author present the method of microbial inactivation by ultrasounds was explained, after that the applications in the food industry for instance in milk, orange juice, wastewater, and whole liquid eggs were well defined, and finally, the advantages, disadvantages and the limitations of this method were examined.

Chapter 6 presents probiotics as an alternative food therapy. Nutraceuticals have therapeutic applications due to the ability of the probiotics to be viable in encapsulated pills and drinks. Due to their ability to exclude carcinogenic microorganisms by limiting the nutrients available and by competing for receptors nutraceuticals are useful against cancers. Nutraceuticals are useful against diabetes by controlling the genes involved in the insulin-signaling pathway. The future perspective for nutraceuticals includes an increase in production, reduction in manufacturing cost and enhanced benefits.

Chapter 7 presents ultrasound pretreatment applications in drying of agricultural products. In this chapter, the authors present the effect of ultrasound pretreatment on drying kinetics and product temperature

Chapter 8 examines information on new emerging technological approach in food preservation. The chapter gives a comprehensive view of nano-antimicrobials of organic and inorganic origin, various mechanisms adopted by these nanoparticles for the destruction of micro-organisms, factors affecting anti-microbial activities of these particles along with their applications in various fields of food technology.

Chapter 9 summarizes recent trends and challengesin sensory and perception science. The author, this chapter aims to provide information about the past and the present of sensory science. Specifically, the chapter is divided into 5 main sections: Sensation and perception (to provide an insight to psychological and physiological approaches), psychophysical laws (to interpret the missing link between the stimulus and perception), sensory, market success and consumer behavior (to highlight the necessity of sensory testing for the industry and consumer), sensory evaluation methods, and finally the recent developments in the field.

Chapter 10 presents novel approach in food processing technology by using. The authors, this chapter will focus on a general review of the advantages of solid-state fermentation over conventional fermentation, bioreactor design for SSF, production of bioactive substances from various foodstuffs, bioconversion of agro-industrial wastes, and biorefining strategy. SSF is a remarkable tool to elevate the nutritional and functional values of the substrate to a large extent.

Hope the book will be useful.

Chapter 1
Non-Thermal Preservation of Dairy Products:
Principles, Recent Advances, and Future Prospects

Alperen Koker
Middle East Technical University, Turkey

İlhami Okur
https://orcid.org/0000-0002-2541-7123
Middle East Technical University, Turkey

Sebnem Ozturkoglu-Budak
Ankara University, Turkey

Hami Alpas
https://orcid.org/0000-0002-7683-8796
Middle East Technical University, Turkey

ABSTRACT

Dairy products include carbohydrates, protein, fatty acids, and different micronutrients, such as minerals and vitamins. Thermal treatment is generally used in dairy products to provide product safety and increase shelf life. But it can also lead to undesirable effects on dairy products such as protein denaturation, maillard reaction, and loss of vitamins. Non-thermal technology is an alternative method in the preservation of food products due to improving product safety and shelf life without any negative effects on food nutritional content. High hydrostatic pressure

DOI: 10.4018/978-1-7998-1924-0.ch001

(HHP), pulsed electric field (PEF), ultrasound, cold plasma (CP), and pulsed light (PL) are the main non-thermal techniques that are used in the food industry. This chapter gives general principles of the non-thermal techniques, current applications in the dairy products, and recent advances in the dairy industry.

INTRODUCTION

Mammals secrete milk from their mammary glands and the primary function of milk is the nutrition of the mammalian neonates. Throughout history, milk and dairy products have been acknowledged as an important source of nutrition and humans domesticated variety of dairy animals such as cow, buffalo, goat, sheep, camel, and many other mammalian species. The major purpose was the production of an adequate amount of milk for the nutritional needs of a human i.e. higher amount of milk that is required for the nourishment of the dairy animal's offspring. Milk is available as a liquid form as pasteurized milk, sterilized milk, and milk of modified composition. Additionally, evaporated milk products, sweetened condensed milk products, and powdered milk products are available in the market. Another milk component, milk cream, is available in the market such as sterilized cream and whipping cream. Cheese, yogurt, butter, kefir, sour cream, fermented buttermilk, and many other fermented dairy products have been important for humans due to their longer shelf life compared with raw and heat-treated milk and nutritious properties developed during the fermentation process. In the class of dairy products, cheeses have the most number of varieties depending on milk type, starter culture, processing, and aging.

Non-thermal technologies are alternative methods in terms of preservation of food products to a certain shelf-life and providing both a healthy and quality product. These technologies have shown not to have any negative effects on the food nutritional content and other quality factors during the process stage (Putnik et al., 2019). Unlike heat treatment applications, non-thermal techniques are known as energy-efficient processes. For these reasons, in recent years non-thermal technologies have gained popularity in the food industry (Santhirasegaram et al., 2016) and there has been much research related to this subject. High hydrostatic pressure (HHP), pulsed electric field (PEF), ultrasound, cold plasma (CP), and pulsed UV-light are the main non-thermal techniques that are used in the industry. They have different microbial inactivation mechanisms and different effects on the physicochemical properties of foods. However, the effects of non-thermal techniques on foods and microorganisms are related to processing parameters, microorganism type and load, and properties of food. In recent years, there has been much research on non-thermal technology applications on milk and other dairy products such as cheese, yogurt, butter, kefir, ice-cream, sour cream, fermented buttermilk, and other fermented dairy products.

This chapter aims to give information about general principles of the non-thermal techniques, current applications in the milk and dairy products industry, recent advances, investigated novel approaches and future expectations from these technologies.

HIGH HYDROSTATIC PRESSURE (HHP)

High Hydrostatic Pressure (HHP), also known as a cold pasteurization technique or pascalization, is a non-thermal technique in which extremely high pressures (between 200 MPa and 800 MPa) are applied to foods that are submerged in a liquid -mostly water- for a desirable period of time (t) and at a desirable temperature (T) (Doona, Kustin, & Feeherry, 2010). It can be said that HHP is a 3-Dimensional process because of having three different parameters as pressure, time and temperature. HHP can destroy vegetative cells of microorganisms, and enzymes (Alpas et al., 1999; Alpas, Lee, Bozoglu, & Kaletunç, 2003). This technique can be applied to all solid and liquid foods except for porous and dry products (Morales-de la Peña, Welti-Chanes, & Martín-Belloso, 2019). Classic heat treatments usually cause the formation of undesirable compounds and caramelization of products due to Maillard reaction and affect color, texture, and flavor of processed foods. Contrary to heat treatment, HHP application retains the nutritional and functional ingredients in the food material (Khan et al., 2018; Misra et al., 2017). Besides, HHP treatment is independent of the mass and geometry of the products (Koutchma, 2014; Misra et al., 2017). Due to these reasons, HHP is one of the most popular non-thermal food processing methods and recently used for the preservation of a wide range of industrial food products like dairy, ready-to-eat, fruit juices and seafoods (Misra et al., 2017). Apart from these, HHP treated food market value is expected to be increased to $ 54.77 billion in 2025 (Huang et al., 2017).

The typical HHP system consists of a high-pressure vessel and its closure, pressure generation device, temperature and pressure control device (Figure 1). Water is used mostly for pressure-transferring medium. Food is put into the pressure vessel with or without packaging. The closed pressure vessel is filled with a pressure transmitting medium and the liquid is compressed by a pump or pressure intensifier. The hydrostatic pressure distributes uniformly throughout the pressure vessel and equally in all directions of the food surfaces (Isostatic Pressure). The temperature of water raises roughly 3°C in each 100 MPa due to adiabatic heating during the

Figure 1. HHP system

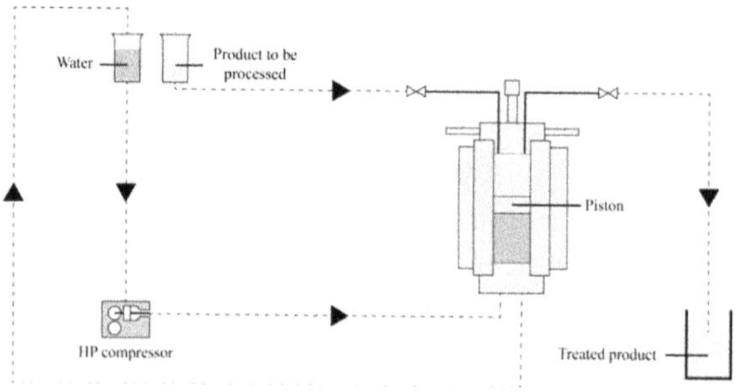

pressure build-up (Bhattacharya, 2015; Durance, 2002; Doona et al., 2010; Morales-de la Peña et al., 2019).

EFFECT OF HHP ON MILK AND DAIRY PRODUCTS

In food science, application of HHP was first introduced by the study of Hite (1899) on raw milk, in which it was found that HHP treatment of milk at 600 MPa for 1 h at room temperature extended shelf life of raw milk. Subsequently, effect of HHP on spoilage and pathogenic microorganisms present in milk and on physicochemical properties of milk have been studied comprehensively.

In terms of extending shelf life, raw milk subjected to 350 MPa for 32 min was found to increase the shelf life of milk to 12, 18 and 25 days at 10, 5 and 0 °C, respectively (Mussa & Ramaswamy, 1997). High-pressure treatment of 300 MPa for 30 min at 20 °C and 500 MPa for 5 min at 20 °C was enough for obtaining shelf life of 10 days at 10 °C (Rademacher, Pfeiffer, & Kessler, 1998). Refrigeration storage (7 °C) of raw milk and application of pressure (400 MPa for 30 min at 25 °C) lead to lower amount of microbial count in pressurized milk after 45 days of storage comparing to microbial count of raw milk after 15 days of storage (García-Risco, Cortés, Carrascosa, & López-fandiño, 1998).

Effect of HHP on pathogenic microorganisms was investigated profoundly in pursuance of achieving food safety conditions for high-pressure treated milk. *Escherichia coli* is an indicator organism for fecal contamination and one of the most important foodborne pathogens for the food industry, particularly serotype *E. coli* O157: H7. Various studies were performed about the application of HHP on milk inoculated with *E.coli*. Patterson and Kilpatrick (1998) achieved 5 log CFU/ml reduction of *E. coli* 0157: H7 NCTC 12079 with the high-pressure treatment of 400 MPa for 15 min at 50 °C. Application of high pressure within a range of 250 MPa and 400 MPa for 0 to 80 min as a holding time at two different temperatures of 3 °C and 21 °C on milk inoculated with *E. coli* ATCC-29055 demonstrated that higher reduction of *E. coli* was achieved at higher pressures, longer holding times and lower temperatures (Pandey, Ramaswamy, & Idziak, 2003). *Listeria monocytogenes* is another foodborne pathogen related to milk and dairy products, which cause listeriosis infection. Styles, Hoover, and Farkas (1991) achieved 6 log CFU/ml reduction in *L. monocytogenes* Scott A number when 345 MPa pressure was applied at 23 °C for 60 min and 80 min in raw and UHT milk, respectively. HHP application of 345 MPa for 5 min at 45 °C and 50 °C lead to 8 log CFU/ml reduction of *L. monocytogenes* CA and *L. monocytogenes* OH, respectively (Alpas, Kalchayanand, Bozoglu, & Ray, 2000). *Staphylococcus aureus* is the most pathogenic staphylococci, which are commonly related to food poisoning due to staphylococcal enterotoxins. *S. aureus* has very high-pressure resistance and pressure resistance of *S. aureus* increases in bovine milk (Patterson & Kilpatrick, 1998). Alpas et al. (2000) studied the reduction degree of *S. aureus, L. monocytogenes, E. coli*, and *Salmonella* spp. in case of HHP treatment and reported that HHP application of 345 MPa for 5 min at 50 °C destroyed *L. monocytogenes, E. coli,* and *Salmonella* spp. completely (8 log CFU/ml reduction), whereas *S. aureus* survived to a degree (5.33 log CFU/ml reduction).

The effect of HHP treatment on milk components, such as proteins and enzymes was also investigated. Casein is the major milk protein found in the cow's milk and due to its primary structure, casein micelles effected by HHP at higher pressure values than 300 MPa. Higher pressure values result in defragmentation of casein micelles as a result of interruption of casein micelle structure's electrostatic and hydrophobic interactions, and colloidal calcium phosphate solubilization (Schrader, Buchheim, & Morr, 1997). It was reported that micellar substructures of casein in HHP treated milk are similar to structures in untreated milk in the view of the electron microscope (Knudsen & Skibsted, 2010).

Main whey proteins present in bovine milk are α-lactoalbumin and β-lactoglobulin. The other whey proteins are found in smaller amounts such as immunoglobulins, bovine serum albumin, and lactoferrin. The effect of pressure treatment of 400 MPa for 15 min on β-lactoglobulin and α-lactoalbumin denaturation were investigated at different temperatures by García-Risco et al. (2000). The percentage loss of β-lactoglobulin was determined as 76 and 95% at the temperatures of 20 °C and 40 °C, respectively. They also found that α-lactoalbumin is more pressure resistant comparing to β-lactoglobulin, and the denaturation ratio was observed as only 3% at 40 °C and no denaturation was detected in α-lactalbumin at 20 °C at the pressure treatment of 400 MPa for 15 min. α-lactoalbumin and β-lactoglobulin denaturation was observed to increase with increasing temperature, treatment time and pH (Trujillo, Ferragut, Juan, Roig-Sagués, & Guamis, 2016).

Effect of HHP on milk enzymes is important considering their potential use of indicator for sufficient processing conditions and their effects on product quality. Rademacher et al. (1998) reported that enzyme activity loss of alkaline phosphatase, γ-glutamyltransferase, and phosphohexose isomerase starts at the HHP treatment of 600, 500 and 400 MPa for 8 min at 20 °C. Rademacher and Hinrichs (2006) proposed that γ-glutamyltransferase could be used as a process indicator for raw milk during HHP treatments as an alternative to alkaline phosphatase due to the efficient inactivation of foodborne pathogens in milk in case of the adequate (500 MPa) inactivation pressure is applied. Otherwise, HHP process indicator may lead to over processing due to the high-pressure inactivation value of alkaline phosphatase (600MPa).

HHP treatments were found to have effects on milk characteristics. It was reported that HHP application decreased the turbidity of milk up to 330 MPa. It was also determined that up to 220 MPa turbidity decreases with increasing processing time from 10 min to 20 min due to changes in casein micelle size (Altuner, Alpas, Erdem, & Bozoglu, 2006). Combination of temperature and high-pressure treatment increases phosphorus and calcium solubility slightly and decreases average particle diameter and lightness (Gaucheron et al., 1997). 40% reduction of level and rate of creaming occurs by the application of HHP higher than 400 MPa, and these creaming properties increases by 70% at pressure values lower than 250 MPa (Huppertz, Fox, & Kelly, 2003). Altuner et al. (2006) reported HHP application to milk leads to the repositioning of hydrophobic areas inside the micelles by the evidence of increasing binding strength and maximum surface allowable for binding. Water binding, gelling, foaming and emulsifying properties of milk is likely to change due to increased exposure of hydrophobic groups (Nakai & Li-chan, 1988). Production of low-fat yogurt from skim milk processed by a combination of heat treatment and HHP (400 MPa to 500 MPa) decreases syneresis and enhances yield stress and elastic modulus (Harte, Luedecke, Swanson, & Barbosa-Cánovas, 2010).

PULSED ELECTRIC FIELD (PEF)

PEF treatment is the application of high-voltage pulses (20–80 kV/cm) during a short time (1-10 μs) to food located between two electrodes. It is usually used for liquid foods (Doona et al., 2010; Shahbaz et al., 2018) and the food material must be homogenous for pasteurization process (Morales-de la Peña et al., 2019). Like HHP treatment, application of PEF also decreases the unacceptable changes in the food materials. The main food products treated with PEF technology are fruit juices, liquid egg, and milk for the purpose of their preservation.

The typical PEF system includes voltage power supply and treatment chamber, which contains electrode, resistor and capacitor (Figure 2). When a voltage is applied to a cell, a sufficiently high transmembrane potential is induced across the cell membrane and this leads to rapture at the membrane (Doona et al., 2010). After damaging a cell, water starts to interrupt to cell and this leads to inactivation of the cell. Unlike HHP which temperature, pressure and time are critical parameters, PEF technology has so many critical parameters to the optimization of the processing system (Ahmed et al., 2010). These are treatment time, electric field strength (E), pulse shape, pulse-frequency, pulse-width, pulse polarity, and temperature. Although there are too many critical parameters to optimize the system, there have been too many researches to find the effects of PEF on food system (Misra et al., 2017).

EFFECT OF PEF ON MILK AND DAIRY PRODUCTS

In terms of extending shelf life, PEF application of 35.5 kV electric field strength for 1000 μs processing time with bipolar 7 μs pulses at 111 Hz on raw milk increased the shelf life of raw milk to 5 days, which is similar to high temperature short time pasteurization process (at 75 °C for 15 s) (Odriozola-Serrano, Bendicho-Porta, & Martín-Belloso, 2010). For the purpose of comparing the effects of conventional heat treatment and the combination of PEF and PEF with (or plus) heat treatment was investigated on the shelf life of skim milk. According to results the shelf life of milk increases to 14 days at 4 °C both for heat treatment at 65 °C for 21 s and PEF treatment of 28, 32 and 36 kV/cm for 30 pulses with a total processing time of 84 μs. Moreover, the combination of PEF and heat treatment increases the shelf life of skim milk to 30 days at the same storage conditions (Fernández-Molina et al., 2005).

Figure 2. PEF system

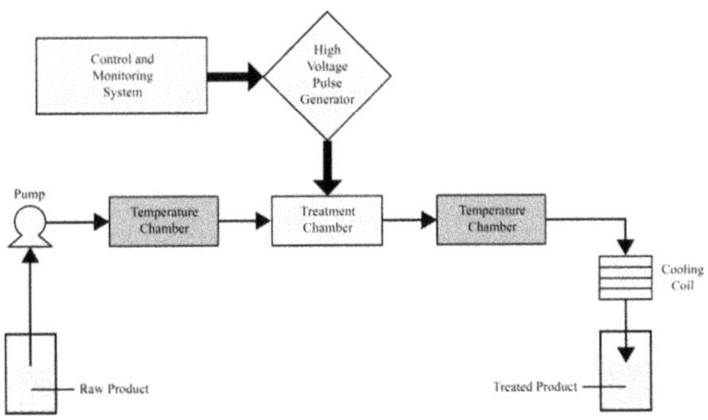

In order to obtain milk without the presence of pathogenic microorganisms, the effects of PEF on foodborne pathogens are crucial. Treatment of milk containing *Escherichia coli* K-12 with preheating unit to increase the initial temperature to 60 °C and PEF application of 40 kVp with frequency of 100 Hz and post holding of milk for 20 s decreases *E. coli* count at least 7 log CFU/ml, and also reduces time required for low temperature long time pasteurization significantly (Ohshima, Tanino, Kameda, & Harashima, 2016). Combination of PEF and electrically induced heat application to pasteurized milk inoculated with *L. innocua* ATCC 51742 at 30 kV/cm, 10 pulses with a pulse width of 2.5 μs at initial temperature of 43 °C leads to 4.3 log CFU/ml reduction, and similar amount of *L. innocua* reductions occur by increasing pulse number and decreasing initial temperature of milk at the same electric field intensity (Guerrero-Beltrán, Sepulveda, Góngora-Nieto, Swanson, & Barbosa-Cánovas, 2010). PEF treatment of skim milk inoculated with *Staphylococcus aureus* ATCC 6538 with electric field strength of 35 kV/cm for 450 μs with bipolar square pulse waves of pulse width of 3.7 μs with 250 Hz pulse rate leads to 3.7 log CFU/ml reduction of *S. aureus* (Evrendilek, Zhang, & Richter, 2004). Heat treatment at 55 °C for 24 s prior to PEF treatment of 23 kV/cm electrical field strength with pulse width 20 μs for 70 μs processing time decreases *E. coli*, *L. innocua* and *S. aureus* below 2 log CFU/ml (Sharma, Bremer, Oey, & Everett, 2014). Comparing PEF resistance of *L. monocytogenes*, *E. coli* and *S. aureus*, with lethal and sublethal injury of these microorganisms, PEF treatment of 15, 20, 25 and 30 kV/cm electric field strength were applied for a treatment time between 0 to 600 μs with square waves of 2 μs width and 200 Hz rate. *L. monocytogenes* was found as the most resistant bacteria to PEF treatment, also inactivation kinetics of all these food-borne pathogens best fit with Hülsheger model (Zhao, Yang, Shen, Zhang, & Chen, 2013).

Application of PEF leads to protein modification and enzyme inactivation in milk as a result of changes in apparent charge of proteins following modification of ionic interactions of proteins (Jaeger, Meneses, & Knorr, 2014). Moreover, free radicals generated by energy absorption of proteins from PEF application can lead to disruption of interactions between proteins and unfolding due to cross-linking of free radicals with proteins (Han, Cai, Cheng, & Sun, 2018).

Loss of β-lactoglobulin, α-lactalbumin and serum albumin in milk after PEF application of 35.5 kV/cm with bipolar pulses of 7 μs pulse width for 1000 μs process time is 20.1%, 40% and 24.5%, respectively, which is similar to loss of whey proteins in milk with the application of traditional heat treatment (Odriozola-Serrano et al., 2010). PEF treatment of reconstituted skim milk (49 kV/cm, 19.36 μs, up to 70 °C) at pH values lower than 7.5 does not affect caseins and have a minor effect on whey protein denaturation. Besides, pH values higher than 7.5 cause dissociations of casein micelles, decrease milk protein sizes and increase casein amount in serum (Liu et al., 2015).

Application of PEF treatment to whey protein isolate at 30-35 kV/cm electric field intensity for a treatment time of 19.2-211 μs at 30-75 °C does not affect emulsification, surface hydrophobicity, and protein aggregation properties, however gel strength of whey protein isolate decreases and gelation time increases (Sui, Roginski, Williams, Versteeg, & Wan, 2011).

Considering milk-fat globule size distributions, the effect of PEF treatments with electric field intensities of 36 kV/cm for 24, 40, 60 pulses and 42 kV/cm for 8, 16, 24 pulses, respectively, gave similar results to thermal treatment at 63 °C for 30 min (Garcia-Amezquita, Primo-Mora, Barbosa-Cánovas, & Sepulveda, 2009). Yu, Ngadi, and Raghavan (2009) studied the effect of PEF on rennet coagulation and they found that increasing electric field intensity, pulse number, and treatment temperature effects coagulation properties of milk and they have concluded that combination of mild heat treatment (50 °C) with PEF (30 kV/cm, 120 pulse) of milk is suitable for cheese production due to increased coagulation activity (or coagulability) of PEF applied milk.

ULTRASOUND (US)

Ultrasound (US) is a novel food preservation technique, in which pressure waves are applied to food material (Morales-de la Peña et al., 2019). The main concept of this technology is to produce acoustic cavitation and this cavitation causes disruption of living cells by micro-mechanical shocks. According to frequency magnitude, US divides into three groups which are power ultrasound (with a frequency of 20-100 kHz), high frequency or extended range for sonochemistry (20 kHz–2 MHz) and diagnostic ultrasound (>1 MHz) (Figure 3) (Awad et al., 2012; Chemat et al., 2011). Furthermore, US types used in the food industry are divided into two groups according to frequency level; a high-frequency ultrasound (HFU) and a low-frequency ultrasound (LFU). A typical ultrasound system is shown in Figure 4.

HFU is used as a non-destructive, non-invasive analytical technique for quality assurance, process monitoring, and control in the food industry. At the low frequency, HFU is enough to lead acoustic cavitation so that minimal physicochemical changes occurred at food material during wave passes. On the other hand, LFU is used for process intensification. Even at the lower frequencies, LFU produces cavitation and it can cause physicochemical changes at food material (Ojha et al., 2017).

Figure 3. The frequency range of application of ultrasound technology reproduced from Ojha et al., (2017)

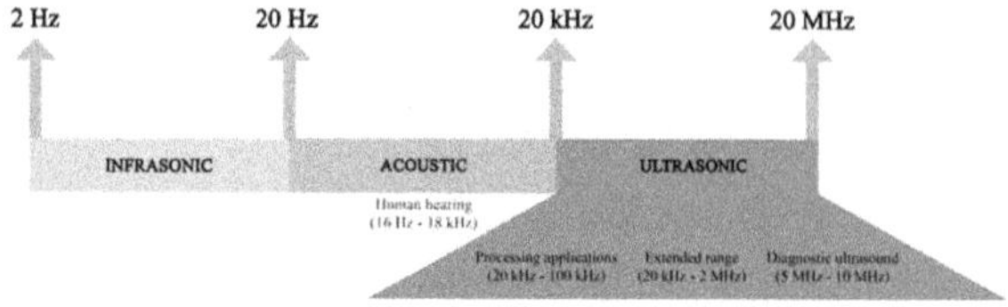

EFFECT OF US ON MILK AND DAIRY PRODUCTS

In terms of extending shelf life, UV treatment of raw milk initially containing 5.31 log CFU/ml coliform and 8.6 log CFU/ml total bacteria lead to 4.01 log CFU/ml and 1.31 log CFU/ml, respectively, in which the treatment conditions of 24 kHz frequency, 100% amplitude, and 15 min was applied (Şengül, Erkaya, Başlar, & Ertugay, 2011). Vijayakumar et al. (2015) studied the effect of thermosonication on coliform and total aerobic bacteria both in skim milk and cream, at 104 W power for 3 min and at 115 W power for 1 min. The authors reported a 2 log CFU/ml reduction in total aerobic bacteria and 3.65 log CFU/ml reduction in the coliform count.

Considering the health risks caused by foodborne pathogens, the effects of ultrasonication on pathogens, which may present in milk, are crucial. Cameron, McMaster, and Britz (2009) inoculated UHT milk with 10^6 CFU/ml *Escherichia coli* and *Listeria monocytogenes*, and obtained 5.34 log CFU/ml and 2.07 log CFU/ml reduction, respectively, after US treatment of 20 kHz for 10 min at 100% amplitude. US treatment (20 kHz, 100% amplitude, 6 min) of raw milk inoculated with 8 log CFU/ml decreased the *L. monocytogenes* ATCC 51414 number to 1.93

Figure 4. Ultrasound system reproduced from Moreno, (2017)

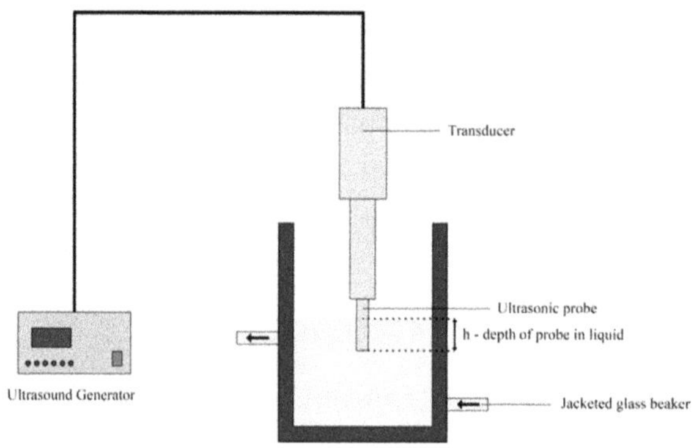

and 3.47 log CFU/ml at 20 °C and 57 ± 1 °C, respectively (D'Amico, Silk, Wu, & Guo, 2006). US treatment of 24 kHz, 100% amplitude and 300 s process time lead to 2.09 log CFU/ml reduction in *Escherichia coli* and 0.55 log CFU/ml reduction in *Staphylococcus aureus*. Sensorial deterioration was also observed during this stage (Marchesini et al., 2015). In an earlier study, Wrigley and Llorca (1992) applied US on skim milk inoculated with *Salmonella typhimurium* ATCC 14028 and observed a decrease of 2.5 and 3.0 log CFU/ml in cell numbers after the treatment of 30 min at 40 °C and 50 °C, respectively.

Effect of ultrasonication on milk proteins and enzymes also investigated by researchers because of their importance in milk and dairy products. Denaturation of whey proteins occurs at higher pH values and lower frequencies. In a research performed by Liu et al. (2014), an increase in protein solubilization, casein release and casein particle negative charge was observed in skim-milk within the US conditions of 20, 400 and 1600 kHz (energy provided for each system adjusted to 286 kJ/kg), at processing temperature lower than 30 °C between pH 6.7 and 8.0. Another study on the effect of US on surface hydrophobicity of reconstituted milk protein concentrate showed that surface hydrophobicity increased from 177.51 to 200.06 and 416.94 with 0.5 and 5 min, respectively. The authors concluded that unfolding of milk protein molecules occurs by US application (Sun et al., 2014). Application of thermo-sonication (115 W average power, 3 min) on skim milk and cream also decreased the total plasmin activity, a heat stable enzyme that can be found in milk, in a ratio of 94% (Vijayakumar et al., 2015).

US treatment of reconstituted milk protein concentrate at 20 kHz frequency and 50% amplitude decreased the particle size from 28.45 μm to 0.13 μm after 30 s of treatment and increased the solubility from 35.8% to 88.3% after 5 min of treatment (Sun et al., 2014). Chandrapala, Martin, Kentish, and Ashokkumar (2014) also examined the effect of US process (20 kHz, 50% amplitude, 10 min) both on solubilization of micellar casein powder and low solubility milk protein concentrate. Similar to the results of Sun et al. (2014), they obtained an increase in solubilization and decrease in average particle sizes. US also leads to a decrease in milk fat globule size due to the homogenization effect caused by the process (Cameron et al., 2009). Zhao et al. (2014) examined effect of US (20 kHz, 0-20 min) on coagulation aspects of goat milk and demonstrated that coagulation time increases 5 to 7 min, when US treatment applied. However, coagulation time again decreases to its original value in the subsequent 10 to 15 minutes, and increases slightly at 20 min. US also increased the coagulum strength, water holding capacity and gel firmness. Effect of US on the particle size of goat milk is similar to cow milk, with increasing processing time average particle size decreases and particle size distribution become more homogeneous.

COLD PLASMA (CP) TECHNOLOGY

Cold Plasma (CP) technology is one of the novel non-thermal technology for food processing that has gained attention from researchers on the world in recent years (Morales-de la Peña et al., 2019). This technology was used for rising surface energy of materials, ameliorating the printing and adhesion properties of polymers, and a variety of usage domains in electronics. However, it was shown that CP is a powerful method for the food industry (Chizoba Ekezie et al., 2017). Plasma is called the fourth state of matter which is an ionized gas including active species such as free radicals, ions, and electrons. Generally, plasma is divided into two groups that are thermal plasma and non-thermal plasma, which is also known as Cold Plasma (CP) (Coutinho et al., 2018; Misra et al., 2017; Misra, Oliver, Schlüter, 2016). In the simplest form, CP technology can be explained that plasma is generated by applying partially ionized state of a gas (usually O_2 and N_2) with free ions, electrons, molecules, atomic species, UV photons, and charged particles. These active species of plasma can react with food and inactivate target microorganism by releasing the stored energy (Misra et al., 2017). The process parameters of this technology are gas fed (type, flow, pressure, type, flow), electric field, time, and media.

EFFECT OF COLD PLASMA (CP) ON MILK AND DAIRY PRODUCTS

CP is an effective method to inactivate microorganisms present in milk (Coutinho et al., 2018; Segat et al., 2016). There are three basic mechanisms that cause inactivation of microorganisms. These mechanisms include;

1. Etching of cell surfaces induced by reactive species formed during plasma generation,
2. Volatilization of compounds and intrinsic photodesorption of ultraviolet (UV) photons,
3. Destruction of genetic material (Laroussi, 2005).

Reactive species such as radicals and chemicals can cause lipid and protein oxidation and DNA damage (Coutinho et al., 2018). UV irradiation in cold plasma causes the breakdown of chemical bonds in bacteria. Then, by-products such as CO and CH_n from the intrinsic atoms of the bacteria occurs. Finally, replicate inhibition and activation of etching happens, which leads to inactivation of bacteria (Niemira, 2012). Changes in membrane integrity by charged particles action makes DNA breakdown interaction with a membrane protein. After this event, cell perforation occurs and DNA is released from the cell and genetic material is destructed (Moreau et al., 2008).

There is little information about the effects of cold plasma on dairy products (Coutinho et al., 2018). Low-Temperature Plasma for inactivation of *E. coli* in milk was examined by Gurol et al. (2012). At this study, three different milk having different fat contents were used. The time-dependent effect of atmospheric corona discharge produced with 9 kV of AC power supply and different exposure time (0, 3, 6, 9,12, 15 and 20 min) was applied. According to results, 4 log reduction occurred after 20 min application. Also, cold plasma treatment did not have any significant effect on the pH and color values of milk samples. Another study performed in terms of microbial inactivation by cold plasma processing is related to microbial safety and quality attributes of milk following treatment with atmospheric pressure encapsulated dielectric barrier discharge (DBD) plasma (Kim et al., 2015). Whole milk was inoculated with *E. coli*, *L. monocytogenes,* and *S. typhimurium*, and plasma were produced by using a plastic container (250 W, 15 kHz) for 10 min. It is stated that roughly 2.4 log reduction occurred after the treatment. Furthermore, pH value and a* of milk were decreased although L* and b* values were increased with respect to the color results. The study suggested that encapsulated DBD plasma treatment for less than 10 min should be done for improving the microbial quality without slight changes in physicochemical quality of milk. Yong et al. (2015) reported pathogen inactivation and quality changes in Cheddar cheese by using flexible thin-layer DBD plasma. *L. monocytogenes*, *E. coli* O157: H7 and *S. typhimurium* were used as pathogen bacteria and plasma was produced by conductive layer (100 W, 15 kHz) for 10 min. 2.1, 3.2, and 5.8 log CFU/g reductions were observed for *L. monocytogenes*, *E. coli* O157: H7, and *S. typhimurium,* respectively after cold plasma treatment. Although sensory attributes such as total color difference (ΔE), and color scores did not show significant changes. Whereas, a significant decrease was determined in flavor and overall acceptance of product after cold plasma treatment due to the increasing off-flavor.

Korachi et al. (2015) stated that there were no significant changes observed for the lipid composition of milk but the quantity of volatile compounds decreased significantly after 20 minCP treatment. Alkaline phosphatase activity has a similar Z-value to heat-resistant pathogens in milk therefore it is used as an indicator to evaluate the efficiency of heat- treatments. Segat et al. (2016) showed that CP technology having high voltages (40, 50 and 60 kV) could inactivate alkaline phosphatase within a few seconds of process time because of the fact that alkaline phosphatase has an α-helix structure and the structure was breakdown under cold plasma process, although the temperature did not change significantly. Moreover, the Weibull model is the best-described model for the inactivation of alkaline phosphatase.

PULSED LIGHT

Pulsed Light (PL) is an emerging technology that is based on the application of using intense and short pulses (100–400 μs) of white light (Li & Farid, 2016; Mahendran et al., 2019). The spectrum of PL consists of wavelengths ranging from ultraviolet to near-infrared region (Li & Farid, 2016). PL was approved by Food and Drug Administration (FDA) in 1996 during food applications up to a cumulative fluence of 12 J cm^{-2}, where emission spectra should be kept between 200 and 1100 nm and pulse duration at ≤ 2 ms (Rowan, 2019). At the beginning of the usage of this technology, it was used for surface and water decontamination. However, in recent years this technology is used for a wide range of food products such as liquid products, fruit, and vegetable surfaces (Morales-de la Peña et al., 2019). The main components of PL are shown in Figure 5, which includes power unit for generation of pulses in the treatment chamber.

The principle of PL relies on the accumulation of high discharge voltage in a capacitor, where the stored energy is delivered in ultra-short pulses through a light source filled with xenon gas. The xenon-light source has a spectrum light flashes with the range of 200–1100 nm and this leads to photochemical modification of microorganism which means disinfection (Garvey & Rowan, 2019; Kramer et al., 2015; Rowan, 2019). The parameters of PL are the fluence (J cm^{-2}), pulse width (τ), number of the pulse (n), frequency (Hz), exposure time (s), and the peak power [W].

Figure 5. Schematic diagram of PL technology system reproduced from Mahendran et al., (2019)

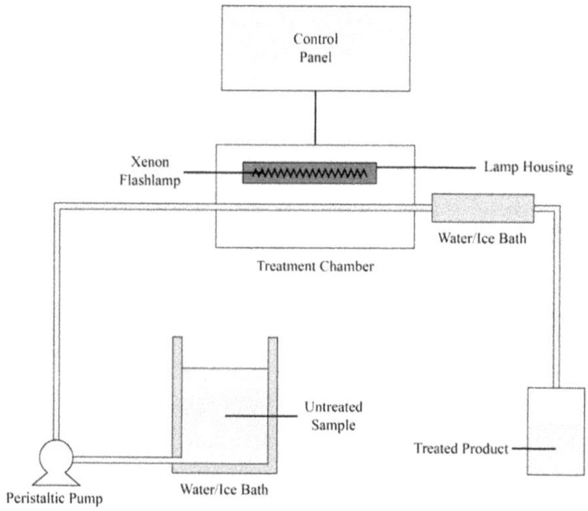

EFFECT OF PULSED LIGHT ON MILK AND DAIRY PRODUCTS

Little information is available on the application of PL to milk and dairy products. Effect of PL on pathogenic microorganisms was found profoundly in pursuance of achieving food safety conditions for PL-treated milk. The inactivation mechanism of PL is the impairment of the DNA of microorganism by forming thymine dimers that cause bacterial death. *Staphylococcus aureus* is one of the most important contagious mastitis foodborne pathogen in dairy products (Kümmel et al., 2016). In terms of inactivation of *S. aureus* in milk, different parameters (distance of the sample from the light strobe (5 to 11 cm), number of passes (1 to 3 passes), and flow rate (20 to 40 mL/min)) were applied into raw milk, which included roughly 8 to 9 log CFU/mL *S. aureus* (Krishnamurthy et al., 2007). Application of 11 cm distance with 2 passes, 11 run order and 30 mL/min flow rate or 8 cm distance with 1 pass, 4 run order and 20 mL/min flow rate showed the highest reduction in *S. aureus* count (roughly 7 $\log_{10}$ CFU/mL). In addition to this, the average bulk temperature of raw milk after PL treatment was found between 22-51°C. This temperature was derived from the distance of the sample to the light strobe (Krishnamurthy et al., 2007).

Inactivation of *E. coli* by using PL treatment was examined by Kasahara et al. (2015). According to the results, it was shown that 6 $\log_{10}$ CFU/mL reduction of *E. coli* was achieved when 10,000 mJ/cm^2 fluence was applied. Furthermore, the physical properties of goat milk such as viscosity, pH, and density were not changed significantly after PL treatment. Similar to physical properties, a chemical composition such as moisture, protein, lipid and ash content, and aroma profiles of the goat milk were not demonstrated any significant change after treatment (Kasahara et al., 2015). Innocente et al. (2014) investigated the effect of PL on the total microbial count and alkaline phosphatase activity of raw milk. In this study, different fluence (0.2626.25 J cm^2) was applied to raw milk. According to results, PL treatments inactivate alkaline phosphatase activity by photochemical and photothermal effects. Moreover, alkaline phosphatase activity was found more photo-resistant than microorganisms.

Effect of PL on pathogenic and spoilage microorganisms for PL-treated cheese surface was performed by Proulx et al., (2015). *Pseudomonas fluorescens* was used as spoilage bacteria, *E. coli* ATCC 25922 was used as a nonpathogenic form of *E. coli* O157: H7, and *L. innocua* was used as a nonpathogenic form of *L. monocytogenes* in this study. 5 or 7 log CFU/slice of bacteria was inoculated on the surface of the cheese and different fluences (1.02 to 12.29 J/cm ^{2}) were applied. Maximum inactivation was obtained at 3.07 J/cm^2 and the most sensitive microorganism to PL treatment was determined as *E. coli* with a nearly 5.4 log reduction. *P. fluorescens* (roughly 3.7 log reduction) and *L. innocua* (approximately 3.3 log reduction) followed *E. coli*, respectively. Proulx et al. (2017) also studied the reduction degree of *E. coli*, *P. fluorescens*, *L. innocua* in case of PL treatment in comparison to PL treatments combined with nisin and natamycin on Cheddar cheese. It was reported that nisin or natamycin before the application of PL treatment reduced the effectiveness of PL t for all bacteria due to decreasing the fluence received by the bacteria. However, the effectiveness of PL treatment increased for all bacteria, when these antimicrobials were applied after the treatment.

PL treatment was reported not to have any negative effects on the quality and sensory characteristics of dairy products. It was shown that no significant changes occurred in color and peroxide value of Cheddar cheese after the PL dose of 9.22 J/ cm^2. In addition to this, PL treatment has a significant effect to delay mold growth during 7 days of refrigeration storage treated with 9.22 J/ cm^2 fluence. It is also reported that PL treatment has an overall liking in terms of flavor, and appearance (Proulx et al., 2016). Protein oxidation in processed cheese slices treated with PL was also examined. Different fluences (0.7, 2.1, 4.2, 8.4 and 11.9 J/cm^2) were applied and it was found that PL treatments up to 4.2 J/cm^2 did not increase protein oxidation whereas higher fluences increased protein oxidation by absorption of light by amino acids or by the involvement of photosensitizers (Fernández et al., 2014).

CONCLUSION AND FUTURE PROSPECTS

Milk and dairy products are important nutrition sources for humans and to preserve the nutritional contents the novel processing method is thermal processing in this sector. Application of thermal treatment leads to the elimination of pathogenic and spoilage microorganisms, together with deterioration in the desirable as well as the undesirable changes of physicochemical properties products. It is possible to eliminate undesirable microorganisms in milk and dairy products without effecting fresh-like properties by application of non-thermal techniques. Microbiological and physicochemical effects of high hydrostatic pressure, pulsed electric field, ultrasound, cold plasma, and pulsed UV-light on milk and dairy products, and advantages of these non-thermal processing technologies comparing with thermal treatments reviewed in this chapter. Future applications of non-thermal processes with or without heat treatment for milk and dairy products will probably advance due to achievement of food safety without undesirable effects caused by thermal treatment, as well as the variety of physicochemical changes induced by non-thermal treatments.

REFERENCES

Ahmed, J., Ramaswamy, H. S., Kasapis, S., & Boye, I. (2010). *J.* Novel food processing effects on rheological and functional properties. CRC Group.

Alpas, H., & Bozoglu, F. (2000, June). The combined effect of high hydrostatic pressure, heat and bacteriocins on inactivation of foodborne pathogens in milk and orange juice. *World Journal of Microbiology & Biotechnology*, *16*(4), 387–392. doi:10.1023/A:1008936607413

Alpas, H., Kalchayanand, N., Bozoglu, F., & Ray, B. (2000). Interactions of high hydrostatic pressure, pressurization temperature and pH on death and injury of pressure-resistant and pressure-sensitive strains of foodborne pathogens. *International Journal of Food Microbiology*, *60*(1), 33–42. doi:10.1016/S0168-1605(00)00324-X PMID:11014520

Alpas, H., Kalchayanand, N., Bozoglu, F., Sikes, A., Dunne, C. P., & Ray, B. (1999). Variation in resistance to hydrostatic pressure among strains of food-borne pathogens. *Applied and Environmental Microbiology*, *65*(9), 4248–4251. PMID:10473446

Alpas, H., Lee, J., Bozoglu, F., & Kaletunç, G. (2003). Evaluation of high hydrostatic pressure sensitivity of Staphylococcus aureus and Escherichia coli O157:H7 by differential scanning calorimetry. *International Journal of Food Microbiology*, *87*(3), 229–237. doi:10.1016/S0168-1605(03)00066-7 PMID:14527795

Altuner, E. M., Alpas, H., Erdem, Y. K., & Bozoglu, F. (2006). Effect of high hydrostatic pressure on physicochemical and biochemical properties of milk. *European Food Research and Technology*, *222*(3–4), 392–396. doi:10.100700217-005-0072-4

Awad, T. S., Moharram, H. A., Shaltout, O. E., Asker, D., & Youssef, M. M. (2012). Applications of ultrasound in analysis, processing and quality control of food: A review. *Food Research International*, *48*(2), 410–427. doi:10.1016/j.foodres.2012.05.004

Bhattacharya, S. (2015). *Conventional and advanced food processing technologies*. Wiley Blackwell.

Cameron, M., McMaster, L. D., & Britz, T. J. (2009). Impact of ultrasound on dairy spoilage microbes and milk components. *Dairy Science & Technology*, *89*(1), 83–98. doi:10.1051/dst/2008037

Chandrapala, J., Martin, G. J. O., Kentish, S. E., & Ashokkumar, M. (2014). Dissolution and reconstitution of casein micelle containing dairy powders by high shear using ultrasonic and physical methods. *Ultrasonics Sonochemistry*, *21*(5), 1658–1665. doi:10.1016/j.ultsonch.2014.04.006 PMID:24798226

Chemat, F., Zill-e-Huma, & Khan, M. K. (2011). Applications of ultrasound in food technology: Processing, preservation, and extraction. *Ultrasonics Sonochemistry, 18*(4), 813–835. doi:10.1016/j.ultsonch.2010.11.023 PMID:21216174

Chizoba Ekezie, F. G., Sun, D. W., & Cheng, J. H. (2017). A review on recent advances in cold plasma technology for the food industry: Current applications and future trends. *Trends in Food Science & Technology, 69*, 46–58. doi:10.1016/j.tifs.2017.08.007

Coutinho, N. M., Silveira, M. R., Rocha, R. S., Moraes, J., Ferreira, M. V. S., Pimentel, T. C., ... Cruz, A. G. (2018). Cold plasma processing of milk and dairy products. *Trends in Food Science & Technology, 74*(February), 56–68. doi:10.1016/j.tifs.2018.02.008

D'Amico, D. J., Silk, T. M., Wu, J., & Guo, M. (2006). Inactivation of microorganisms in milk and apple cider treated with ultrasound. *Journal of Food Protection, 69*(3), 556–563. doi:10.4315/0362-028X-69.3.556 PMID:16541685

Doona, C. J., Kustin, K., & Feeherry, F. (2010). *Case studies in novel food processing technologies*. Woodhead Publishing; doi:10.1533/9780857090713

Durance, T. (2002). *Handbook of food preservation* (Vol. 35). Food Research International; doi:10.1016/S0963-9969(00)00143-5

Evrendilek, G. A., Zhang, Q. H., & Richter, E. R. (2004). Application of pulsed electric fields to skim milk inoculated with Staphylococcus aureus. *Biosystems Engineering, 87*(2), 137–144. doi:10.1016/j.biosystemseng.2003.11.005

Fernández, M., Ganan, M., Guerra, C., & Hierro, E. (2014). Protein oxidation in processed cheese slices treated with pulsed light technology. *Food Chemistry, 159*, 388–390. doi:10.1016/j.foodchem.2014.02.165 PMID:24767071

Fernández-Molina, J. J., Fernández-gutiérrez, S. A., Altunakar, B., Bermúdez-Aguirre, D., Swanson, B. G., & Barbosa-Cánovas, G. V. (2005). The combined effect of pulsed electric fields and conventional heating on the microbial quality and shelf life of skim milk. *Journal of Food Processing and Preservation, 29*(5–6), 390–406. doi:10.1111/j.1745-4549.2005.00036.x

Garcia-Amezquita, L. E., Primo-Mora, A. R., Barbosa-Cánovas, G. V., & Sepulveda, D. R. (2009). Effect of nonthermal technologies on the native size distribution of fat globules in bovine cheese-making milk. *Innovative Food Science & Emerging Technologies, 10*(4), 491–494. doi:10.1016/j.ifset.2009.03.002

García-Risco, M. R., Cortés, E., Carrascosa, A. V., & López-Fandiño, R. (1998). Microbiological and chemical changes in high-pressure-treated milk during refrigerated storage. *Journal of Food Protection*, *61*(6), 735–737. doi:10.4315/0362-028X-61.6.735 PMID:9709260

García-Risco, M. R., Olano, A., Ramos, M., & López-Fandiño, R. (2000). Micelar changes induced by high pressure. influence in the proteolytic activity and organoleptic properties of milk. *Journal of Dairy Science*, *83*(10), 2184–2189. doi:10.3168/jds.S0022-0302(00)75101-0 PMID:11049057

Garvey, M., & Rowan, N. J. (2019). Pulsed UV as a potential surface sanitizer in food production processes to ensure consumer safety. *Current Opinion in Food Science*, *26*, 65–70. doi:10.1016/j.cofs.2019.03.003

Gaucheron, F., Famelart, M. H., Mariette, F., Raulot, K., Michel, F., & Le Graetf, Y. (1997). Combined effects of temperature and high-pressure treatments on physicochemical characteristics of skim milk. *Food Chemistry*, *59*(3), 439–447. doi:10.1016/S0308-8146(96)00301-9

Guerrero-Beltrán, J. Á., Sepulveda, D. R., Góngora-Nieto, M. M., Swanson, B., & Barbosa-Cánovas, G. V. (2010). Milk thermization by pulsed electric fields (PEF) and electrically induced heat. *Journal of Food Engineering*, *100*(1), 56–60. doi:10.1016/j.jfoodeng.2010.03.027

Gurol, C., Ekinci, F. Y., Aslan, N., & Korachi, M. (2012). Low temperature plasma for decontamination of E. coli in milk. *International Journal of Food Microbiology*, *157*(1), 1–5. doi:10.1016/j.ijfoodmicro.2012.02.016 PMID:22622128

Han, Z., Cai, M., Cheng, J., & Sun, D. (2018). Effects of electric fields and electromagnetic wave on food protein structure and functionality: A review. *Trends in Food Science & Technology*, *75*(March), 1–9. doi:10.1016/j.tifs.2018.02.017

Harte, F., Luedecke, L., Swanson, B., & Barbosa-Cánovas, G. V. (2010). Low-fat set yogurt made from milk subjected to combinations of high hydrostatic pressure and thermal processing. *Journal of Dairy Science*, *86*(4), 1074–1082. doi:10.3168/jds.S0022-0302(03)73690-X PMID:12741531

Hite, B. H. (1899). The effects of pressure in the preservation of milk. *Bulletin of the West Virginia University Agricultural Experimental Station Morgantown*, (58), 15–35.

Huang, H. W., Wu, S. J., Lu, J. K., Shyu, Y. T., & Wang, C. Y. (2017). Current status and future trends of high-pressure processing in food industry. *Food Control*, *72*(12), 1–8. doi:10.1016/j.foodcont.2016.07.019

Huppertz, T., Fox, P. F., & Kelly, A. L. (2003). High pressure-induced changes in the creaming properties of bovine milk, *4*(July), 349–359. doi:10.1016/S1466-8564

Innocente, N., Segat, A., Manzocco, L., Marino, M., Maifreni, M., Bortolomeoli, I., ... Nicoli, M. C. (2014). Effect of pulsed light on total microbial count and alkaline phosphatase activity of raw milk. *International Dairy Journal*, *39*(1), 108–112. doi:10.1016/j.idairyj.2014.05.009

Jaeger, H., Meneses, N., & Knorr, D. (2014). Pulsed electric field technology. In Y. Motarjemi, G. Moy, & E. Todd (Eds.), *Encyclopedia of food safety* (Vol. 3, pp. 239–244). Elsevier. doi:10.1016/B978-0-12-378612-8.00260-2

Kasahara, I., Carrasco, V., & Aguilar, L. (2015). Inactivation of Escherichia coli in goat milk using pulsed ultraviolet light. *Journal of Food Engineering*, *152*, 43–49. doi:10.1016/j.jfoodeng.2014.11.012

Khan, M. K., Ahmad, K., Hassan, S., Imran, M., Ahmad, N., & Xu, C. (2018). Effect of novel technologies on polyphenols during food processing. *Innovative Food Science and Emerging Technologies*, *45*(December 2017), 361–381. doi:10.1016/j.ifset.2017.12.006

Kim, H. J., Yong, H. I., Park, S., Kim, K., Choe, W., & Jo, C. (2015). Microbial safety and quality attributes of milk following treatment with atmospheric pressure encapsulated dielectric barrier discharge plasma. *Food Control*, *47*, 451–456. doi:10.1016/j.foodcont.2014.07.053

Knudsen, J. C., & Skibsted, L. H. (2010). High pressure effects on the structure of casein micelles in milk as studied by cryo-transmission electron microscopy. *Food Chemistry*, *119*(1), 202–208. doi:10.1016/j.foodchem.2009.06.017

Korachi, M., Ozen, F., Aslan, N., Vannini, L., Guerzoni, M. E., Gottardi, D., & Ekinci, F. Y. (2015). Biochemical changes to milk following treatment by a novel, cold atmospheric plasma system. *International Dairy Journal*, *42*, 64–69. doi:10.1016/j.idairyj.2014.10.006

Koutchma, T. (2014). *Adapting high hydrostatic pressure (HPP) for food processing operations. Adapting high hydrostatic pressure (HPP) for food processing operations*. doi:10.1016/C2013-0-12997-2

Kramer, B., Wunderlıch, J., & Muranyı, P. (2015). Pulsed light decontamination of endive salad and mung bean sprouts and impact on color and respiration activity. *Journal of Food Protection*, *78*(2), 340–348. doi:10.4315/0362-028X.JFP-14-262 PMID:25710149

Krishnamurthy, K., Demirci, A., & Irudayaraj, J. M. (2007). Inactivation of Staphylococcus aureus in milk using flow-through pulsed UV-light treatment system. *Journal of Food Science*, *72*(7), M233–M239. doi:10.1111/j.1750-3841.2007.00438.x PMID:17995646

Kümmel, J., Stessl, B., Gonano, M., Walcher, G., Bereuter, O., Fricker, M., ... Ehling-Schulz, M. (2016). Staphylococcus aureus entrance into the dairy chain: Tracking S. aureus from dairy cow to cheese. *Frontiers in Microbiology*, *7*(OCT), 1–11. doi:10.3389/fmicb.2016.01603 PMID:27790200

Laroussi, M. (2005). Low temperature plasma-based sterilization: Overview and state-of-the-art. *Plasma Processes and Polymers*, *2*(5), 391–400. doi:10.1002/ppap.200400078

Li, X., & Farid, M. (2016). A review on recent development in non-conventional food sterilization technologies. *Journal of Food Engineering*, *182*, 33–45. doi:10.1016/j.jfoodeng.2016.02.026

Liu, Z., Hemar, Y., Tan, S., Sanguansri, P., Niere, J., Buckow, R., & Augustin, M. A. (2015). Pulsed electric field treatment of reconstituted skim milks at alkaline pH or with added EDTA. *Journal of Food Engineering*, *144*, 112–118. doi:10.1016/j.jfoodeng.2014.06.033

Liu, Z., Juliano, P., Williams, R. P., Niere, J., & Augustin, M. A. (2014). Ultrasound effects on the assembly of casein micelles in reconstituted skim milk. *The Journal of Dairy Research*, *81*(2), 146–155. doi:10.1017/S0022029913000721 PMID:24351847

Mahendran, R., Ramanan, K. R., Barba, F. J., Lorenzo, J. M., López-Fernández, O., Munekata, P. E. S., ... Tiwari, B. K. (2018, December). Recent advances in the application of pulsed light processing for improving food safety and increasing shelf life. *Trends in Food Science & Technology*, *88*, 67–79. doi:10.1016/j.tifs.2019.03.010

Marchesini, G., Fasolato, L., Novelli, E., Balzan, S., Contiero, B., Montemurro, F., ... Segato, S. (2015). Ultrasonic inactivation of microorganisms: A compromise between lethal capacity and sensory quality of milk. *Innovative Food Science & Emerging Technologies*, *29*, 215–221. doi:10.1016/j.ifset.2015.03.015

Misra, N. N., Koubaa, M., Roohinejad, S., Juliano, P., Alpas, H., Inácio, R. S., ... Barba, F. J. (2017). Landmarks in the historical development of twenty first century food processing technologies. *Food Research International*, *97*(May), 318–339. doi:10.1016/j.foodres.2017.05.001 PMID:28578057

Misra, N. N., Oliver, K., & Schlüter, P. J. C. (2016). *Cold plasma in food and agriculture*. Academic Press; doi:10.1016/b978-0-12-801365-6.09991-1

Morales-de la Peña, M., Welti-Chanes, J., & Martín-Belloso, O. (2019). Novel technologies to improve food safety and quality. *Current Opinion in Food Science*, *30*, 1–7. doi:10.1016/j.cofs.2018.10.009

Moreau, M., Orange, N., & Feuilloley, M. G. J. (2008). Non-thermal plasma technologies: New tools for bio-decontamination. *Biotechnology Advances*, *26*(6), 610–617. doi:10.1016/j.biotechadv.2008.08.001 PMID:18775485

Moreno, S. E. H. (2017). Distribución de pesticidas en sistemas de abejas, polen, cera y miel. Análisis por LC-MS/MS. Retrieved from http://repositorio.ual.es/bitstream/handle/10835/6453/16628_Memoria_TFG FINAL2.pdf?sequence=1&isAllowed=y

Mussa, D. M., & Ramaswamy, H. S. (1997). Ultra high pressure pasteurization of milk: Kinetics of microbial destruction and changes in Physico-chemical characteristics. *Lebensmittel-Wissenschaft + Technologie*, *30*(6), 551–557. doi:10.1006/fstl.1996.0223

Nakai, S., & Li-chan, E. (1988). *Hydrophobic interactions in food systems authors*. Boca Raton, FL: CRC Press.

Niemira, B. A. (2012). Cold plasma decontamination of foods. *Annual Review of Food Science and Technology*, *3*(1), 125–142. doi:10.1146/annurev-food-022811-101132 PMID:22149075

Odriozola-Serrano, I., Bendicho-Porta, S., & Martín-Belloso, O. (2010). Comparative study on shelf life of whole milk processed by high-intensity pulsed electric field or heat treatment. *Journal of Dairy Science*, *89*(3), 905–911. doi:10.3168/jds.S0022-0302(06)72155-5 PMID:16507684

Ohshima, T., Tanino, T., Kameda, T., & Harashima, H. (2016). Engineering of operation condition in milk pasteurization with PEF treatment. *Food Control*, *68*, 297–302. doi:10.1016/j.foodcont.2016.03.047

Ojha, K. S., Mason, T. J., O'Donnell, C. P., Kerry, J. P., & Tiwari, B. K. (2017). Ultrasound technology for food fermentation applications. *Ultrasonics Sonochemistry*, *34*, 410–417. doi:10.1016/j.ultsonch.2016.06.001 PMID:27773263

Pandey, P. K., Ramaswamy, H. S., & Idziak, E. (2003). High pressure destruction kinetics of in raw milk at two temperatures. *Journal of Food Process Engineering*, *26*(3), 265–283. doi:10.1111/j.1745-4530.2003.tb00601.x

Patterson, M. F., & Kilpatrick, D. J. (1998). The combined effect of high hydrostatic pressure and mild heat on inactivation of pathogens in milk and poultry. *Journal of Food Protection*, *61*(4), 432–436. doi:10.4315/0362-028X-61.4.432 PMID:9709206

Proulx, J., Agustin, M., Sullivan, G., VanWees, S., Jian, J., Hilton, S. T., & Moraru, C. I. (2016). Short communication: Influence of pulsed light treatment on the quality and sensory characteristics of Cheddar cheese. *Journal of Dairy Science, 100*(2), 1004–1008. doi:10.3168/jds.2016-11579 PMID:28012618

Proulx, J., Hsu, L. C., Miller, B. M., Sullivan, G., Paradis, K., & Moraru, C. I. (2015). Pulsed-light inactivation of pathogenic and spoilage bacteria on cheese surface. *Journal of Dairy Science, 98*(9), 5890–5898. doi:10.3168/jds.2015-9410 PMID:26162787

Proulx, J., Sullivan, G., Marostegan, L. F., VanWees, S., Hsu, L. C., & Moraru, C. I. (2017). Pulsed light and antimicrobial combination treatments for surface decontamination of cheese: Favorable and antagonistic effects. *Journal of Dairy Science, 100*(3), 1664–1673. doi:10.3168/jds.2016-11582 PMID:28109595

Putnik, P., Kresoja, Ž., Bosiljkov, T., Režek Jambrak, A., Barba, F. J., Lorenzo, J. M., ... Bursać Kovačević, D. (2018, July). Comparing the effects of thermal and non-thermal technologies on pomegranate juice quality: A review. *Food Chemistry, 279*, 150–161. doi:10.1016/j.foodchem.2018.11.131 PMID:30611474

Rademacher, B., & Hinrichs, J. (2006). Effects of high-pressure treatment on indigenous enzymes in bovine milk: Reaction kinetics, inactivation and potential application. *International Dairy Journal, 16*(6), 655–661. doi:10.1016/j.idairyj.2005.10.021

Rademacher, B., Pfeiffer, B., & Kessler, H. G. (1998). *Inactivation of microorganisms and enzymes in pressure-treated raw milk. High pressure food science, bioscience, and chemistry*. The Royal Society of Chemistry; doi:10.1533/9781845698379.3.145

Rowan, N. J. (2019). Pulsed light as an emerging technology to cause disruption for food and adjacent industries – Quo vadis? *Trends in Food Science & Technology, 88*(March), 316–332. doi:10.1016/j.tifs.2019.03.027

Santhirasegaram, V., Razali, Z., & Somasundram, C. (2016). *Safety improvement of fruit juices by novel thermal and nonthermal processing. Food hygiene and toxicology in ready-to-eat foods*. Elsevier Inc.; doi:10.1016/B978-0-12-801916-0.00012-1

Schrader, K., Buchheim, W., & Morr, C. V. (1997). High pressure effects on the colloidal calcium phosphate and the structural integrity of micellar casein in milk. Part 1 phosphate in heated milk systems. *Food / Nahrung, 41*(3), 133–138.

Segat, A., Misra, N. N., Cullen, P. J., & Innocente, N. (2016). Effect of atmospheric pressure cold plasma (ACP) on activity and structure of alkaline phosphatase. *Food and Bioproducts Processing, 98*, 181–188. doi:10.1016/j.fbp.2016.01.010

Şengül, M., Erkaya, T., Başlar, M., & Ertugay, M. F. (2011). Effect of photosonication treatment on inactivation of total and coliform bacteria in milk. *Food Control, 22*(11), 1803–1806. doi:10.1016/j.foodcont.2011.04.015

Shahbaz, H. M., Kim, J. U., Kim, S.-H., & Park, J. (2018). *Advances in nonthermal processing technologies for enhanced microbiological safety and quality of fresh fruit and juice products. food processing for increased quality and consumption.* Elsevier; doi:10.1016/b978-0-12-811447-6.00007-2

Sharma, P., Bremer, P., Oey, I., & Everett, D. W. (2014). Bacterial inactivation in whole milk using pulsed electric field processing. *International Dairy Journal, 35*(1), 49–56. doi:10.1016/j.idairyj.2013.10.005

Styles, M. F., Hoover, D. G., & Farkas, D. F. (1991). Response of Listeria monocytogenes and Vibrio parahaemolyficus to high hydrostatic pressure. *Journal of Food Science, 56*(5), 1404–1407. doi:10.1111/j.1365-2621.1991.tb04784.x

Sui, Q., Roginski, H., Williams, R. P. W., Versteeg, C., & Wan, J. (2011). Effect of pulsed electric field and thermal treatment on the physicochemical and functional properties of whey protein isolate. *International Dairy Journal, 21*(4), 206–213. doi:10.1016/j.idairyj.2010.11.001

Sun, Y., Chen, J., Zhang, S., Li, H., Lu, J., Liu, L., ... Jiaping, L. (2014). Effect of power ultrasound pre-treatment on the physical and functional properties of reconstituted milk protein concentrate. *Journal of Food Engineering, 124*, 11–18. doi:10.1016/j.jfoodeng.2013.09.013

Trujillo, A. J., Ferragut, B., Juan, B., Roig-Sagués, A. X., & Guamis, B. (2016). Processing of dairy products utilizing high pressure. In V. M. Balasubramaniam, G. V. Barbosa-Cánovas, & H. L. M. Lelieveld (Eds.), *High pressure processing of food* (pp. 553–590). New York: Springer; doi:10.1007/978-1-4939-3234-4_25

Vijayakumar, S., Grewell, D., Annandarajah, C., Benner, L., & Clark, S. (2015). Quality characteristics and plasmin activity of thermosonicated skim milk and cream. *Journal of Dairy Science, 98*(10), 6678–6691. doi:10.3168/jds.2015-9429 PMID:26233461

Wrigley, D. M., & Llorca, N. G. (1992). Decrease of Salmonella typhimurium in skim milk and egg by heat and ultrasonic wave treatment. *Journal of Food Protection, 55*(9), 678–680. doi:10.4315/0362-028X-55.9.678 PMID:31084132

Yong, H. I., Kim, H. J., Park, S., Kim, K., Choe, W., Yoo, S. J., & Jo, C. (2015). Pathogen inactivation and quality changes in sliced cheddar cheese treated using flexible thin-layer dielectric barrier discharge plasma. *Food Research International*, *69*, 57–63. doi:10.1016/j.foodres.2014.12.008

Yu, L. J., Ngadi, M., & Raghavan, G. S. V. (2009). Effect of temperature and pulsed electric field treatment on rennet coagulation properties of milk. *Journal of Food Engineering*, *95*(1), 115–118. doi:10.1016/j.jfoodeng.2009.04.013

Zhao, L., Zhang, S., Uluko, H., Liu, L., Lu, J., Xue, H., ... Lv, J. (2014). Effect of ultrasound pretreatment on rennet-induced coagulation properties of goat's milk. *Food Chemistry*, *165*, 167–174. doi:10.1016/j.foodchem.2014.05.081 PMID:25038663

Zhao, W., Yang, R., Shen, X., Zhang, S., & Chen, X. (2013). Lethal and sublethal injury and kinetics of Escherichia coli, Listeria monocytogenes and Staphylococcus aureus in milk by pulsed electric fields. *Food Control*, *32*(1), 6–12. doi:10.1016/j.foodcont.2012.11.029

Chapter 2
Recent Trends in Cold Storage Systems

Kapilan N.
Nagarjuna College of Engineering and Technology, India

ABSTRACT

The cold storage is a big room or commercial building used to store perishable items like vegetables, fruits, fish, milk products, meat, etc. under controlled conditions where temperature, humidity, etc. are controlled for a longer duration. Different types of cold storages are available and selection of cold storage is based on the application. In recent years, solar hybrid cold storage systems are getting popular due to lower operating cost and reduced greenhouse gas emissions. Also, thermal energy storage systems with phase change materials are being used in cold storage systems. The new technologies are developed to reduce the energy consumption of cold storage systems. The quality of the perishable items stored in the cold storage may be retained by applying nano coating on the fruits. This chapter reviews cold storage systems, elements of cold storage, design of cold storage, application of phase change and nano particles in cold storage system.

INTRODUCTION

The domestic refrigeration system is used for storing fruits, vegetables, milk products, and other perishable items for a longer duration. Similarly, cold storage systems are used for refrigerating a huge amount of vegetables, fruits, perishable items, etc. The cold storage is a typical room or building where temperature, humidity, etc. are controlled to maintain required condition using various refrigeration methods. In cold storage, insulations are provided on walls, ceiling, and floors of the cold storage

DOI: 10.4018/978-1-7998-1924-0.ch002

room to reduce heat loss. The cold storage systems are used to preserve perishable nutrient products, perishable foods like fruits, meats, vegetables, and dairy products, fish, medicines, flowers, pharmaceutical products, etc. The important functions of the cold storage systems are as follows:

- To increase the shelf life of vegetables, fruits, etc.
- To increase the taste of the beverages.
- To slow the deterioration of the perishable products.
- To maintain product quality.
- To reduce the wastage of commodities.

The climatic change increases ambient temperature and this may affect the development of cold chain in near future. Also, an increase in ambient temperature will affect the quality of food stored and in extreme cases may result in food spoilage unless quality of the cold-chain is improved. The cold-chain affects CO_2 emission and accounts for 1% of CO_2 emission in the world. The CO_2 production may increase the global temperatures significantly. However, it is possible to reduce CO_2 emission by adopting environmental-friendly cold storage systems (James & James, 2010).

MAJOR ELEMENTS OF COLD STORAGE SYSTEM

The major elements of cold storage systems are as follows. Figure 1 shows the typical cold storage system.

1. Compressor

The compressor may be called as heart of the system as it is the driving force. The compressor of the cold storage is used for circulation of the refrigerant and also to increase the pressure of the system. It is compact and available in various capacities to meet the requirements. The selection of the compressor is based on applications. The recent advances in compressor technology reduce power consumption and also help in reliable operation.

2. Condenser

The compressor is used for condensing the refrigerant and it may be air- or water-cooled. The air-cooled condenser speed is low and hence, produces low noise. It has higher air volume and also higher efficiency. It is easy to operate and maintain the air-cooled condensers. The condensers are made of high-grade, seamless pipe, hence its corrosion rate is less. The coil out-frame is made of high-quality steel and this steel has higher corrosion resistance and higher strength.

3. Evaporator or Ceiling unit cooler

The evaporator or ceiling unit cooler is made of thin-walled brass and the second flanging punching creates a highly efficient aluminum fins. These materials have higher thermal conductivity and hence, have higher heat transfer efficiency. This unit is provided with an anti-moisture fan which produces higher air volume, low noise, and low temperature resistance, and provides reliable operation. It is provided with stainless steel tube with electric defrosting system. The coil tube is made of high-quality, corrosion-resistant steel and uses electrical heater for defrosting.

4. Wind curtain

The wind curtain is installed inside the cold room at the top of the door. It blows the air such that it divides the air in and out of the room to reduce cold air loss. It is an essential component with a slick design and is also reliable.

5. Door

The door is made of low thermal conductivity materials. The external and internal review of the cold storage door is based on the needs of the customer like painted color steel or an aluminum or galvanized panel, etc. The door shall have lower

Figure 1. Typical cold storage system

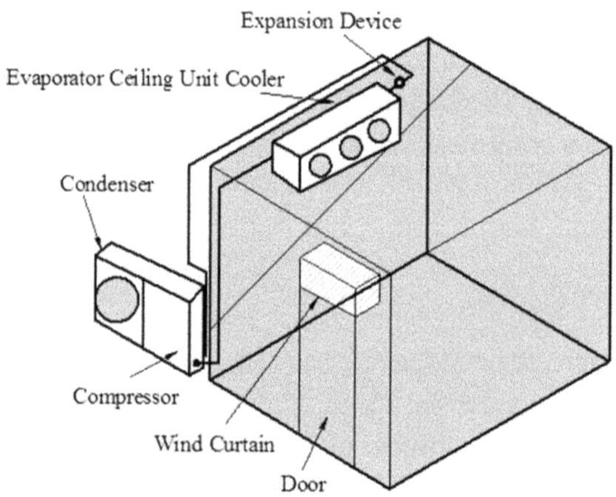

moisture absorption, corrosion resistance, a light weight, better insulation properties, and an aesthetic look.

TYPES OF COLD STORAGE SYSTEM

The cold storage system may be classified into different types based on a number of variables and applications (https://www.winnesota.com/news/coldstorage). Hence, the cold storage systems are available from individual units to entire dedicated facilities based on the demand and refrigeration requirements.

1. Control Atmosphere and Cold Storage:

This type of system is preferred for long-term storage of fruits. The cold storage room temperature, and concentration of gases such as oxygen, carbon dioxide, ethylene, etc., affect the quality of stored fruits. However, the requirement may vary depending on the storage products.

2. Temperature Controlled Cold Storage:

This type of system is selected to store the dry fruits and grains. In this cold storage, temperature is maintained constant throughout the cold room.

3. High Humidity Cold Storage:

This kind of storage system is used for storing vegetables. The cold storage temperature and humidity are controlled and maintained as per the type of vegetables that is stored.

4. Refrigerated Containers:

This is the simplest form of cold storage and is used to store small quantities of temperature-sensitive products. Also, it can be used for mobile application and hence, it gives extra flexibility.

5. Blast Freezers and Chillers:

This type of system is preferred for quick cooling and storage of food before it reaches its end consumer. It is most commonly used in larger restaurants and catering companies.

6. Pharmaceutical Grade Cold Storage:

This is a special type of cold storage used in hospitals and research institutions. It is provided with additional features which make them suitable for blood storage, pharmaceutical applications, and vaccines storage.

7. Plant-attached Cold Storage:

This type is preferred when the user wants to place cold storage in house. In this system, products can be transported from a manufacturing location to a cold room facility on-site.

STORAGE OF FOODS AND STORAGE CONDITIONS

The microorganisms consist of yeasts and bacteria, and spoil the food items. The microorganisms' activities can be controlled by maintaining low temperature in a room and storing foods and many other items inside the room. The low temperature reduces the microorganisms' activities and does not kill the microorganisms. Hence, we can preserve perishable foods and other commodities for a shorter period in their natural state. The preservation temperature depends on the storage time and product type (http://agritech.tnau.ac.in/agricultural_marketing/agrimark_cold%20 storage.html).

In general, there are three groups of products:

- Food items such as vegetables, fruits, etc., which are alive at the time of delivery.
- Processed food items such as meat, chicken, fish, etc., which are no longer alive.
- Commodities such as cold drinks, beer, tobacco, khandsari, etc., which need cooling before delivery.

The fruits and vegetables are living foods and these items have few natural protections against microorganism activities. Hence, for preserving these items, it is necessary to keep these items alive and retard the ripening rate by retarding the natural enzyme activity.

The non-living foods are susceptible to spoilage and hence, preservation of these items is complicated. This is due to decay of dead tissues. The meat and fish are frozen and stored below -15°C for long-term storage. However, most of the fruits and vegetables cannot be frozen and hence, fruits and vegetables are carefully stored as per the recommended storage temperature and humidity. Any deviation in the recommended condition may affect the quality of the stored items and spoil them.

The fruits and vegetables like tomatoes, oranges, etc. cannot be frozen, but can be stored at low temperature for long-term storage. The dairy products are nonliving foodstuffs and can be stored at low temperature for a short duration. However, these items undergo oxidation during a long storage period which breaks down fats, causing rancidity. This problem can be overcome by excluding air in the package. The absence of oxygen helps to store these foods for a longer period.

COLD STORAGE TECHNOLOGY

The suitable refrigeration system is used in cold storage rooms to maintain the required indoor condition for the commodities to be stored. The working of the refrigeration system is based on two principles and the systems are called:

1. Vapour absorption system
2. Vapour compression system

The vapour absorption system (VAS) requires lower operating cost, and adequately compensates the higher initial cost. The waste industrial waste or solar heat can be used to reduce the operating cost. Hence, this type of system can be selected to conserve energy and operational cost. However, it is not preferred when storage temperature is below 10^{0}C, and few fruits need storage temperature below 10^{0}C for long storage.

Figure 2. Vapour absorption system

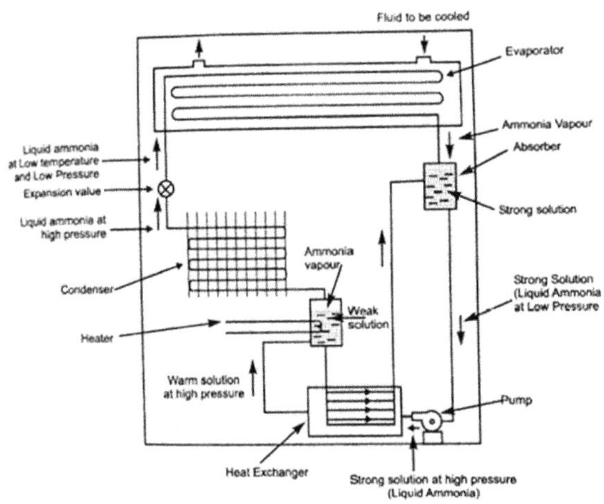

Figure 3. Vapour compression system

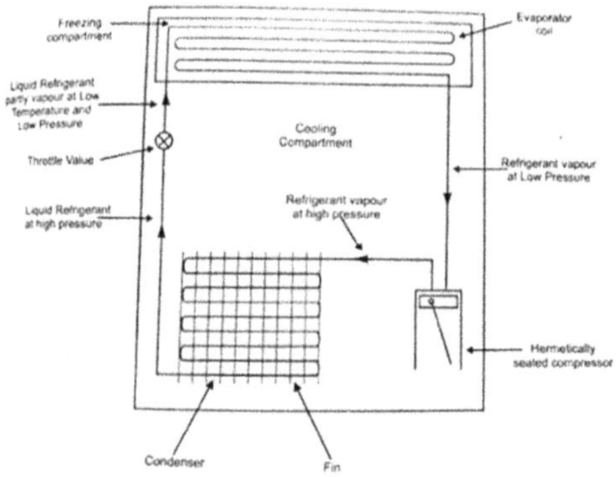

The vapour compression system (VCS) is cheap as compared to VAS. The various types of VCS systems such as diffuser type, bunker type, and fin coil type, are available commercially and selection of VAS depends upon the indoor cooling requirements of cold storage units. Among the VCS, diffuser type VCS is costlier and is preferred when storage room heights are low. Also, this system's operating cost is higher. The cheapest VCS type is Bunker type and it is preferred when cold storage unit heights normally exceed 11.5 m. The operating cost of this VCS is low as compared to other VCS types. The operational cost of fin coil type is low and it is highly energy efficient; however it is 5% higher than the bunker type and has higher storage space. It is preferred for room heights of 5.4 m onwards.

In refrigeration system, refrigerant absorbs heat from evaporator unit and in condenser it rejects the heat to the surrounding. The most commonly used refrigerant in VCS is Freon and use of this refrigerant degrades the environment. However, it is replaced by alternative refrigerants which are environmental friendly. Carbon dioxide, ammonia, etc. are used as alternative refrigerant, and ammonia is preferred for horticultural produce in cold storage units.

HEAT LOAD CALCULATION IN COLD STORAGE DESIGN

The following factors should be considered in heat load calculation:

1. Heat gain due to conduction through wall, floor, and ceiling.
2. Heat gain from solar radiation through wall and ceiling.

3. Heat load due to frequent door opening and closing and also during fresh air charge.
4. Heat gain due to workers working in cold storage.
5. Heat gain due to incoming goods.
6. Heat gain due to respiration of stored product.
7. Heat gain due to lighting and fans.
8. Miscellaneous heat loads.

STORAGE CONDITIONS REQUIRED

The cold storage condition may vary based on type of fruits and vegetables. Table 1 shows the preferable storage conditions for vegetables and fruits (https://www.winnesota.com/news/coldstorage).

Energy Efficiency

The operating cost of the cold storage can be reduced by increasing the energy efficiency through reducing heat loss. Hence, good-quality doors and insulation should be provided to reduce the heat loss and maintain required cooling temperature. The temperature distribution, charging and discharging rate, and capacity are key indexes to evaluate the cold storage performance. It was during a tank charging-chiller delivering load operation, using bypass is unfavourable for thermal storage effectiveness. However, during a tank-discharging, chiller-delivering load operation, chiller starting at a later stage with bypass is favorable for the system efficiency (Xiaolin Wang et al., 2019).

The variety of cooling load of the internal mass is linear dependent on the variety of room temperature (Shen Tian et al., 2019). It is reported that the volume of the storage tank in ice storage-full mode, cold water storage-partial mode, and cold water storage-full mode cases is higher than volume of the storage tank in ice storage-partial mode. For full cold water storage, the current energy consumption is reduced by 72% compared to the direct cooling mode without the storage tank. The system with the partial storage of cold water has a lower initial cost than a non-storage system. Furthermore, less energy and current costs will be achieved by the partial cold water storage system (Ali Reza Shaibani et al., 2019).

A study carried out on the flooded tube for cold thermal storage with stirred water, stagnant water, and recirculated water shows that the recirculated water gives best performance as compared to other types of water source (Hani Hussain Sait, 2018). The deposition of frost on the surface of the evaporator will reduce performance of the cold storage and cooling capacity. The periodic defrosting reduces the evaporator blockage. Hence, reversed cycle defrosting (RCD) and electric heater defrosting (EHD) methods are preferred in cold storages. A novel defrosting system based on EHD and RCD results in higher energy utilization and minimizes temperature fluctuation. Also, it improves the compressor operational stability (Dong Wang et al., 2017).

Phase Change Materials

The use of phase-change material (PCM) with conventional insulation improves the cold storage performance. This is because of latent heat energy transfer in PCM. Hence, PCM plays a key role in energy storage and minimizes the operating cost at peak load demand as the peak load energy charges are higher than the non-peak load. In the distribution process, the use of PCM can ensure the quality of agricultural products. The quality of agricultural products can be maintained for longer duration if lower phase transition temperature is maintained. The transportation of agricultural products in the frozen state maintains the freshness of the agricultural products (Bing Bai et al., 2019). The high latent heat density of water/ice PCM makes it a promising material in cold storage applications. During peak load, the cold storage system uses water/ice PCM so electricity load reduces, cutting operating cost. Also, water/ice PCM enhances the insulation effect and minimizes the requirement of cold energy. It is reported that the return on investment of water/ice PCM is approximately 4.1 years in laboratory scale (Changjiang Wang et al., 2017). The high thermal storage capacity PCM is considered as promising materials for the cold storage and it can be used during frequent door openings, power failure, compressor's on-off cycling, and also to maintain the product temperature within safe limits (Wei Lu et al., 2019).

The microencapsulated PCMs have large surface per volume and hence, provides better heat transfer performance to the surroundings. Also, the restricted phase separation in microscopic distances improves its cycling stability. The addition of nanoparticles to the PCM improves the heat transfer rate and is also for better storage process (Gang Li et al., 2013). The composite building material for thermal storage mixed with the composite phase change material (CPCM) and the cement, early strength agent, and water-reducer, was prepared at a constant temperature of 70^0 C. It is reported that this type of composite building materials with CPCM provides higher thermal conductivity and good stability, higher heat storage capacity, and higher compression resistance capability, and also meets the requirement of ordinary mortar building material (Qunli Zhang et al., 2017).

A paraffin with the melting point of 22°C can be used as the PCM to cool hot air during the daytime in summer and to store cold during the night-time. The result shows that the PCM usage minimizes the fluctuation in temperature and also reduces the requirement of cooling and air-conditioning system addition. The higher temperature difference between inlet outlet and higher cold power helps the cold storage to work efficiently for cooling and also for ventilating applications (Stritih & Butala, 2017). It is reported that the cold storage unit with three-stage cascade system gives better thermal performance than a single-stage cold storage system (Shaon Talukdar et al., 2019).

The falling film type of cold energy regenerator with a PCM results in higher heat transfer rate and improves the cold storage characteristics. Also, it requires lower pumping power. The cold storage performance can be improved by lower film Reynolds number (Xiwen Cheng et al., 2018). The improvement in maximum melting time of the PCM at optimal insulation thickness keeps the product at a low temperature in cold storage unit for a longer period (Jun-Feng Jiang et al., 2018). The stable COP for ejector cooling unit can be obtained with PCM cold storage unit integrated with ejector. Also, the solar heating or industrial waste heat may be used as the heat source for this system (Yoram Kozak et al., 2017).

Solar Cold Storage

A hybrid cold storage system which uses both solar PV system and solar thermal energy was tested successfully and it is reported that this hybrid system has a return on investment in less than four years (Xiangjie Chen et al., 2014). The PCM can be used in solar cold storage applications to meet the peak power demand or when there is no sunshine. The PCM can be used in solar cold storage applications to meet the peak power demand or when there is no sunshine. A PCM-based latent heat thermal energy storage system which uses a heat exchanger made up from evaporator tube along with rectangular metal fins increases heat transfer rate during phase change of PCM. The PCM units with metal fins improve solidification, enhance thermal energy storage capacity, and results in higher heat flux during melting (Dipankar et al., 2016). The application of microcapsule phase change materials (MEPCM) in solar powered air conditioning cold storage system increases the system efficiency and reduces the electrical power requirement. The heat transfer rate of MEPCM is better than pure water (Shaon Talukdar et al., 2019).

The work carried out on solar-powered air conditioning system integrated with PCM cold storage shows that the comfort temperature was exceeded more than 26% of the time, without cold storage. However, with cold storage, the periods of high indoor temperatures reduced significantly. The study gives the maximum COP and solar thermal ratio of 0.193 and 0.097 respectively (Lin Zheng et al., 2017).

EFFECT OF COLD STORAGE ON STORED ITEMS

The psychrotrophic microorganisms affect the quality of the food items stored as these microorganisms reproduce rapidly, developing lipases and proteases with high thermal stability. Different techniques are used to detect these microorganisms in foods due to introduction of rapid and automatic technologies. However, this detection system should be simple and easy to use in the food industry and also suitable with online microbial detection (Yosr Allouche et al., 2017). A study carried out with Listeria monocytogenes (Lm) on blueberries shows that inoculation method affects the survival of Lm. It also shows that the Dip-inoculated Lm is resistant to intervention of sanitizer as compared to spot-inoculated Lm on blueberries (Qingyi Wei et al., 2019). The survival of probiotic bacteria during longer storage period in cold storage and the gastrointestinal tract affects the quality of probiotic products. The total dietary fiber content of EPS is similar to its total carbohydrate content. However, the protective effect of exopolysaccharide (EPS) reduces when the concentration was reduced or by partial degradation with power ultrasound. In all bacterial strains in bile juice, EPS provides significant protective effect; hence the EPS has significant potential to use as the medicinal fungus (Lina Sheng et al., 2019).

The Campylobacter affects the chicken liver stored and hence the undercooked chicken livers are not advisable to eat. The gamma irradiation can be used in the contaminated chicken livers to reduce Campylobacter jejuni. Campylobacter jejuni radiation resistance can be increased by storing chicken livers at −20 °C before the treatment of radiation. By combining cold storage and irradiation in chicken livers, campylobacter jejuni level can be reduced significantly (Ang-Xin Song et al., 2019). The conventional freezing system requires higher energy as compared to constant volume system freezing. However, the energy requirement can be reduced further changing the thermodynamic state of food stored, from constant pressure process to constant volume process, without making major changes in the existing refrigeration system. There is an increase in thiobarbituric acid reactive substance, peroxide value, and total oxidation when the mussels meat is kept in cold storage at 4 °C. However, the percentage of polyunsaturated fatty acid reduces which shows the lipid oxidation (Nereus et al., 2019).

The postharvest disease in different fruits can be reduced with the help of ozone treatment. During storage, the ozone treatment reduces the range of fluctuations which generally occurs in untreated apples. The ozone treatment does not affect the concentration of ursolic acid in the apple peel, but there is a change in oleanolic acid value. The total phenols concentration is reduced when peel of both cultivars were treated by both gaseous ozone and ozonated water treatment. However, in the apple flesh, the ozone treatment does not affect the concentration of total phenols (Xin Zhou et al., 2019). The refrigeration of Inoculated slices of raw salmon at 4 °C did not inhibit growth of A. salmonicida. The pH value reduces in the salmon when it was contacted with acidified rice and this is due to reduction in bacterial viability. During cold storage, the microbiological hazard was observed in retail sushi products. However, the growth of pathogenic aeromonas species can be reduced with the combination of rice acidification and low storage temperature (Yanrong et al., 2019).

A research work was carried out to study the impacts of ultrasonic treatment on the foaming and physicochemical properties of egg white from shell egg stored for several days. The results show that the ultrasound treatment increases the foaming ability significantly. During 60 days storage, the highest foaming ability was observed with reference to untreated egg white. This enhancement may be due to increase in free sulfhydryl content and surface hydrophobicity. This shows that egg white protein became easier to adsorb at the air-water interface and it has flexible structure. These results show that during storage, the protein properties can be affected by ultrasound and hence ultrasound can be used in the food storage industry (SunnivaHoel et al., 2018).

The original microwave-extracted pectin prior to cold storage was compared with the properties of pectin, acid-extracted by microwave heating under pressure from orange and lime albedo which was stored in cold storage for 14 years. From the analysis, it is observed that the galacturonic acid content was significantly changed during cold storage. However, the degree of esterification reduced by 50% and there was an increase in total sugar content during cold storage for both lime pectin and orange. It is suggested that the higher protein was associated with higher molar mass pectin (Yinxia Chen et al., 2019). The storage time, storage temperature, and food matrix affect the growth of L. monocytogenes. The desired temperature for the storage of the food items with low growth of L. monocytogenes is 5 °C. The growth of L. monocytogenes is observed in most of the salad's samples, however no growth was observed on corn, celeriac, and salad of carrot products (Marshall et al., 2019).

The bamboo shoots-subjected oxalic acid treatment increases the membrane integrity, reduces respiration, weight, losses of total sugar content, and disease incidence, inhibits enzymatic browning, and retards lignification during cold storage. Hence, edible quality of bamboo shoots can be maintenance due to the above said characteristics. The postharvest of bamboo shoots can be done with the use of oxalic acid (Matthias Ziegler et al., 2019). During cold storage, quality traits, and bioactive compounds of kiwifruit affected by aminoethoxy vinyl glycine (AVG) and modified atmosphere packaging (MAP) treatments. The firmness of fruit treated with MAP is higher than the AVG, however it results in lower respiration rate. The fruit treated with MAP has higher vitamin C content than the control. The fruit treated with AVG was observed with phenolic components. Hence, during cold storage, the MAP can be used to reduce the fruit quality losses and bioactive components of kiwifruit (Jian Zheng et al., 2019).

The psychrotrophs-removed commercial food products subjected to various temperature and volatile organic compounds show the prominent effect of genus pseudomonas on food products followed by Psychrobacter, Brochothrix, Serratia, and Stenotrophomonas. However, this dominance is based on the origin of the bacteria. The psychrotrophs grow in frozen food, livestock products, and aquatic products. The genus Pantoea grows in fruits and Pseudomonas present in cereal grain, dairy products, and bean (Burhan Ozturk et al., 2019). The fruit aroma decreases during cold storage and it is reported that Micro RNAs reduce the aroma of pear during cold storage (Yuxiang Zhang et al., 2019).

The heat stress applied during cold storage significantly inactivates the campylobacter jejuni. The consecutive stresses enhance the inactivation of C. jejuni during storage (Fei Shi et al., 2019). The fungal infection on grapes may decay it after harvesting. The study carried out with grapes dipped into water or kombucha for 15 min and at 4 °C in cold storage shows that the deterioration in fruit quality is reduced significantly and also delays the reduction in fruit soluble solid, vitamin C, and hardness, and inhibits the MDA accumulation. Hence, the kombucha application can inhibit the decay of table grapefruit after postharvest and also stored in cold storage unit (Benjamin Duqué et al., 2019).

The postharvest quality of various fruits is affected by the stages of fruit harvesting and storage in cold storage. The citric acid is the prominent individual organic acid in lemon fruit juice. During green to yellow maturity stage, levels of fumaric acid, malic acid, and citric acid can be reduced significantly during cold storage. The extension of cold storage period increases malic acid and reduces fumaric acid levels significantly. During cold storage and harvest maturity stages, the vitamin C content in the juice was not affected significantly. However, there is a reduction in mean total antioxidant capacity in the juice with increase in storage period. The lemon fruit harvested during its yellow and green stage can be stored for 30 days and 90 days respectively (Xian Zhou et al., 2019). The loquat fruit has lower shelf life at ambient condition and it is affected by chilling injury during cold storage. However, the coating of chitosan nano silica on loquat fruit improves the chilling tolerance and enhances the shelf life with slight variation in internal and external quality (Yongdong Sun et al., 2019). The microbial contamination and water losses were observed with minimum processed pomegranate, however the coating of combination of carboxymethyl cellulose and nano ZnO on pomegranate arils reduced the yeast and mold during cold storage of this fruit in 12 days. Also, it is observed that this coating reduces mesophilic bacteria level during cold storage period of six days. Also, the coatings reduce weight loss and soluble solids content during cold storage. The coated arils have higher vitamin C, antioxidant and total anthocyanin (Huwei Song et al., 2016).

CONCLUSION

Most of the underdeveloped and developing countries of the world are facing food crops wastage due to lack of cold storage facilities. The cold storage is the effective technique to preserve the vegetables, fruits, etc. for longer duration. Hence, cold storage is gaining popularity in various applications and is being installed in different locations. In recent years, energy efficiency cold storage units are being developed and used commercially. Various types of cold storage systems are available and selection of a particular type depends upon the demand of the consumers. The quality of the food stored in the cold storage is affected by the cold storage temperature and storage time. However, different techniques were used to reduce these problems. New types of cold storage units with phase change materials and nano particles are used in the cold storage to enhance the performance of the cold storage units.

REFERENCES

Allouche, Y., Varga, S., Bouden, C., & Oliveira, A. C. (2017). Dynamic simulation of an integrated solar-driven ejector-based air conditioning system with PCM cold storage. *Applied Energy*, *190*, 600–611. doi:10.1016/j.apenergy.2017.01.001

Bai, B., Zhao, K., & Li, X. (2019). Application research of nano-storage materials in cold chain logistics of e-commerce fresh agricultural products. *Results in Physics*, *13*, 102049. doi:10.1016/j.rinp.2019.01.083

Basu, D. N., & Ganguly, A. (2016). Solar thermal–photovoltaic powered potato cold storage – Conceptual design and performance analyses. *Applied Energy*, *165*, 308–317. doi:10.1016/j.apenergy.2015.12.070

Chen, X., Worall, M., Omer, S., Su, Y., & Riffat, S. (2014). Experimental investigation on PCM cold storage integrated with ejector cooling system. *Applied Thermal Engineering*, *63*(1), 419–427. doi:10.1016/j.applthermaleng.2013.11.029

Chen, Y., Sheng, L., Gouda, M., & Ma, M. (2019). Impact of ultrasound treatment on the foaming and physicochemical properties of egg white during cold storage. *Lwt*, *113*, 108303. doi:10.1016/j.lwt.2019.108303

Cheng, X., & Zhai, X. (2018). Thermal performance analysis of a cascaded cold storage unit using multiple PCMs. *Energy*, *143*, 448–457. doi:10.1016/j.energy.2017.11.009

Duqué, B., Haddad, N., Rossero, A., Membré, J.-M., & Guillou, S. (2019). Influence of cell history on the subsequent inactivation of Campylobacter jejuni during cold storage under modified atmosphere. *Food Microbiology*, *84*, 103263. doi:10.1016/j.fm.2019.103263 PMID:31421767

Fishman, M. L., Chau, H. K., Hotchkiss, A. T. Jr, White, A., Garcia, R. A., & Cooke, P. H. (2019). Effect of long-term cold storage and microwave extraction time on the physical and chemical properties of citrus pectin. *Food Hydrocolloids*, *92*, 104–116. doi:10.1016/j.foodhyd.2018.12.047

Gunther, N. W. IV, Abdul-Wakeel, A., Scullen, O. J., & Sommers, C. (2019). The evaluation of gamma irradiation and cold storage for the reduction of Campylobacter jejuni in chicken livers. *Food Microbiology*, *82*, 249–253. doi:10.1016/j.fm.2019.02.014 PMID:31027780

Hoel, S., Vadstein, O., & Jakobsen, A. N. (2018). Growth of mesophilic Aeromonas salmonicida in an experimental model of nigiri sushi during cold storage. *International Journal of Food Microbiology*, *285*, 1–6. doi:10.1016/j.ijfoodmicro.2018.07.008 PMID:30005315

http://agritech.tnau.ac.in/agricultural_marketing/agrimark_cold%20storage.html(browsed on 20-6-2019)

https://www.winnesota.com/news/coldstorage

https://www.winnesota.com/news/coldstorage(browsed on 20-6-2019)

James, S., & James, C. (2010). The food cold-chain and climate change. *Food Research International*, *43*(7), 1944–1956. doi:10.1016/j.foodres.2010.02.001

Jiang, J.-F., Li, S.-F., & Liu, Z.-H. (2018). Study on heat transfer and cold storage characteristics of a falling film type of cold energy regenerator with PCM. *Applied Thermal Engineering*, *143*, 676–687. doi:10.1016/j.applthermaleng.2018.07.127

Kozak, Y., Farid, M., & Ziskind, G. (2017). Experimental and comprehensive theoretical study of cold storage packages containing PCM. *Applied Thermal Engineering*, *115*, 899–912. doi:10.1016/j.applthermaleng.2016.12.127

Li, G., Hwang, Y., Radermacher, R., & Chun, H.-H. (2013). Review of cold storage materials for subzero applications. *Energy*, *51*, 1–17. doi:10.1016/j.energy.2012.12.002

Lu, W., Liu, G., Xing, X., & Wang, H. (2019). Investigation on ternary salt-water solutions as phase change materials for cold storage. *Energy Procedia*, *158*, 5020–5025. doi:10.1016/j.egypro.2019.01.662

Lv, Y., Tahir, I. I., & Olsson, M. E. (2019). Effect of ozone application on bioactive compounds of apple fruit during short-term cold storage. *Scientia Horticulturae, 253*, 49–60. doi:10.1016/j.scienta.2019.04.021

Ozturk, B., Uzun, S., & Karakaya, O. (2019). Combined effects of aminoethoxyvinylglycine and MAP on the fruit quality of kiwifruit during cold storage and shelf life. *Scientia Horticulturae, 251*, 209–214. doi:10.1016/j.scienta.2019.03.034

Saba, M. K., & Amini, R. (2017). Nano-ZnO/carboxymethyl cellulose-based active coating impact on ready-to-use pomegranate during cold storage. *Food Chemistry, 232*, 721–726. doi:10.1016/j.foodchem.2017.04.076 PMID:28490133

Saba, M. K., & Amini, R. (2017). Nano-ZnO/carboxymethyl cellulose-based active coating impact on ready-to-use pomegranate during cold storage. *Food Chemistry, 232*, 721–726. doi:10.1016/j.foodchem.2017.04.076 PMID:28490133

Sait, H. H. (2018). Design and analysis of a flooded tube for cold energy storage. *International Journal of Refrigeration, 94*, 151–160. doi:10.1016/j.ijrefrig.2018.07.018

Shaibani, A. R., Keshtkar, M. M., & Sardari, P. T. (2019). Thermo-economic analysis of a cold storage system in full and partial modes with two different scenarios: A case study. *Journal of Energy Storage, 24*, 100783. doi:10.1016/j.est.2019.100783

Sheng, L., Tsai, H.-C., Zhu, H., & Zhu, M.-J. (2019). Survival of Listeria monocytogenes on blueberries post-sanitizer treatments and subsequent cold storages. *Food Control, 100*, 138–143. doi:10.1016/j.foodcont.2019.01.019

Shi, F., Zhou, X., Yao, M.-M., Tan, Z., Zhou, Q., Zhang, L., & Ji, S.-J. (2018). miRNAs play important roles in aroma weakening during the shelf life of 'Nanguo' pear after cold storage. *Food Research International, 116*, 942-952. doi:10.1101/247932

Song, A.-X., Mao, Y.-H., Siu, K.-C., Tai, W. C. S., & Wu, J.-Y. (2019). Protective effects of exopolysaccharide of a medicinal fungus on probiotic bacteria during cold storage and simulated gastrointestinal conditions. *International Journal of Biological Macromolecules, 133*, 957–963. doi:10.1016/j.ijbiomac.2019.04.108 PMID:31028812

Song, H., Yuan, W., Jin, P., Wang, W., Wang, X., Yang, L., & Zhang, Y. (2016). Effects of chitosan/nano-silica on postharvest quality and antioxidant capacity of loquat fruit during cold storage. *Postharvest Biology and Technology, 119*, 41–48. doi:10.1016/j.postharvbio.2016.04.015

Stritih, U., & Butala, V. (2007). Energy saving in building with PCM cold storage. *International Journal of Energy Research*. doi:10.1002/er.1318

Sun, Y., Singh, Z., Tokala, V. Y., & Heather, B. (2019). Harvest maturity stage and cold storage period influence lemon fruit quality. *Scientia Horticulturae*, *249*, 322–328. doi:10.1016/j.scienta.2019.01.056

Talukdar, S., Afroz, H. M. M., Hossain, M. A., Aziz, M., & Hossain, M. M. (2019). Heat transfer enhancement of charging and discharging of phase change materials and size optimization of a latent thermal energy storage system for solar cold storage application. *Journal of Energy Storage*, *24*, 100797. doi:10.1016/j.est.2019.100797

Tian, S., Shao, S., & Liu, B. (2019). Investigation on transient energy consumption of cold storages: Modeling and a case study. *Energy*, *180*, 1–9. doi:10.1016/j.energy.2019.04.217

Wang, C., He, Z., Li, H., Wennerstern, R., & Sun, Q. (2017). Evaluation on performance of a phase change material based cold storage house. *Energy Procedia*, *105*, 3947–3952. doi:10.1016/j.egypro.2017.03.820

Wang, D., Jiang, J., Tao, L., Kou, Z., & Yao, L. (2017). Experimental investigation on a novel cold storage defrosting device based on electric heater and reverse cycle. *Applied Thermal Engineering*, *127*, 1267–1273. doi:10.1016/j.applthermaleng.2017.08.122

Wang, X., Dennis, M., Jiang, J., Zhou, L., Zhai, X., & Lipiński, W. (2019). Performance of a novel cold thermal storage material in an emulated air conditioning system using different storage strategies. *International Journal of Refrigeration*, *104*, 259–269. doi:10.1016/j.ijrefrig.2019.05.038

Wei, Q., Wang, X., Sun, D.-W., & Pu, H. (2019). Rapid detection and control of psychrotrophic microorganisms in cold storage foods: A review. *Trends in Food Science & Technology*, *86*, 453–464. doi:10.1016/j.tifs.2019.02.009

Zhang, Q., Rao, Y., Jiao, Y., Li, L., Li, Y., & Jin, L. (2017). preparation and performance of composite building materials with phase change material for thermal storage. *Energy Procedia*, *143*, 125–130. doi:10.1016/j.egypro.2017.12.659

Zhang, Y., Wei, J., Yuan, Y., & Yue, T. (2019). Diversity and characterization of spoilage-associated psychrotrophs in food in cold chain. *International Journal of Food Microbiology*, *290*, 86–95. doi:10.1016/j.ijfoodmicro.2018.09.026 PMID:30317110

Zheng, J., Li, S., Xu, Y., & Zheng, X. (2019). Effect of oxalic acid on edible quality of bamboo shoots (Phyllostachys prominens) without sheaths during cold storage. *Lwt*, *109*, 194–200. doi:10.1016/j.lwt.2019.04.014

Zheng, L., Zhang, W., & Liang, F. (2017). Experiment study on thermal conductivity of microcapsule phase change suspension applied to solar powered air conditioning cold storage system. *Procedia Engineering*, *205*, 1237–1244. doi:10.1016/j.proeng.2017.10.364

Zhou, X., Tan, J., Gou, Y., Liao, Y., Xu, F., Li, G., ... Chen, Z. (2019). The biocontrol of postharvest decay of table grape by the application of kombucha during cold storage. *Scientia Horticulturae*, *253*, 134–139. doi:10.1016/j.scienta.2019.04.025

Zhou, X., Zhou, D.-Y., Liu, Z.-Y., Yin, F.-W., Liu, Z.-Q., Li, D.-Y., & Shahidi, F. (2019). Hydrolysis and oxidation of lipids in mussel Mytilus edulis during cold storage. *Food Chemistry*, *272*, 109–116. doi:10.1016/j.foodchem.2018.08.019 PMID:30309519

Ziegler, M., Kent, D., Stephan, R., & Guldimann, C. (2019). Growth potential of Listeria monocytogenes in twelve different types of RTE salads: Impact of food matrix, storage temperature and storage time. *International Journal of Food Microbiology*, *296*, 83–92. doi:10.1016/j.ijfoodmicro.2019.01.016 PMID:30851644

Chapter 3
Non-Thermal Food Preservation Methods in the Meat Industry

Basak Gokce Col
https://orcid.org/0000-0002-7627-9867
Istanbul Gelisim University, Turkey

Sergen Tuggum
https://orcid.org/0000-0002-0519-7039
Tekirdag Namik Kemal University, Turkey

Seydi Yıkmış
https://orcid.org/0000-0001-8694-0658
Tekirdağ Namık Kemal University, Turkey

ABSTRACT

The most commonly used meat preservation methods include cooling, freezing, drying, vacuum packing, and curing. Meat quality is impaired by a wide range of changes including physical, chemical, microbiological, and enzymatic reactions. Food manufacturers focus on processes that require fewer chemical additives to meet the increased demand of consumers and to obtain more natural, healthy, and nutritious meat products. Non-thermal food preservation methods are one of the new trends to minimise thermal effects on texture, nutritional value, and flavor losses of meats. The chapter focuses on two novel approaches; non-thermal (Pulsed Electric Field) and Atmospheric Pressure Cold Plasma (APCP) Technologies.

DOI: 10.4018/978-1-7998-1924-0.ch003

INTRODUCTION

Containers or cases made of special materials such as metal, glass, plastic, which protect the products against external factors and facilitate the marketing and consumption of foods are referred to as food packages. The main purpose of food packaging is to ensure food safety by preserving the overall quality during the production, shelf life and consumption of products (Cutter, 2006).

The UK Packaging Institute defines packaging in three different ways (Gawith & Robertson, 2000):

1. Preparation of products for transportation, distribution, storage, retailing and final use in a coordinated manner,
2. Safe and cost-efficient delivery way of products to the final consumers,
3. Technological and economic function of the goal of minimizing delivery costs while maximizing sales and profits.

Food packaging is being developed day by day, upon the demands of the consumers and the novel trends applied in food industry. Four important functions should be considered when developing a food package: storage, protection, convenience and communication. In other words, package should be able to protect the product against external factors such as water, gas, odor, microorganisms, dust and pressure. They also have to contain information about the product and should be constantly improved to adapt to varying living conditions (Gawith & Robertson, 2000). Since milk and dairy products are particularly prone to physical, chemical and biological changes in a short time, the packaging technologies have been being developed in order to extend the shelf life of them.

The novel methods used in packaging technology can be listed as follows (Patel, Prajapati, & Balakrishnan, 2015):

1. Nanotechnology
2. Modified Atmosphere Packaging
3. Active Packaging
4. Intelligent/Smart Packaging

Nanotechnology

Nanotechnology is an applied science that provides the control of occurrences at atomic or molecular level below 100 nm (Anonymous, 2019). It is implemented in many food fields such as increasing food safety, reducing agricultural inputs and preventing the nutritional factors (Schnettler et al., 2013). In food science, food

Figure 1. The advantages and disadvantages of nanotechnological applications (Chau, Wu, & Yen, 2007; Buzby, 2010; Gruere, Narrod, & Abbott, 2011; Momin, Jayakumar, & Prajapati, 2013)

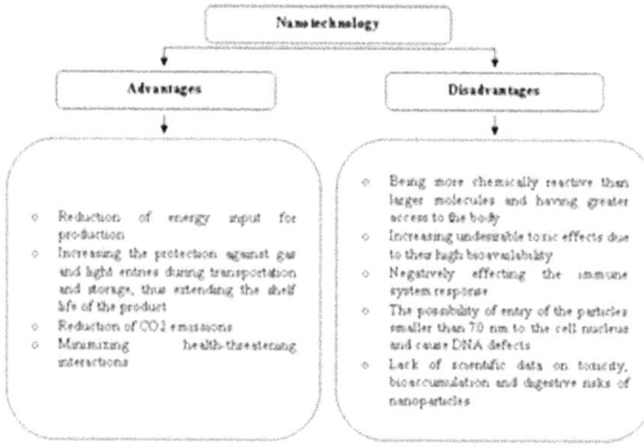

packaging is known as the most common field where the nanotechnology is applied (Sürengil & Kılınç, 2011) and dairy products are not exception.

Nanotechnology in food/dairy packaging can be used in three different ways (Duncan, 2011):

1. Producing synthetic polymer and biopolymer based packaging materials with improved barrier and mechanical properties.
2. Developing active packaging materials having properties antimicrobial or oxygen absorption such as Ag, ZnO, TiO_2.
3. Monitoring the storage conditions in which food products are exposed by use of different nanoparticles such as Fe_2O_3, TiO_2 in intellegent packaging technology and to produce markers that inform the manufacturer, seller and consumer.

Nanotechnological applications, in food/dairy technology, have various advantages and disadvantages as indicated in Figure 1.

In addition to the disadvantages mentioned in Figure 1, major concerns with the nanotechnological applications in food/dairy products are the lack of scientific data. Therefore, nanotechnological applications in food/dairy packaging are considered to be more reliable than the applications in food products.

In a survey that examined the consumers' view about nanotechnological applications in the food industry, panelists stated that they would prefer to buy neither nanotechnological foods nor the products packaged with nanotechnological treatments (Siegrist, Cousin, Kastenholz, & Wiek, 2007). However, consumers also believe that the usage of nanotechnology in food packaging is more beneficial than the usage in foods. Siegrist, Stampfli, Kastenholz, & Keller (2008) also investigated the preference of 337 consumers in 19 nanotechnology products. Results showed that nanotechnological packaging materials were determined to be more reliable by consumers than the foods produced by nanotechnological approaches (Siegrist et al., 2008).

Modified Atmosphere Packaging

In 1930s, the great losses caused by the rapid deterioration of the products in the transportation especially of fruits and vegetables, led to new searches to increase the shelf life of food products. When CO_2 was supplied to the food stores, it was seen that the products could remain stable for a longer period and this has been the starting point for the development of Modified Atmosphere Packaging (MAP) technology (Demir, 1999).

MAP is based on the principle of replacing the air in the food package with a mixture of different gas or gases. This change can be implemented in two different ways: active MAP and passive MAP technology. In the Active MAP technology, the air in the package is directly replaced with the desired gas or gases. However, passive MAP technology is based on the occurrence of naturally desired composition in the atmosphere of package over time, depending on the respiration of the food and the permeability of the special packaging material used (Lee, Arul, Lencki, & Castaigne, 1996).

Oxygen (O_2), nitrogen (N_2) and carbon dioxide (CO_2) gases are the main gases used in MAP technology. In addition to these gases, carbon monoxide, ozone, ethylene oxide, nitrous oxide, sulfur dioxide, helium, neon, argon, propylene oxide, ethanol, hydrogen and chlorine are the gases that can be used in MAP technology. However, they are not preferred because they are costly and impair the sensory properties of the products (Farber, 1991; Sivertsvik et al., 2002).

During the storage of the product, O_2 is consumed and CO_2 is produced by food respiration. N_2 is an inert gas used to prevent shrinkage of packages due to CO_2 absorption (Sandhya, 2010). N_2 is a tasteless and a low solubility gas that is also insoluble in water and oil, which causes its not to be absorbed by foods, including dairy products. Although this gas alone does not have any antimicrobial effect, it is indirectly inhibits the development of aerobic microorganisms when used instead of O_2 to retard the oxidative rancidity in oxygen sensitive products (Farber, 1991; Sivertsvik et al., 2002).

It is stated that the development of aerobic bacteria will be supported by using high amounts of O_2 in MAP technology, on one hand the reactions of enzymatic discoloration and anaerobic fermentation can be prevented on the other hand (Van der Steen, Jacxsens, Devlieghere, & Debevere, 2002). In MAP technology, O_2 is mainly used for packaging fresh red meat products, by this way meat maintains its bright red color (Farber, 1991).

CO_2, one of the main gases used in MAP technology, is the only gas with a significant direct antimicrobial effect. Easy solubility of CO_2 in water and oil readily inhibits the microbial growth, and affects the lag phase (negatively or positively) leads to, maximum growth rate and maximum population density of microorganisms (Devlieghere & Debevere, 2000). Although the mechanism of action of CO_2 application is not known exactly, this gas penetrates into microbial cell and decrease the pH of stoplazm hence inhibiting the growth of microorganisms. It is reported that the use of CO_2 also prevents the bad odor that may occur during storage and transportation of the product (Mullan, 2002). CO_2 is also known as having a prevention effect on food respiration (Farber, 1991).

The degree of main gases effect used in MAP technology varies depending on the type of microorganism, temperature, water activity and characteristics of product (Oliveira et al., 2015). Therefore, combinations and usage rates of double or triple gas mixtures used in packaging change according to the characteristics of the food/dairy product (such as pH, water activity, type and amount of fat) to be maintained during storage. In this way, the product is effectively protected against microbiological, chemical and enzymatic changes (Devlieghere, Gil, & Debevere, 2002). As a result, the shelf life of a food product packaged by using MAP technology depends on the type and initial quality of the product, storage temperature, gas mixture used inside the package, gas / product volume ratio and the preservation properties of the packaging materials (Sivertsvik et al., 2002; Sivertsvik, Rosnes, & Jeksrud, 2004).

As mentioned above, within the scope of MAP technology, different concentrations of gas mixtures are used during the packaging of different types of foods. If the deterioration is mostly due to microbial spoilage, such as in milk and dairy products, the most important deterioration parameter is generally regarded as high water activity. In this case, the CO_2 value of the gas mixture used in MAP application recommended to be high. In foods where oxidative rancidity is observed, such as milk and dairy products, all atmosphere in the package must be replaced with nitrogen gas or CO_2/N_2 mixture, and O_2 should not be used in the packaging. In addition to the gas mixture applied in MAP technology, the packaging material also has a crucial effect on the shelf life of foods. Packaging materials such as polyester, polyethylene, nylon, polyvinyldichloride and polypropylene are frequently used in atmosphere modified packaging (Sivertsvik et al., 2002; Kılınç & Çaklı, 2004). When low permeability packaging material is used, the gas remains in the package for a longer period of time and if the permeability of the package increases, this gas escapes and lose its protective effectiveness (Batu, 1994).

Spoilage in milk and dairy products varies according to product properties. For example, it is usual to observe mould in products with low water activity such as hard cheeses. However, mainly yeast, bacterial spoilage, oxidative rancidity and physical separation are observed in products with higher water activity such as soft type cheeses and cream. It is possible to extend shelf life of products by applying the MAP technology on milk and dairy products (Sivertsvik et al., 2002; Velu et al., 2013). It is stated that the shelf life of cheeses which usually can be remain undeteriorated in refrigerator conditions for 3 weeks, can be extended as far as 8 weeks (Anonymous 1998: Batu et al., 2008). Since, especially some types of moulds can grow despite vacuum packaging, application of MAP technology is more effective on milk and dairy products in which mould development is observed. Similarly, the existence of gases used in MAP technology is effective in terms of preservation of integrity of products such as sliced cheese (Taniwaki, Hocking, Pitt, & Fleet, 2001). It is suggested to use the 30/70 proportion of CO_2/N_2 for the sliced cheeses (Farber, 1991). CO_2/N_2 composition is effective on the prevention of mould development, and it is usually used in packaging of hard cheeses. CO_2 proportion used in this stage is desired to be at least 30%, while this proportion can increase as far as 70%. On the other hand, N_2 proportion is suggested to be in the range of 30-100% (Farber, 1991; Fierheller, 1991). Since this gas composition is effective on the prevention of bacterial deterioration and oxidative rancidity, it is similarly used in soft cheeses. MAP technology can also be applied easily on crumbly cheeses which can be deformed when vacuum packaging is applied. It is inconvenient to apply MAP technology on mould-ripened cheeses since the gas composition prevents the mold growth which is desired to occur in the product (Fierheller, 1991). This packaging method is also used on headspace of fresh milk products such as pasteurized milk, yoghurt, ice cream, cream and sour cream by way of CO_2 application in different amounts for each product type (Hotchkiss & Chen, 1996; Batu et al., 2008).

Active Packaging

Active packaging is considered as an alternative method to chemical preservatives and MAP technology. With this method, it is made possible to protect the product characteristics and to extend the shelf life of product by adding specific substances into headspace, inner side between the different layers or surface of packaging material (Labuza & Breene, 1989; Dobrucka & Cierpiszewski, 2014). Active packaging is a packaging type which has an extra function providing a protective barrier for foods against external factors according to European Union Guidance to the Commission Regulation (EUGCR). The package used in this type of packaging absorbs the unwanted components from food and headspace surrounding the food or on the contrary it enables the release of components, preservatives and antioxidants which are intended to be supplied to food or the air surrounding the food (Anonymous, 2009). The fact that developed packaging machinery and packaging material of high barrier characteristics are the important advantages of this method over MAP technology (Gutiérrez et al., 2011).

Different systems are used in active packaging technology depending on the food/dairy product, which can be evaluated under two main headings: 1. Active absorber-scavenger systems, 2. Active release-emitter systems (Üçüncü, 2011). The most commonly used active absorber-scavenger systems are oxygen, carbon dioxide, ethylene, moisture scavenger systems, and the most commonly used active release-emitter systems are carbon dioxide, ethanol, antioxidant and antimicrobial emitter systems (Hurme, Sipilainen-Malm, & R. Ahvenainen, 2002). More than 15 active packaging products have been being used commercially, especially in Japan, Australia and the United States, with the majority of oxygen scavengers. Among these products, oxygen scavengers are followed by ethylene scavengers, antimicrobial emitters and moisture scavengers (Biji, Ravishankar, Mohan, & Gopal, 2015).

Oxygen Scavengers

O_2 present in the packages of foods such as dairy products can not be completely removed even with vacuum packaging and MAP technology, thus causes an increase in microbial load, aroma losses, unwanted odor, color changes and nutrition losses in time. The amount of O_2 in the package can be reduced to 0.01% in active packaging systems which O_2 scavengers are used. For this purpose, O_2 scavengers such as iron powder, ascorbic acid, enzymes and photosensitive dyes are used to absorb O_2 in the pack after packaging (Hurme et al. 2002; Göncü & Özkal, 2017). Another method used in this system is to absorb O_2 in the package by the use of sachets and pads placed inside the packaging material (Huff, 2008). Using oxygen scavengers, it is possible to prevent rancidity, discoloration and especially mould growth in cheese during storage of milk and dairy products (Kartal, 2010).

Among the O_2-scavenging components, iron-based ones are the most commonly used ones (Vermeiren et al., 1999). In these systems, the iron-based component reacts with O_2 and is oxidized to the iron oxide after packaging. Iron-based O_2 scavengers are effective in many food/dairy products with low, medium and high moisture content (Özdemir & Floros, 2004). When used for cheese-spreads, Gomes et al. (2009) showed that the active packaging using iron-based oxygen scavengers decreased the rancidity of samples and that all the product properties were preserved for a longer time period.

In addition to iron-based components, some enzymes such as glucose oxidase are also used as active O_2 scavengers during active packaging (Vermeiren et al., 1999). In these systems, the enzymes react with the O_2 capture substrate present in the package, thereby reducing the amount of O_2 in the package. However, due to the enzymes used, these systems are more expensive than iron based O_2 scavenging systems. In addition to being costly, these systems are highly sensitive to other factors such as temperature, pH, water activity and solvent/substrate balance in the package, and thus its use is very limited (Özdemir & Floros, 2004).

Carbon Dioxide Scavengers and Emitters

The amount of CO_2 required in the packaging material varies entirely depending on the type of product in the package. In some products, the amount of CO_2 in the package is desired to decrease while in some others to increase. Therefore, in the active packaging technology, it is possible to use CO_2 scavenger or CO_2 emitting components depending on the purpose.

CO_2 is produced as a result of spoilage or respiration by some foods, and may result in reduction in shelf life of the product, deformation or even the explosion of the package (Vermeiren et al., 1999). The use of CO_2 scavengers such as silica gel, calcium hydroxide, sodium hydroxide, potassium hydroxide, calcium oxide is effective in such products (Fang, Zhao, Warner, & Johnson, 2017).

In some cases, it is aimed to increase the CO_2 level in the package. It is because CO_2 has a direct inhibitory effect on many aerobic bacteria and fungi, and this effect varies according to the type of microorganism. For example, it is possible to inhibit *Pseudomonas* spp. with about 20% CO_2 while only a small proportion of pathogens such as *Clostridium perfringens*, *C. botulinum* and *Listeria monocytogenes* are inhibited with the presence of less than 50% CO_2 (Fang et al., 2017). CO_2 level in the package should be adjusted according to the type of microorganism which has a risk of occurrence in the product, by using various components in the packages. In such cases, it is aimed to spontaneously consume the O_2 present in the package and to produce CO_2, after packaging. Iron carbonate and ascorbic acid/sodium bicarbonate mixtures are generally used for this purpose. In addition to the above-mentioned effects on microorganisms, O_2 scavengers or CO_2 emitters are also used in active packaging technology to maintain a high volume in food packages and to preserve the appearance of bulk packages (Vermeiren et al., 1999).

Ethylene Scavengers

The use of ethylene scavengers in milk and dairy products where active packaging technology is applied is not very common. Ethylene scavengers are more important for fruits and vegetables as ethylene acts like a plant hormone accelerates the respiration rate and ripening. Furthermore, it causes softening in some fruit varieties and leads to yellowing of green vegetables. Therefore, the removal of ethylene from the package atmosphere is very important for fruits and vegetables, and for this purpose various ethylene scavengers are used in the active packaging technology.

The most commonly used one is potassium permanganate and the minerals such as zeolite, silica gel and activated carbon are also used either alone or in combination to remove ethylene in the package (Vermeiren et al., 1999; Dainelli et al., 2008; Pereira de Abreu, Cruz, & Paseiro Losada, 2012) Ergun, 2016).

Moisture Absorbers

As in many foods, the water content is one of the most important causes for microbial spoilage in milk and dairy products. In the presence of high moisture, microbial growth is accelerated. In order to increase the shelf life of the product moisture content of the packages should be kept under control To do this, various moisture scavengers are used in the active packaging technology. Moisture absorbing pads and sheets are generally used where the moisture content is required to be kept under control., Active clay, silica gel and calcium oxide are commonly used components due to their high moisture-absorbing properties (Brody, Strupinsky, & Kline, 2001; Dobrucka, 2013; Biji et al., 2015).

Antimicrobial Emitters

Antimicrobial active packaging is a very important packaging method in the packaging of foods such as milk and dairy products containing many nutrients which are necessary for the development of microorganisms. The antimicrobial components used in this technique have an inhibitory effect on microorganisms by extending the lag phase and decreasing the expansion section of the microorganisms (Patel et al., 2015). The antimicrobial components are added to the packaging material or the product coating, and the mechanism of action of these components can be occur in two different ways. These components can migrate through food migration or prevent microbial growth that can be seen on the surface of the food without migration (Irkin & Esmer, 2015; Ergun, 2016). Examples of antimicrobial agents used in antimicrobial active packaging include alcohol, bacteriocin, nisin, natamycin and various metals such as silver, copper (Nicholson, 1998; Suppakul, Miltz, Sonneveld, & Bigger, 2003). In a study using antimicrobial active packaging in butter, it was determined that the product showed longer resistance to fungi and yeasts (Moraes et al., 2007).

Antioxidant Emitters

Lipid oxidation is one of the most important factors in shortening the shelf life of products such as milk and dairy products. The odor, aroma and color of the products change and also the nutritional value of food is lost with oxidation, due to the formation of toxic aldehydes and degradation of polyunsaturated fatty acid (PUFA) (Gomez-Estaca et al., 2014). To prevent the changes in the product occurred by oxidation and to prolong the shelf life of the foods, the antioxidants are usually incorporated into the packaging materials and these antioxidants pass into the food or air space around the food during storage. Waxed papers, butylated hydroxy toluene impregnated packaging materials and tocopherols, essential fatty acids and plant extracts obtained from plants such as rosemary, oregano and tea have been being used in antioxidant spreading systems. The effectiveness of α-tocopherol against lipid oxidation has already been shown in active packaged whole-fat milk Vitamins E and C appeared to be promising in these systems in recent years due to their antioxidative effects (Wessling, Nielsen, Leufvén, & Jägerstad, 1998; Tian, Decker, & Goddard, 2012; Yang, Lee, Won, & Song, 2016; Fang et al., 2017).

Ethanol Emitters

Ethanol emitters are especially used for active packaging of products with moderate moisture content against microbial spoilage and are particularly effective against mould growth. These components are also stated to be effective against yeasts and reduce staling and oxidative changes when used in high concentrations (Hurme et al. 2002; Suppakul et al., 2003; Dainelli et al., 2008). Ethanol emitters are reported to be particularly effective on bakery products (Dainelli et al., 2008).

The use of ethanol in active packaging technology can be applied in two ways: 1. Direct injection of ethanol into the package, 2. Use of packages with encapsulated ethanol that can release ethanol vapor (Hempel, O'Sullivan, Papkovsky, & Kerry, 2013).

Intelligent/Smart Packaging

Intelligent food packaging is defined as materials that can monitor the conditions of food and the surrounding environment. This technology, also known as smart labels, provides information to the producer, seller and consumer about changes in food during transport and storage (de Kruijf et al., 2002; Majid, Nayik, Dar, & Nanda, 2018). For this purpose, various sensors and indicators are integrated into the package or directly adhered to the package so that they provide information about some of the product characteristics such as freshness, shelf life and usage conditions. The working principle of these sensors and indicators is generally based on temperature-time measurements and the measurement of changes in chemical and microbiological properties (Yam, Takhistov, & Miltz, 2006; Dobrucka, 2013). Intelligent packaging technology provides useful information on the product in a short time, thus providing an alternative to time-consuming and costly analysis.

Sensors

Sensors are units that generally detect changes in the atmosphere of the product or packaging itself and transfer it to the manufacturer, seller and consumer. The sensors consist essentially of a receptor and a transducer. It is possible to examine the sensors used in smart packaging under different headings as biosensors, gas sensors, chemical sensors and pathogen sensors (Biji et al., 2015).

Biosensors

Devices that detect, record and transmit the biological reactions occur in food packages are called as biosensors. Biosensors, like other sensors, consist of a bioreceptor and energy converting devices (transducers). Bioreceptors are responsible for detecting the target parameter, usually organic materials such as various enzymes, antigens, hormones and nucleic acids. Transducers consist of electrochemical, optical or calorimetric systems that convert biological signals into measurable electrical messages (Smolander, 2003; Yam et al., 2006; Otles & Yalcin, 2008).

Gas Sensors

It is important that the gas mixture in the package does not change until it reaches to the consumer, particularly in products packaged using MAP systems, in order to maintain the quality of the product. Gas sensors are systems that detect and transmit the presence or absence of gases used in modified atmosphere packaging, packaging integrity and leaks (Otles & Yalcin, 2008; Robertson, 2012; Heising, Dekker, Bartels, & Van Boekel, 2014). O_2 and CO_2 sensors are the most commonly used gas sensors (Ergun, 2016).

Chemical Sensors

Chemical sensors detect the presence of a specific chemical or gas in the packaged product or in the headspace of the package. The substance detected by chemical sensors is converted into signals by transducers and enables the consumer to perceive the presence of the substance (Vanderroost, Ragaert, Devlieghere, & Meulenaer, 2014).

Pathogen Sensors

They are used for the detection of pathogenic microorganisms that infect the packaged products after production. Antibodies are generally used for this purpose and visual warning appears in the package as result of the reaction of these antibodies with the pathogens present in the product (Smolander, 2003).

Indicators

Indicators used in intelligent packaging technology indicate the presence, concentration or deficiency of a particular substance, particularly by color changes. They are generally classified as freshness indicators, time-temperature indicators (TTI), integrity indicators and Radio frequency information device (RFID) (Hogan & Kerry, 2008).

Freshness Indicators

The freshness indicators are generally used in products packaged with MAP technology, which informs the consumer on the change in gas composition in the package through the label printed on the package. If the necessary conditions are not met during the transportation and storage of foods, some microbiological spoilage may occur in the product and as a result of this deterioration, metabolites such as CO_2, SO_2, NH_3 and ethanol are formed. Freshness indicators identify these metabolites and change the color of the label on the packaging, thus providing information about the quality of food (Smolander, 2003; Smolander, 2008). When classified according to working principles, it is possible to list the most frequently used freshness indicators as indicators; sensitive to pH change, volatile compounds, hydrogen sulfide (H_2S) and various microbial metabolites (Smolander, 2003; Gök, 2007).

Time-Temperature Indicators (TTI)

Time-temperature indicators (TTI) detect physical, chemical, enzymatic and microbial deteriorations in the product when food product is exposed to temperatures that they should not be exposed to during transport and storage. They are particularly effective in controlling temperature changes in refrigerated or frozen foods such as milk and dairy products. In case of temperature change, the barcode on the packaging turns into a dark color, thus it prevents the data transfer when the barcode is scanned and prevents the sale of the product. It is possible to classify the most commonly used time-temperature indicators as polymer, diffusion and enzymatic based ones (Riva, Piergiovanni, & Schiraldi, 2001; Taoukis & Labuza, 2003; Gök, 2007; Lee & Rahman, 2014; Taoukis & Tsironi, 2016). Time-temperature indicators can be used for all types of food products, including milk and dairy products (Ergun, 2016).

Integrity Indicators

The package may be punctured or torn particularly during transport of the products. Such situations can be detected and transferred to the seller and the consumer through integrity indicators. Among the integrity indicators, the most widely used ones are based on O_2 measurement. Integrity indicators detect O_2 entering the package when the package is punctured or torn and stain the active substance with redox effect (Mattila-Sandholm, Ahvenainen, Hurme, & Jarvi-Kaariainen, 1995; Davies & Gardner, 1996), indicating the integrity of the packaging material has been impaired.

Radio Frequency Information Device (RFID)

Radio frequency information device (RFID) is an advanced technology that make possible to read labels using radio waves without human intervention. In this system, unlike the other methods described above, various changes in food are not detected by physical interaction, but by radio waves through the microchips inserted into the product (Tajima, 2007; Lee & Rahman, 2014). RFID tags carry the information of all the changes in the food starting from the packaging of the product and allow remote monitoring of this information via radio waves thereafter (Karagöz & Demirdöven, 2017). RFID indicators can be found in many different forms such as disk, glass, capsule and label can also be combined with other indicators and can be used for all foods including milk and dairy products (Yuksel & Zaim, 2009; Ruiz-Garcia & Lunadei, 2011).

REFERENCES

Anonymous. (2009). *EU Guidance to the commission regulation (EC) No 450/2009 of 29 May 2009 on active and intelligent materials and articles intended to come into contact with food. Version 1.0. European Commission Health and Consumers Directorate-General Directorate E-Safety of the Food chain. E6- Innovation and sustainability.* Retrieved from https://ec.europa.eu/food/sites/food/files/safety/docs/cs_fcm_legis_active-intelligent_guidance.pdf

Anonymous. (2019). *Nanotechnology.* European Food Safety Authority. Retrieved from https://www.efsa.europa.eu/en/topics/topic/nanotechnology

Batu, A. (1994). Properties of modified atmosphere packaging films and application of fruits vegetables. *Gida, 19*(6), 397–403.

Batu, A., Caglar, A., & Kara, H. H. (2008). Afyon kaymagının raf ömrünün uzatılmasında modifiye atmosferde paketleme önerisi. *Gıda Teknolojileri Elektronik Dergisi, 2008*(2), 43-46.

Biji, K. B., Ravishankar, C. N., Mohan, C. O., & Gopal, T. S. (2015). Smart packaging systems for food applications: A review. *Journal of Food Science and Technology, 52*(10), 6125–6135. doi:10.100713197-015-1766-7 PMID:26396360

Brody, A. L., Strupinsky, E. R., & Kline, L. R. (2001). *Active packaging for food applications*. Boca Raton, FL: CRC Press. doi:10.1201/9781420031812

Buzby, J. C. (2010). Nanotechnology for food applications. More questions than answers. *The Journal of Consumer Affairs, 44*(3), 528–545. doi:10.1111/j.1745-6606.2010.01182.x

Chau, C. F., Wu, S. H., & Yen, G. C. (2007). The development of regulations for food nanotechnology. *Trends in Food Science & Technology, 18*(5), 169–280. doi:10.1016/j.tifs.2007.01.007

Cutter, C. N. (2006). Opportunities for bio-based packaging technologies to improve the quality and safety of fresh and further processed muscle foods. *Meat Science, 74*(1), 131–142. doi:10.1016/j.meatsci.2006.04.023 PMID:22062722

Dainelli, D., Gontard, N., Spyropoulos, D., Zondervan-van den Beuken, E., & Tobback, P. (2008). Active and intelligent food packaging: Legal aspects and safety concerns. *Trends in Food Science & Technology, 19*, 103–112. doi:10.1016/j.tifs.2008.09.011

Davies, E. S., & Gardner, C. D. (1996). *UK Patent No. GB-2298273.* Oxygen indicating composition, The Victoria University of Manchester.

de Kruijf, N. N., van Beest, M., Rijk, R., Sipilainen-Malm, T., Paseiro, L. P., & De Meulenaer, B. (2002). Active and intelligent packaging: Applications and regulatory aspects. *Food Additives and Contaminants, 19,* Suppl, 144-162.

Demir, M. (1999). *Modified atmosphere packaging*. Retrieved from http://www.apack.com.tr/images/userfiles/1467054400909499 18.pdf

Devlieghere, F., & Debevere, J. (2000). Influence of dissolved carbon dioxide on the growth of spoilage bacteria. *Lebensmittel-Wissenschaft + Technologie, 33*(8), 531–537. doi:10.1006/fstl.2000.0705

Devlieghere, F., Gil, M. I., & Debevere, J. (2002). Modified atmosphere packaging (MAP). In C. J. K. Henry, & C. Chapman (Eds.), *The nutrition handbook for food processors* (pp. 342–370). England: Woodhead Publishing. doi:10.1533/9781855736658.2.342

Dobrucka, R. (2013). The future of active and intelligent packaging industry. *LogForum*, *9*(2), 103–110.

Dobrucka, R., & Cierpiszewski, R. (2014). Active and intelligent packaging food - Research and development - A review. *Polish Journal of Food and Nutrition Sciences*, *64*(1), 7–15. doi:10.2478/v10222-012-0091-3

Duncan, T. V. (2011). Applications of nanotechnology in food packaging and food safety: Barrier materials, antimicrobials, and sensors. *Journal of Colloid and Interface Science*, *363*(1), 1–24. doi:10.1016/j.jcis.2011.07.017 PMID:21824625

Ergun, M. (2016). Taze meyve ve sebzeler için aktif, zeki veya akıllı paketleme teknolojileri. *Alatarım*, *15*(2), 51–60.

Fang, Z., Zhao, Y., Warner, R. D., & Johnson, S. K. (2017). Active and intelligent packaging in meat industry. *Trends in Food Science & Technology*, *61*, 60–71. doi:10.1016/j.tifs.2017.01.002

Farber, J. M. (1991). Microbiological aspects of modified-atmosphere packaging technology - A review. *Journal of Food Protection*, *54*(1), 58–70. doi:10.4315/0362-028X-54.1.58 PMID:31051584

Fierheller, M. G. (1991). Modified atmosphere packaging of miscellaneous products. In B. Ooraikul, & M. E. Stiles (Eds.), *Modified atmosphere packaging of food* (pp. 246–260). Boston, MA: Springer. doi:10.1007/978-1-4615-2117-4_8

Gawith, J. A., & Robertson, T. R. (2000). Wrapping up packaging technology. *Journal of the Home Economics Institute of Australia*, *7*(2), 6–14.

Gök, V. (2007). Gıda paketleme sanayinde akıllı paketleme teknolojisi. *Gıda Teknolojileri Elektronik Dergisi, 2007*(1), 45-58.

Gomes, C., Elena Castell-Perez, M., Chimbombi, E., Barros, F., Sun, S., Liu, J. D., ... Wright, A. O. (2009). Effect of oxygen-absorbing packaging on the shelf life of a liquid-based component of military operational rations. *Journal of Food Science*, *74*(4), 167–176. doi:10.1111/j.1750-3841.2009.01120.x PMID:19490321

Gomez-Estaca, J., Lopez-de-Dicastillo, C., Hernandez-Munoz, P., Catala, R., & Gavara, R. (2014). Advances in antioxidant active food packaging. *Trends in Food Science & Technology*, *35*(1), 42–51. doi:10.1016/j.tifs.2013.10.008

Göncü, A., & Özkal, S. G. (2017). Ekmeklerde aktif paketleme uygulamaları. *Türk Tarım - Gıda Bilim ve Teknoloji Dergisi, 5*(11), 1264-1273.

Granda-Restrepo, D. M., Soto-Valdez, H., Peralta, E., Troncoso-Rojas, R., Vallejo-Córdoba, B., Gámez-Meza, N., & Graciano-Verdugo, A. Z. (2009). Migration of α-tocopherol from an active multilayer film into whole milk powder. *Food Research International*, *42*(10), 1396–1402. doi:10.1016/j.foodres.2009.07.007

Gruere, G. P., Narrod, C. A., & Abbott, L. (2011). Agriculture, food, and water nanotechnologies for the poor: Opportunities and constraints, Policy briefs 19, International Food Policy Research Institute (IFPRI). Retrieved from http://cdm15738.contentdm.oclc.org/utils/getfile/collection/p15738coll2/id/124891/filename/124892.pdf

Gutiérrez, L., Batlle, R., Andújar, S., Sánchez, C., & Nerín, C. (2011). Evaluation of antimicrobial active packaging to increase shelf life of gluten-free sliced bread. *Packaging Technology & Science*, *24*(8), 485–494. doi:10.1002/pts.956

Heising, J. K., Dekker, M., Bartels, P. V., & Van Boekel, M. A. J. S. (2014). Monitoring the quality of perishable foods: Opportunities for intelligent packaging. *Critical Reviews in Food Science and Nutrition*, *54*(5), 645–654. doi:10.1080/10408398.2011.600477 PMID:24261537

Hempel, A. W., O'Sullivan, M. G., Papkovsky, D. B., & Kerry, J. P. (2013). Use of smart packaging technologies for monitoring and extending the shelf-life quality of modified atmosphere packaged (MAP) bread: Application of intelligent oxygen sensors and active ethanol emitters. *European Food Research and Technology*, *237*(2), 117–124. doi:10.100700217-013-1968-z

Hogan, S. A., & Kerry, J. P. (2008). Smart packaging of meat and poultry products. In J. Kerry & P. Butler (Eds.), *Smart packaging technologies for fast moving consumer goods* (pp. 33–59). West Sussex: John Wiley & Sons. doi:10.1002/9780470753699.ch3

Hotchkiss, J. H., & Chen, J. H. (1996). Microbiological effects of the direct addition of CO_2 to pasteurized milk. *Journal of Dairy Science*, *79*(Supplement 1), 87. PMID:8675787

Huff, K. (2008). *Active and intelligent packaging: Innovations for the future*. Retrieved from http://www.iopp.org/files/public/VirginiaTechKarleighHuff.pdf

Hurme, E., Sipilainen-Malm, T., & Ahvenainen, R. V. T. T. (2002). Active and intelligent packaging. In T. Ohlsson, & N. Bengtsson (Eds.), Minimal processing technologies in the food industry (pp. 87-123). Cambridge: CRC Press.

Irkin, R., & Esmer, O. K. (2015). Novel food packaging systems with natural antimicrobial agents. *Journal of Food Science and Technology*, *52*(10), 6095–6111. doi:10.100713197-015-1780-9 PMID:26396358

Karagöz, Ş., & Demirdöven, A. (2017). Gıda ambalajlamada güncel uygulamalar: Modifiye atmosfer, aktif, akıllı ve nanoteknolojik ambalajlama uygulamaları. *Gaziosmanpaşa Bilimsel Araştırma Dergisi*, *6*(1), 9–21.

Kartal, S. (2010). *Çileğin raf ömrünün mikroperfore filmler ve oksijen tutucular kullanılarak denge modifiye atmosfer ile arttırılması*. (Unpublished master's thesis). Çanakkale Onsekiz Mart University, Çanakkale, Turkey.

Kılınç, B., & Çaklı, Ş. (2004). Su ürünlerinin modifiye atmosferde paketlenmesi. *E.U. Su Ürünleri Dergisi*, *21*(3-4), 349–353.

Labuza, T. P., & Breene, W. M. (1989). Applications of "active packaging" for improvement of shelf-life and nutritional quality of fresh and extended shelf-life foods. *Journal of Food Processing and Preservation*, *13*(1), 1–69. doi:10.1111/j.1745-4549.1989.tb00090.x

Lee, L., Arul, J., Lencki, R., & Castaigne, F. (1996). A review on modified atmosphere packaging and preservation of fresh fruits and vegetables: Physiological basis and practical aspects-Part II. *Packaging Technology & Science*, *9*(1), 1–17. doi:10.1002/(SICI)1099-1522(199601)9:1<1::AID-PTS349>3.0.CO;2-W

Lee, S. J., & Rahman, A. T. M. M. (2014). Intelligent packaging for food products. In J. H. Han (Ed.), *Innovations in food packaging* (pp. 171–209). Academic Press. doi:10.1016/B978-0-12-394601-0.00008-4

Majid, I., Nayik, G. A., Dar, S. M., & Nanda, V. (2018). Novel food packaging technologies: Innovations and future prospective. *Journal of the Saudi Society of Agricultural Sciences*, *17*(4), 454–462. doi:10.1016/j.jssas.2016.11.003

Mattila-Sandholm, T., Ahvenainen, R., Hurme, E., & Jarvi-Kaariainen, T. (1995). *Finnish Patent No. FI-94802*. Leakage Indicator, VTT Biotechnology and Food Research.

Momin, J. K., Jayakumar, C., & Prajapati, J. B. (2013). Potential of nanotechnology in functional foods. *Emirates Journal of Food and Agriculture*, *25*(1), 10–19. doi:10.9755/ejfa.v25i1.9368

Moraes, A. R. F., Gouveia, L. E. R., Soares, N. F. F., Santos, M. M. S., & Gonçalves, M. P. J. C. (2007). Development and evaluation of antimicrobial film on butter conservation. *Food Science and Technology (Campinas)*, *27*, 33–36. doi:10.1590/S0101-20612007000500006

Mullan, W. M. A. (2002). *Science and technology of modified atmosphere packaging*. Retrieved from https://www.dairyscience.info/index.php/packaging/117-modified-atmosphere-packaging.html

Nicholson, M. D. (1998). The role of natural antimicrobials in food/packaging biopreservation. *Journal of Plastic Film & Sheeting*, *14*(3), 234–241. doi:10.1177/875608799801400306

Oliveira, M., Abadias, M., Usall, J., Torres, R., Teixido, N., & Vinas, I. (2015). Application of modified atmosphere packaging as a safety approach to fresh-cut fruits and vegetables - A review. *Trends in Food Science & Technology*, *46*(1), 13–26. doi:10.1016/j.tifs.2015.07.017

Otles, S., & Yalcin, B. (2008). Intelligent food packaging. *LogForum*, *4*(3), 1–9.

Özdemir, M., & Floros, J. D. (2004). Active food packaging technologies. *Critical Reviews in Food Science and Nutrition*, *44*(3), 185–193. doi:10.1080/10408690490441578 PMID:15239372

Patel, R., Prajapati, J. P., & Balakrishnan, S. (2015). *Recent trends in packaging of dairy and food products*. Paper presented at the meeting National seminar on Indian Dairy Industry - Opportunities and Challenges. Gujarat, India.

Pereira de Abreu, D. A., Cruz, J. M., & Paseiro Losada, P. (2012). Active and intelligent packaging for the food industry. *Food Reviews International*, *28*(2), 146–187. doi:10.1080/87559129.2011.595022

Riva, M., Piergiovanni, L., & Schiraldi, A. (2001). Performances of time–temperature indicators in the study of temperature exposure of packaged fresh food. *Packaging Technology & Science*, *14*(1), 1–9. doi:10.1002/pts.521

Robertson, G. L. (2012). *Food packaging, principles and practice*. London, UK: CRC Press.

Ruiz-Garcia, L., & Lunadai, L. (2011). The role of RFID in agriculture: Applications, limitations, and challenges. *Computers and Electronics in Agriculture*, *79*(1), 42–50. doi:10.1016/j.compag.2011.08.010

Sandhya. (2010). Modified atmosphere packaging of fresh produce: Current status and future needs. *Food Science and Technology, 43*, 381-392.

Schnettler, B., Crisostomo, G., Mora, M., Lobos, G., Miranda, H., & Grunert, K. G. (2013). Acceptance of nanotechnology applications and satisfaction with food-related life in southern Chile. *Food Science and Technology (Campinas)*, *34*(1), 157–163. doi:10.1590/S0101-20612014005000001

Siegrist, M., Cousin, M. E., Kastenholz, H., & Wiek, A. (2007). Public acceptance of nanotechnology foods and food packaging: The influence of affect and trust. *Appetite, 49*(2), 459–466. doi:10.1016/j.appet.2007.03.002 PMID:17442455

Siegrist, M., Stampfli, N., Kastenholz, H., & Keller, C. (2008). Perceived risks and perceived benefits of different nanotechnology foods and nanotechnology food packaging. *Appetite, 51*(2), 283–290. doi:10.1016/j.appet.2008.02.020 PMID:18406006

Sivertsvik, M., Rosnes, J. T., & Bergslien, H. (2002). Modified atmosphere packaging. In T. Ohlsson & N. Bengtsson (Eds.), *Minimal processing technologies in the food industry* (pp. 61–86). Cambridge: CRC Press. doi:10.1533/9781855736795.61

Sivertsvik, M., Rosnes, J. T., & Jeksrud, W. K. (2004). Solubility and absorption rate of carbon dioxide into non-respiring foods. Part 2: Raw fish fillets. *Journal of Food Engineering, 63*(4), 451–458. doi:10.1016/j.jfoodeng.2003.09.004

Smolander, M. (2003). The use of freshness ındicators in packaging. In R. Ahvenainen (Ed.), *Novel food packaging techniques* (pp. 127–143). Cambridge: Woodhead Publishing. doi:10.1533/9781855737020.1.127

Smolander, M. (2008). Freshness indicators and food packaging. In J. Kerry, & P. Butler (Eds.), *Smart packaging technologies for fast moving consumer goods* (pp. 111–127). West Sussex: John Wiley & Sons. doi:10.1002/9780470753699.ch7

Suppakul, P., Miltz, J., Sonneveld, K., & Bigger, S. W. (2003). Active packaging technologies with an emphasis on antimicrobial packaging and its applications. *Journal of Food Science, 68*(2), 408–420. doi:10.1111/j.1365-2621.2003.tb05687.x

Sürengil, G., & Kılınç, B. (2011). Gıda - Ambalaj sektöründe nanoteknolojik uygulamalar ve su ürünleri açısından önemi. *Journal of Fisheries Sciences Com., 5*(4), 317–325.

Tajima, M. (2007). Strategic value of RFID in supply chain management. *Journal of Purchasing and Supply Management, 13*(4), 261–273. doi:10.1016/j.pursup.2007.11.001

Taniwaki, M. H., Hocking, A. D., Pitt, J. I., & Fleet, G. H. (2001). Growth of fungi and mycotoksin production on cheese under modified atmospheres. *International Journal of Food Microbiology, 68*(1-2), 125–133. doi:10.1016/S0168-1605(01)00487-1 PMID:11545212

Taoukis, P., & Tsironi, T. (2016). Smart packaging for monitoring and managing food and beverage shelf life. In P. Subramaniam, & P. Wareing (Eds.), *The stability and shelf life of food* (pp. 141–168). Woodhead Publishing. doi:10.1016/B978-0-08-100435-7.00005-8

Taoukis, P. S., & Labuza, T. P. (2003). Time-temperature indicators. In R. Ahvenainen (Ed.), *Novel food packaging techniques* (pp. 103–126). Cambridge: Woodhead Publishing. doi:10.1533/9781855737020.1.103

Tian, F., Decker, E. A., & Goddard, J. M. (2012). Development of an iron chelating polyethylene film for active packaging applications. *Journal of Agricultural and Food Chemistry*, *60*(8), 2046–2052. doi:10.1021/jf204585f PMID:22288894

Üçüncü, M. (2011). *Gıda ambalajlanma teknolojisi*. İstanbul, Turkey: Ambalaj Sanayiciler Derneği.

Van der Steen, C., Jacxsens, L., Devlieghere, F., & Debevere, J. (2002). Combining high oxygen atmospheres with low oxygen modified atmosphere packaging to improve the keeping quality of strawberries and raspberries. *Postharvest Biology and Technology*, *26*(1), 49–58. doi:10.1016/S0925-5214(02)00005-4

Vanderroost, M., Ragaert, P., Devlieghere, F., & Meulenaer, B. D. (2014). Intelligent food packaging: The next generation. *Trends in Food Science & Technology*, *39*(1), 47–62. doi:10.1016/j.tifs.2014.06.009

Velu, S., Abu Bakar, F., Mahyudin, N. A., Saari, N., & Zaman, M. Z. (2013). Effect of modified atmosphere packaging on microbial flora changes in fishery products. *International Food Research Journal*, *20*(1), 17–26.

Vermeiren, L., Devlieghere, F., Van Beest, M., De Kruijf, N., & Debevere, J. (1999). Developments in the active packaging of foods. *Trends in Food Science & Technology*, *10*(3), 77–86. doi:10.1016/S0924-2244(99)00032-1

Wessling, C., Nielsen, T., Leufvén, A., & Jägerstad, M. (1998). Mobility of α-tocopherol and BHT in LDPE in contact with fatty food simulants. *Food Additives and Contaminants*, *15*(6), 709–715. doi:10.1080/02652039809374701 PMID:10209582

Yam, K. L., Takhistov, P. T., & Miltz, J. (2006). Intelligent packaging: concepts and applications. *Journal of Food Science*, *70*(1), 1–10. doi:10.1111/j.1365-2621.2005.tb09052.x

Yang, H. J., Lee, J. H., Won, M., & Song, K. B. (2016). Antioxidant activities of distiller dried grains with solubles as protein films containing tea extracts and their application in the packaging of pork meat. *Food Chemistry*, *196*, 174–179. doi:10.1016/j.foodchem.2015.09.020 PMID:26593480

Yüksel, M. E., & Zaim, A. H. (2009). *Yeni nesil teknoloji olarak RFID, RFID sistem yapıları ve bir RFID sistem tasarımı yaklaşımı.* Paper presented at the meeting 5th International Advanced Technologies Symposium. Karabük, Turkey.

Chapter 4
Novel Packaging Technologies in Dairy Products:
Principles and Recent Advances

Nazli Turkmen
Ankara University, Turkey

Sebnem Ozturkoglu-Budak
Ankara University, Turkey

ABSTRACT

The packaging process is an important step in maintaining the quality characteristics of foods. Packaging foods protects products from external effects and provides product information to consumers. Due to the various changes occurring during the distribution and storage of the products, some significant quality characteristics can be lost. In recent years, novel packaging technologies have been developed to supply long shelf life, safety, and 'fresh-like' characteristics to the products. These novel technologies include nanotechnology, modified atmosphere packaging, active packaging, and intelligent/smart packaging. Since dairy products are generally vulnerable to biological, physical, and chemical changes, they lose their quality characteristics within a short term. Therefore, the use of these novel techniques in dairy products is greatly important. This chapter informs about general principles of the novel packaging techniques and their current applications in dairy technology.

INTRODUCTION

Containers or cases made of special materials such as metal, glass, and plastic

DOI: 10.4018/978-1-7998-1924-0.ch004

which protect the products against external factors and facilitate the marketing and consumption of foods are called food packaging. To ensure the food safety, the stages of packaging, transportation, and storage until the product reaches consumers are as important as the production of foods (Çelik & Tümer, 2016). The main purpose of food packaging is to ensure food safety by preserving the overall quality throughout the time interval of production and consumption (Cutter, 2006).

The UK Packaging Institute defines packaging in three different ways (Gawith & Robertson, 2000):

1. Preparation of products for transportation, distribution, storage, retailing, and final use in a coordinated manner.
2. Safe and cost-efficient delivery way of products to the final consumers.
3. Technological and economic function of the goal of minimizing delivery costs while maximizing sales and profits.

Food packaging is being developed day by day upon the demands of the consumers and the novel trends applied in food industry. There are four important functions to be considered when developing a food package: storage, protection, convenience, and communication. In other words, package should be able to protect the product against external factors such as water, gas, odor, microorganisms, dust, and pressure. In addition, food packages need to contain information about the product and should be constantly improved to adapt to varying living conditions (Gawith & Robertson, 2000). Particularly in milk and dairy products which are prone to physical, chemical, and biological changes in a short time, the packaging technologies have been constantly developed io extend the shelf life of products.

The novel methods used in packaging technology can be listed as follows (Patel, Prajapati, & Balakrishnan, 2015):

1. Nanotechnology
2. Modified Atmosphere Packaging
3. Active Packaging
4. Intelligent/Smart Packaging

NANOTECHNOLOGY

Nanotechnology is an applied science that controls occurrences at atomic or molecular level below 100 nm (Anonymous, 2019). Nanotechnology is implemented in many food fields such as increasing food safety, reducing agricultural inputs, and preventing the nutritional factors (Schnettler et al., 2013). In food science, food packaging is known as the most common field where nanotechnology is applied (Sürengil & Kılınç, 2011).

Nanotechnology can be used in food packaging in three ways (Duncan, 2011):

1. To produce synthetic polymer and biopolymer-based packaging materials for the purpose of developing packages with better barrier and mechanical properties.
2. To develop active packaging materials by using nanoparticles having antimicrobial properties or oxygen absorption such as Ag, ZnO, and TiO_2.
3. To detect the storage conditions in which food products are exposed by use of different nanoparticles such as Fe_2O_3 and TiO_2 in intelligent packaging technology and to produce markers that inform the manufacturer, seller, and consumer.

Nanotechnological applications, which are used in many fields in food technology, have advantages and disadvantages.

The disadvantages and the concerns arising from the lack of scientific data on the subject are mostly related to nanotechnological applications in food products. Therefore, nanotechnological applications in food packaging are considered to be more reliable than the applications in food products.

In a survey that examined the consumers' view about nanotechnological applications in the food industry, panelists stated they would prefer to buy neither nanotechnological foods nor the products packaged with nanotechnological treatments (Siegrist, Cousin, Kastenholz, & Wiek, 2007). However, consumers also believe that the usage of nanotechnology in food packaging is more beneficial than the usage in foods. Siegrist, Stampfli, Kastenholz, & Keller (2008) also investigated the preference of 337 consumers in 19 nanotechnology products. Results showed that nanotechnological packaging materials were determined to be more reliable by consumers than the foods produced by nanotechnological approaches (Siegrist et al., 2008).

Modified Atmosphere Packaging

In the 1930s, the great losses caused by the rapid deterioration of the products in the transportation, especially of fruits and vegetables, led to new searches to increase the shelf life of food products. When CO_2 was supplied to the stores where food products are stored, it was seen that the products could remain stable for a longer period and this has been the starting point for the development of Modified Atmosphere Packaging (MAP) technology (Demir, 1999).

MAP is based on the principle of replacing the air in the food package with a mixture of different gas or gases. This change can be implemented in two ways: active MAP and passive MAP technology. In the Active MAP technology, the air in the package is directly replaced with the desired gas or gases. However, passive MAP technology is based on the occurrence of naturally desired composition in the atmosphere of package over time, depending on the respiration of the food and the permeability of the special packaging material used (Lee, Arul, Lencki, & Castaigne, 1996).

Oxygen (O_2), nitrogen (N_2) and carbon dioxide (CO_2) gases are the main gases used in MAP technology. In addition to these gases, carbon monoxide, ozone, ethylene oxide, nitrous oxide, sulfur dioxide, helium, neon, argon, propylene oxide, ethanol, hydrogen, and chlorine are the gases that can be used in MAP technology. However, they are not preferred because they are costly and impair the sensory properties of the products (Farber, 1991; Sivertsvik et al., 2002).

During the storage of the product, O_2 is consumed and CO_2 is produced by food respiration. N_2 is an inert gas used to prevent shrinkage of packages due to CO_2 absorption (Sandhya, 2010). N_2 is a tasteless and a low solubility gas that is also insoluble in water and oil, which causes it not to be absorbed by foods. Although this gas alone does not have any antimicrobial effect, it indirectly inhibits the development of aerobic microorganisms when used instead of O_2 to delay the oxidative rancidity in oxygen sensitive products (Farber, 1991; Sivertsvik et al., 2002).

It is stated that the development of aerobic bacteria will be supported by using high amounts of O_2 in MAP technology. On the contrary, the reactions of enzymatic discoloration and anaerobic fermentation can be prevented (Van der Steen, Jacxsens, Devlieghere, & Debevere, 2002). In MAP technology, O_2 is mainly used for packaging fresh red meat products so they maintain their bright red color (Farber, 1991).

CO_2, one of the main gases used in MAP technology, is the only gas with a significant direct antimicrobial effect. Easy solubility of CO_2 in water and oil causes the inhibition of microbial growth and also affects the lag phase, maximum growth rate, and maximum population density of microorganisms (Devlieghere & Debevere, 2000). Although the mechanism of action of CO_2 application is not known exactly, this gas penetrates into microbial cell and decreases the pH of microorganisms which inhibits the growth of microorganisms. Using CO_2 also prevents the bad odor that may occur during storage and transportation of the product (Mullan, 2002). CO_2 is also known to have a prevention effect on food respiration (Farber, 1991).

The degree of main gases effect used in MAP technology varies depending on the type of microorganism, temperature, water activity, and characteristics of product (Oliveira et al., 2015). Therefore, combinations and usage rates of double or triple gas mixtures used in packaging change according to the characteristics of the food product (such as pH, water activity, type, and amount of fat) to be maintained during storage. In this way, the product is effectively protected against microbiological, chemical, and enzymatic changes (Devlieghere, Gil, & Debevere, 2002). As a result, the shelf life of a food product packaged by using MAP technology depends on the type and initial quality of the product, storage temperature, gas mixture used inside the package, gas / product volume ratio and the preservation properties of the packaging materials (Sivertsvik et al., 2002; Sivertsvik, Rosnes, & Jeksrud, 2004).

As mentioned above, within the scope of MAP technology, different concentrations of gas mixtures are used during the packaging of different types of foods. If the deterioration is mostly due to microbial spoilage, such as in milk and dairy products, the most important deterioration parameter is generally regarded as high water activity. In this case, the CO_2 value of the gas mixture used in MAP application is recommended to be high. In foods where oxidative rancidity is observed, such as milk and dairy products, all atmosphere in the package must be replaced with nitrogen gas or CO_2/N_2 mixture, and O_2 should not be used in the packaging. In addition to the gas mixture applied in MAP technology, the packaging material also has a crucial effect on the shelf life of foods. Packaging materials such as polyester, polyethylene, nylon, polyvinyldichloride and polypropylene are frequently used in atmosphere modified packaging (Sivertsvik et al., 2002; Kılınç & Çaklı, 2004). When low permeability packaging material is used, the gas remains in the package for a longer period of time and if the permeability of the package increases, this gas escapes and loses its protective effectiveness (Batu, 1994).

Spoilage in milk and dairy products varies according to product properties. For example, it is usual to observe mould in products with low water activity such as hard cheeses. However, mainly yeast, bacterial spoilage, oxidative rancidity, and physical separation are observed in products with higher water activity such as soft type cheeses and cream. It is possible to extend shelf life of products by applying the MAP technology on milk and dairy products (Sivertsvik et al., 2002; Velu et al., 2013). It is stated that the shelf life of cheeses which usually can remain undeteriorated in refrigerator conditions for three weeks, can be extended as far as eight weeks (Anonymous 1998: Batu et al., 2008). Since some types of moulds can grow despite vacuum packaging, application of MAP technology is more effective on milk and dairy products in which mould development is observed. Similarly, the existence of gases used in MAP technology is effective in terms of preservation of integrity of products such as sliced cheese (Taniwaki, Hocking, Pitt, & Fleet, 2001). It is suggested to use the 30/70 proportion of CO_2/N_2 for the sliced cheeses (Farber, 1991). CO_2/N_2 composition is effective on the prevention of mould development, and it is usually used in packaging of hard cheeses. CO_2 proportion used in this stage is desired to be at least 30%, while this proportion can increase as far as 70%. On the other hand, N_2 proportion is suggested to be in the range of 30-100% (Farber, 1991; Fierheller, 1991). Since this gas composition is effective on the prevention of bacterial deterioration and oxidative rancidity, it is similarly used in soft cheeses. MAP technology can also be applied easily on crumbly cheeses which can be deformed when vacuum packaging is applied. It is inconvenient to apply MAP technology on mould-ripened cheeses since the gas composition prevents the mold growth which is desired to occur in the product (Fierheller, 1991). This packaging method is also used on headspace of fresh milk products such as pasteurized milk, yoghurt, ice cream, cream, and sour cream by way of CO_2 application in different amounts for each product type (Hotchkiss & Chen, 1996; Batu et al., 2008).

Active Packaging

Active packaging is accepted as an alternative method to chemical preservatives and MAP technology, which are used for extending the shelf life of foods. With this method, it is made possible to protect the product characteristics and to extend the shelf life of product by specific substances added into headspace, inner side between the different layers or surface of packaging material (Labuza & Breene, 1989; Dobrucka & Cierpiszewski, 2014). Active packaging is a packaging type which has an extra function providing a protective barrier for foods against external factors according to European Union Guidance to the Commission Regulation (EUGCR). The package used in this type of packaging absorbs the unwanted components from food and headspace surrounding the food or on the contrary it enables the release of components, preservatives, and antioxidants which are intended to be supplied to food or the air surrounding the food (Anonymous, 2009). Developed packaging machinery and packaging material of high barrier characteristics are the important advantages of this method over MAP technology (Gutiérrez et al., 2011).

Different systems are used in active packaging technology depending on the product to be protected. These systems can be evaluated under two main headings: 1. Active absorber-scavenger systems, 2. Active release-emitter systems (Üçüncü, 2011). The most commonly used active absorber-scavenger systems are oxygen, carbon dioxide, ethylene, moisture scavenger systems, and the most commonly used active release-emitter systems are carbon dioxide, ethanol, antioxidant, and antimicrobial emitter systems (Hurme, Sipilainen-Malm, & R. Ahvenainen, 2002). More than 15 active packaging products have been used commercially, especially in Japan, Australia, and the United States, with the majority of oxygen scavengers. Among these products, oxygen scavengers are followed by ethylene scavengers, antimicrobial emitters, and moisture scavengers (Biji, Ravishankar, Mohan, & Gopal, 2015).

Oxygen Scavengers

O_2 present in the packages of foods such as dairy products cannot be completely removed even with vacuum packaging and MAP technology, thus it causes an increase in microbial load, aroma losses, unwanted odor, color changes, and nutrition losses in time. The amount of O_2 in the package can be reduced to 0.01% in active packaging systems which O_2 scavengers are used. For this purpose, O_2 scavengers such as iron powder, ascorbic acid, enzymes, and photosensitive dyes are used to absorb O_2 in the pack after packaging (Hurme et al., 2002; Göncü & Özkal, 2017). Another method used in this system is to absorb O_2 in the package by the use of sachets and pads placed inside the packaging material (Huff, 2008). It is possible to prevent rancidity, discoloration and especially mould growth in cheese during storage of milk and dairy products (Kartal, 2010).

Among the O_2-scavenging components, iron-based ones are the most commonly used ones (Vermeiren et al., 1999). In these systems, the iron-based component used during packaging reacts using the O_2 contained in the packaging and is oxidized to the iron oxide after packaging. Iron-based O_2 scavengers are effective in many food products with low, medium, and high moisture content (Özdemir & Floros, 2004). In a study performed by Gomes et al. (2009) active packaging was made for cheese-spread samples using iron-based oxygen scavengers and it was observed that rancidity of samples reduces and all the product properties were preserved for a longer time period.

In addition to iron-based components, some enzymes such as glucose oxidase are also used as active O_2 scavengers during active packaging (Vermeiren et al., 1999). In enzymatic O_2 capture systems, the enzymes react with the O_2 capture substrate present in the package, thereby reducing the amount of O_2 in the package. However, due to the enzymes used, these systems are more expensive than iron-based O_2 scavenging systems. In addition to being costly, these systems are highly sensitive to factors such as temperature, pH, water activity, and solvent/substrate balance in the package, and the use of enzyme-based O_2 scavenging systems is very limited (Özdemir & Floros, 2004).

Carbon Dioxide Scavengers and Emitters

The amount of CO_2 required in the packaging material varies entirely depending on the type of product in the package. In some products, the amount of CO_2 in the package is desired to decrease and in some others it is desired to increase. Therefore, in the active packaging technology, it is possible to use CO_2 scavenger or CO_2 emitting components depending on the purpose.

CO_2 occurs as a result of spoilage and respiration in some foods, and may result in reduced shelf life of the product, deformation of the package, or even explosion (Vermeiren et al., 1999). The use of CO_2 scavengers such as silica gel, calcium hydroxide, sodium hydroxide, potassium hydroxide, and calcium oxide is effective in such products (Fang, Zhao, Warner, & Johnson, 2017).

In some cases, it is aimed to increase the amount of CO_2 in the package because CO_2 has a direct inhibitory effect on many aerobic bacteria and fungi. The rate of this effect varies according to the type of microorganism. For example, while it is possible to inhibit *Pseudomonas* spp. in the presence of about 20% CO_2, only a small proportion of pathogens such as *Clostridium perfringens*, *C. botulinum*, and *Listeria monocytogenes* are inhibited with the presence of less than 50% CO_2 (Fang et al., 2017). It is possible to adjust the amount of CO_2 in the package to desired ratio, according to the type of microorganism which has a risk of occurrence in the product, by using various components in the packages. In such cases, it is aimed to spontaneously consume the O_2 in the package and to produce CO_2, after packaging. Iron carbonate and ascorbic acid/sodium bicarbonate mixtures are generally used for this purpose. In addition to the above-mentioned effects on microorganisms, O_2 scavengers or CO_2 emitters are also used in active packaging technology to increase volume in food packages and to preserve the appearance of bulk packages (Vermeiren et al., 1999).

Ethylene Scavengers

Ethylene has similar effects on fruits and vegetables as a plant hormone, and accelerates respiration and provides ripening. Furthermore, it causes softening in some fruit varieties and leads to yellowing of green vegetables. Therefore, the removal of ethylene from the package atmosphere is very important for fruits and vegetables, and for this purpose various ethylene scavengers are used in the active packaging technology. The most commonly used one is potassium permanganate and the minerals such as zeolite, silica gel, and activated carbon are also used either alone or in combination to remove ethylene in the package (Vermeiren et al., 1999; Dainelli et al., 2008; Pereira de Abreu, Cruz, & Paseiro Losada, 2012; Ergun, 2016). However, the use of ethylene scavengers in milk and dairy products where active packaging technology is applied is not very common.

Moisture Absorbers

As in many foods, high humidity is one of the most important causes of microbial spoilage in milk and dairy products. Since water activity is high in the presence of moisture, microbial growth accelerates in such environments. To increase the shelf life of the product by keeping the moisture content of the packages under control, various moisture scavengers are used in the active packaging technology. Moisture absorbing pads and sheets are generally used in the active packaging of the products in which moisture content is desired to be kept under control. Moreover, active clay, silica gel, and calcium oxide substances are commonly used components due to their moisture-absorbing properties (Brody, Strupinsky, & Kline, 2001; Dobrucka, 2013; Biji et al., 2015).

Antimicrobial Emitters

Antimicrobial active packaging is a very important packaging method in the packaging of foods such as milk and dairy products containing many nutrients which are necessary for the development of microorganisms. The antimicrobial components used in this technique have an inhibitory effect on microorganisms by extending the lag phase and decreasing the expansion section of the microorganisms (Patel et al., 2015). The antimicrobial components are added to the packaging material or the product coating, and the mechanism of action of these components can occur in two ways. These components can migrate through food migration or prevent microbial growth that can be seen on the surface of the food without migration (Irkin & Esmer, 2015; Ergun, 2016). Examples of antimicrobial agents used in antimicrobial active packaging include alcohol, bacteriocin, nisin, natamycin, and various metals such as silver and copper (Nicholson, 1998; Suppakul, Miltz, Sonneveld, & Bigger, 2003). In a study using antimicrobial active packaging in butter, it was determined that the product showed longer resistance to fungi and yeasts (Moraes et al., 2007).

Antioxidant Emitters

Lipid oxidation is one of the most important factors which shortens the shelf life of products such as milk and dairy products. The odor, aroma, and color of the products change and also the nutritional value of food is lost with oxidation, due to the formation of toxic aldehydes and degradation of polyunsaturated fatty acids (PUFA) (Gomez-Estaca et al., 2014). To prevent the changes in the product occurred by oxidation and to prolong the shelf life of the foods, the antioxidants are usually incorporated into the packaging materials and these antioxidants pass into the food or air space around the food during storage. Waxed papers, butylated hydroxy toluene impregnated packaging materials and tocopherols, essential fatty acids, and plant extracts obtained from plants such as rosemary, oregano, and tea have been used in antioxidant spreading systems. Vitamins E and C may be used in these systems in recent years due to their antioxidative effects (Wessling, Nielsen, Leufvén, & Jägerstad, 1998; Tian, Decker, & Goddard, 2012; Yang, Lee, Won, & Song, 2016;

Fang et al., 2017). In a study by Granda-Restrepo et al. (2009), it was determined that lipid oxidation decreased in whole-fat milk powder that was active-packaged by using α-tocopherol.

Ethanol Emitters

Ethanol emitters are especially used for active packaging of products with moderate moisture content against microbial spoilage and are particularly effective against mould growth. These components are also stated to be effective against yeasts and reduce staling and oxidative changes when used in high concentrations (Hurme et al., 2002; Suppakul et al., 2003; Dainelli et al., 2008). Ethanol emitters are reported to be particularly effective on bakery products (Dainelli et al., 2008).

The use of ethanol in active packaging technology can be applied in two ways: 1. Direct injection of ethanol into the package, 2. Use of packages with encapsulated ethanol that can release ethanol vapor (Hempel, O'Sullivan, Papkovsky, & Kerry, 2013).

Intelligent/Smart Packaging

Intelligent food packaging is defined as materials that can monitor the conditions of food and the surrounding environment. This technology, also known as smart labels, provides information to the producer, seller, and consumer about changes in food during transport and storage of the product (de Kruijf et al., 2002; Majid, Nayik, Dar, & Nanda, 2018). For this purpose, various sensors and indicators are used by being integrated into the package or directly adhered to the package and they provide information about the product characteristics such as freshness, shelf life and usage conditions. The working principle of these sensors and indicators is generally based on temperature-time measurements and the measurement of changes in chemical and microbiological properties (Yam, Takhistov, & Miltz, 2006; Dobrucka, 2013). Intelligent packaging technology provides information on the product in a short time, thus providing an alternative to time-consuming and costly analysis.

Sensors

Sensors are units that generally detect changes in the atmosphere within the product or packaging itself and transfer it to the manufacturer, seller, and consumer. The sensors consist essentially of a receptor and a transducer. It is possible to examine the sensors used in smart packaging under different headings as biosensors, gas sensors, chemical sensors, and pathogen sensors (Biji et al., 2015).

Biosensors

Devices that detect, record, and transmit the biological reactions occurring in food packages are called biosensors. Biosensors, like other sensors, consist of a bioreceptor and energy-converting devices (transducers). Bioreceptors are responsible for detecting the target parameter, usually organic materials such as various enzymes, antigens, hormones, and nucleic acids. Transducers consist of electrochemical, optical, or calorimetric systems and convert biological signals into measurable electrical messages (Smolander, 2003; Yam et al., 2006; Otles & Yalcin, 2008).

Gas sensors

It is important that the gas mixture in the package does not change until it reaches the consumer, particularly in products packaged using modified atmosphere packaging systems, to maintain the quality of the product. Gas sensors are systems that detect and transmit the presence or absence of gases used in modified atmosphere packaging, packaging integrity, and leaks (Otles & Yalcin, 2008; Robertson, 2012; Heising, Dekker, Bartels, & Van Boekel, 2014). O_2 and CO_2 sensors are the most commonly used gas sensors (Ergun, 2016).

Chemical sensors

Chemical sensors detect the presence of a specific chemical or gas in the packaged product or in the headspace of the package. The substance detected by chemical sensors is converted into signals by transducers and enables the consumer to perceive the presence of the substance (Vanderroost, Ragaert, Devlieghere, & Meulenaer, 2014).

Pathogen sensors

They are used for the detection of pathogenic microorganisms that infect the packaged products after production. Antibodies are generally used for this purpose and visual warning appears in the package as result of the reaction of these antibodies with the pathogens present in the product (Smolander, 2003).

Indicators

Indicators used in intelligent packaging technology indicate the presence, concentration, or deficiency of a particular substance, particularly by color changes. They are generally classified as freshness indicators, time-temperature indicators

(TTI), integrity indicators, and Radio frequency information device (RFID) (Hogan & Kerry, 2008).

Freshness indicators

The freshness indicators are generally used in products packaged with modified atmosphere packaging technology, which informs the consumer of the change in gas composition in the package through the label printed on the package. If the necessary conditions are not met during the transportation and storage of foods, some microbiological spoilage may occur in the product and as a result of this deterioration, metabolites such as CO_2, SO_2, NH_3 and ethanol are formed. Freshness indicators identify these metabolites and change the color of the label on the packaging, thus providing information about the quality of food (Smolander, 2003; Smolander, 2008). When classified according to working principles, it is possible to list the most frequently used freshness indicators as indicators; sensitive to pH change, volatile compounds, hydrogen sulfide (H_2S), and various microbial metabolites (Smolander, 2003; Gök, 2007).

Time-Temperature indicators (TTI)

Time-temperature indicators (TTI) detect physical, chemical, enzymatic, and microbial deteriorations in the product due to the exposure of the products to temperatures that they should not be exposed to during transport and storage. They are particularly effective in controlling temperature changes in refrigerated or frozen foods such as milk and dairy products. In case of temperature change, the barcode on the packaging turns into a dark color, thus it prevents the data transfer when the barcode is scanned and prevents the sale of the product. It is possible to classify the most commonly used time-temperature indicators as polymer, diffusion, and enzymatic-based ones (Riva, Piergiovanni, & Schiraldi, 2001; Taoukis & Labuza, 2003; Gök, 2007; Lee & Rahman, 2014; Taoukis & Tsironi, 2016). Time-temperature indicators can be used for all types of food products, including milk and dairy products (Ergun, 2016).

Integrity indicators

The package may be punctured or torn particularly during transport of the products. Such situations can be detected and transferred to the seller and the consumer through integrity indicators. Among the integrity indicators, the most widely used ones are based on O_2. Integrity indicators detect O_2 entering the package when the package is punctured or torn and stain the active substance with redox effect (Mattila-Sandholm, Ahvenainen, Hurme, & Jarvi-Kaariainen, 1995; Davies &

Gardner, 1996). Therefore, it can be determined that the integrity of the packaging material is impaired by color change.

Radio frequency information device (RFID)

Radio frequency information device (RFID) is a technology that makes it possible to read labels using radio waves without human intervention. In this system, as in the other methods described above, various changes in food are not detected by physical interaction, but by radio waves, through the microchips inserted into the product (Tajima, 2007; Lee & Rahman, 2014). RFID tags carry the information of all the changes in the food as from the packaging of the product and also allow remote monitoring of this information via radio waves (Karagöz & Demirdöven, 2017). RFID indicators which can be found in many different forms such as disk, glass, capsule, and label can also be combined with other indicators and can be used for all foods including milk and dairy products (Yuksel & Zaim, 2009; Ruiz-Garcia & Lunadei, 2011).

REFERENCES

Anonymous. (2009). *EU Guidance to the commission regulation (EC) No 450/2009 of 29 May 2009 on active and intelligent materials and articles intended to come into contact with food. Version 1.0. European Commission Health and Consumers Directorate-General Directorate E-Safety of the Food chain. E6- Innovation and sustainability.* Retrieved from https://ec.europa.eu/food/sites/food/files/safety/docs/cs_fcm_legis_active-intelligent_guidance.pdf

Anonymous. (2019). *Nanotechnology.* European Food Safety Authority. Retrieved from https://www.efsa.europa.eu/en/topics/topic/nanotechnology

Batu, A. (1994). Properties of modified atmosphere packaging films and application of fruits vegetables. *Gida, 19*(6), 397–403.

Batu, A., Caglar, A., & Kara, H. H., (2008). Afyon kaymagının raf ömrünün uzatılmasında modifiye atmosferde paketleme önerisi. *Gıda Teknolojileri Elektronik Dergisi, 2008*(2), 43-46.

Biji, K. B., Ravishankar, C. N., Mohan, C. O., & Gopal, T. S. (2015). Smart packaging systems for food applications: A review. *Journal of Food Science and Technology, 52*(10), 6125–6135. doi:10.100713197-015-1766-7 PMID:26396360

Brody, A. L., Strupinsky, E. R., & Kline, L. R. (2001). *Active packaging for food applications*. Boca Raton, FL: CRC Press. doi:10.1201/9781420031812

Buzby, J. C. (2010). Nanotechnology for food applications. More questions than answers. *The Journal of Consumer Affairs*, *44*(3), 528–545. doi:10.1111/j.1745-6606.2010.01182.x

Çelik, İ., & Tümer, G. (2016). Gıda ambalajlamada son gelişmeler. *Akademik Gıda*, *14*(2), 180–188.

Chau, C. F., Wu, S. H., & Yen, G. C. (2007). The development of regulations for food nanotechnology. *Trends in Food Science & Technology*, *18*(5), 169–280. doi:10.1016/j.tifs.2007.01.007

Cutter, C. N. (2006). Opportunities for bio-based packaging technologies to improve the quality and safety of fresh and further processed muscle foods. *Meat Science*, *74*(1), 131–142. doi:10.1016/j.meatsci.2006.04.023 PMID:22062722

Dainelli, D., Gontard, N., Spyropoulos, D., Zondervan-van den Beuken, E., & Tobback, P. (2008). Active and intelligent food packaging: Legal aspects and safety concerns. *Trends in Food Science & Technology*, *19*, 103–112. doi:10.1016/j.tifs.2008.09.011

Davies, E. S., & Gardner, C. D. (1996). *UK Patent No. GB-2298273*. Oxygen indicating composition, The Victoria University of Manchester.

de Kruijf, N. N., van Beest, M., Rijk, R., Sipilainen-Malm, T., Paseiro, L. P., & De Meulenaer, B. (2002). Active and intelligent packaging: Applications and regulatory aspects. *Food Additives and Contaminants, 19* Suppl, 144-162.

Demir, M. (1999). *Modified atmosphere packaging*. Retrieved from http://www.apack.com.tr/images/userfiles/146705440090949918.pdf

Devlieghere, F., & Debevere, J. (2000). Influence of dissolved carbon dioxide on the growth of spoilage bacteria. *Lebensmittel-Wissenschaft + Technologie*, *33*(8), 531–537. doi:10.1006/fstl.2000.0705

Devlieghere, F., Gil, M. I., & Debevere, J. (2002). Modified atmosphere packaging (MAP). In C. J. K. Henry, & C. Chapman (Eds.), *The nutrition handbook for food processors* (pp. 342–370). England: Woodhead Publishing. doi:10.1533/9781855736658.2.342

Dobrucka, R. (2013). The future of active and intelligent packaging industry. *LogForum*, *9*(2), 103–110.

Dobrucka, R., & Cierpiszewski, R. (2014). Active and intelligent packaging food - Research and development - A review. *Polish Journal of Food and Nutrition Sciences*, *64*(1), 7–15. doi:10.2478/v10222-012-0091-3

Duncan, T. V. (2011). Applications of nanotechnology in food packaging and food safety: Barriermaterials, antimicrobials and sensors. *Journal of Colloid and Interface Science*, *363*(1), 1–24. doi:10.1016/j.jcis.2011.07.017 PMID:21824625

Ergun, M. (2016). Taze meyve ve sebzeler için aktif, zeki veya akıllı paketleme teknolojileri. *Alatarım*, *15*(2), 51–60.

Fang, Z., Zhao, Y., Warner, R. D., & Johnson, S. K. (2017). Active and intelligent packaging in meat industry. *Trends in Food Science & Technology*, *61*, 60–71. doi:10.1016/j.tifs.2017.01.002

Farber, J. M. (1991). Microbiological aspects of modified-atmosphere packaging technology - A review. *Journal of Food Protection*, *54*(1), 58–70. doi:10.4315/0362-028X-54.1.58 PMID:31051584

Fierheller, M. G. (1991). Modified atmosphere packaging of miscellaneous products. In B. Ooraikul, & M. E. Stiles (Eds.), *Modified atmosphere packaging of food* (pp. 246–260). Boston, MA: Springer. doi:10.1007/978-1-4615-2117-4_8

Gawith, J. A., & Robertson, T. R. (2000). Wrapping up packaging technology. *Journal of the Home Economics Institute of Australia*, *7*(2), 6–14.

Gök, V. (2007). Gıda paketleme sanayinde akıllı paketleme teknolojisi. *Gıda Teknolojileri Elektronik Dergisi, 2007*(1), 45-58.

Gomes, C., Elena Castell-Perez, M., Chimbombi, E., Barros, F., Sun, S., Liu, J. D., ... Wright, A. O. (2009). Effect of oxygen-absorbing packaging on the shelf life of a liquid-based component of military operational rations. *Journal of Food Science*, *74*(4), 167–176. doi:10.1111/j.1750-3841.2009.01120.x PMID:19490321

Gomez-Estaca, J., Lopez-de-Dicastillo, C., Hernandez-Munoz, P., Catala, R., & Gavara, R. (2014). Advances in antioxidant active food packaging. *Trends in Food Science & Technology*, *35*(1), 42–51. doi:10.1016/j.tifs.2013.10.008

Göncü, A., & Özkal, S. G. (2017). Ekmeklerde aktif paketleme uygulamaları. *Türk Tarım - Gıda Bilim ve Teknoloji Dergisi, 5*(11), 1264-1273.

Granda-Restrepo, D. M., Soto-Valdez, H., Peralta, E., Troncoso-Rojas, R., Vallejo-Córdoba, B., Gámez-Meza, N., & Graciano-Verdugo, A. Z. (2009). Migration of α-tocopherol from an active multilayer film into whole milk powder. *Food Research International*, *42*(10), 1396–1402. doi:10.1016/j.foodres.2009.07.007

Gruere, G. P., Narrod, C. A., & Abbott, L. (2011). Agriculture, food, and water nanotechnologies for the poor: Opportunities and constraints, Policy briefs 19, International Food Policy Research Institute (IFPRI). Retrieved from http://cdm15738.contentdm.oclc.org/utils/getfile/collection/p15738coll2/id/124891/filename/124892.pdf

Gutiérrez, L., Batlle, R., Andújar, S., Sánchez, C., & Nerín, C. (2011). Evaluation of antimicrobial active packaging to increase shelf life of gluten-free sliced bread. *Packaging Technology & Science*, *24*(8), 485–494. doi:10.1002/pts.956

Heising, J. K., Dekker, M., Bartels, P. V., & Van Boekel, M. A. J. S. (2014). Monitoring the quality of perishable foods: Opportunities for intelligent packaging. *Critical Reviews in Food Science and Nutrition*, *54*(5), 645–654. doi:10.1080/10408398.2011.600477 PMID:24261537

Hempel, A. W., O'Sullivan, M. G., Papkovsky, D. B., & Kerry, J. P. (2013). Use of smart packaging technologies for monitoring and extending the shelf-life quality of modified atmosphere packaged (MAP) bread: Application of intelligent oxygen sensors and active ethanol emitters. *European Food Research and Technology*, *237*(2), 117–124. doi:10.100700217-013-1968-z

Hogan, S. A., & Kerry, J. P. (2008). Smart packaging of meat and poultry products. In J. Kerry, & P. Butler (Eds.), *Smart packaging technologies for fast moving consumer goods* (pp. 33–59). West Sussex: John Wiley & Sons. doi:10.1002/9780470753699.ch3

Hotchkiss, J. H., & Chen, J. H. (1996). Microbiological effects of the direct addition of CO_2 to pasteurized milk. *Journal of Dairy Science*, *79*(Supplement 1), 87. PMID:8675787

Huff, K. (2008). *Active and intelligent packaging: Innovations for the future*. Retrieved from http://www.iopp.org/files/public/VirginiaTechKarleighHuff.pdf

Hurme, E., Sipilainen-Malm, T., & Ahvenainen, R. V. T. T. (2002). Active and intelligent packaging. In T. Ohlsson, & N. Bengtsson (Eds.), Minimal processing technologies in the food industry (pp. 87-123). Cambridge: CRC Press.

Irkin, R., & Esmer, O. K. (2015). Novel food packaging systems with natural antimicrobial agents. *Journal of Food Science and Technology*, *52*(10), 6095–6111. doi:10.100713197-015-1780-9 PMID:26396358

Karagöz, Ş., & Demirdöven, A. (2017). Gıda ambalajlamada güncel uygulamalar: Modifiye atmosfer, aktif, akıllı ve nanoteknolojik ambalajlama uygulamaları. *Gaziosmanpaşa Bilimsel Araştırma Dergisi*, *6*(1), 9–21.

Kartal, S. (2010). *Çileğin raf ömrünün mikroperfore filmler ve oksijen tutucular kullanılarak denge modifiye atmosfer ile arttırılması.* (Unpublished master's thesis). Çanakkale Onsekiz Mart University, Çanakkale, Turkey.

Kılınç, B., & Çaklı, Ş. (2004). Su ürünlerinin modifiye atmosferde paketlenmesi. *E.U. Su Ürünleri Dergisi, 21*(3-4), 349–353.

Labuza, T. P., & Breene, W. M. (1989). Applications of "active packaging" for improvement of shelf-life and nutritional quality of fresh and extended shelf-life foods. *Journal of Food Processing and Preservation, 13*(1), 1–69. doi:10.1111/j.1745-4549.1989.tb00090.x

Lee, L., Arul, J., Lencki, R., & Castaigne, F. (1996). A review on modified atmosphere packaging and preservation of fresh fruits and vegetables: Physiological basis and practical aspects-Part II. *Packaging Technology & Science, 9*(1), 1–17. doi:10.1002/(SICI)1099-1522(199601)9:1<1::AID-PTS349>3.0.CO;2-W

Lee, S. J., & Rahman, A. T. M. M. (2014). Intelligent packaging for food products. In J. H. Han (Ed.), *Innovations in food packaging* (pp. 171–209). Academic Press. doi:10.1016/B978-0-12-394601-0.00008-4

Majid, I., Nayik, G. A., Dar, S. M., & Nanda, V. (2018). Novel food packaging technologies: Innovations and future prospective. *Journal of the Saudi Society of Agricultural Sciences, 17*(4), 454–462. doi:10.1016/j.jssas.2016.11.003

Mattila-Sandholm, T., Ahvenainen, R., Hurme, E., & Jarvi-Kaariainen, T. (1995). *Finnish Patent No. FI-94802*. Leakage Indicator, VTT Biotechnology, and Food Research.

Momin, J. K., Jayakumar, C., & Prajapati, J. B. (2013). Potential of nanotechnology in functional foods. *Emirates Journal of Food and Agriculture, 25*(1), 10–19. doi:10.9755/ejfa.v25i1.9368

Moraes, A. R. F., Gouveia, L. E. R., Soares, N. F. F., Santos, M. M. S., & Gonçalves, M. P. J. C. (2007). Development and evaluation of antimicrobial film on butter conservation. *Food Science and Technology (Campinas), 27*, 33–36. doi:10.1590/S0101-20612007000500006

Mullan, W. M. A. (2002). *Science and technology of modified atmosphere packaging*. Retrieved from https://www.dairyscience.info/index.php/packaging/117-modified-atmosphere-packaging.html

Nicholson, M. D. (1998). The role of natural antimicrobials in food/packaging biopreservation. *Journal of Plastic Film & Sheeting*, *14*(3), 234–241. doi:10.1177/875608799801400306

Oliveira, M., Abadias, M., Usall, J., Torres, R., Teixido, N., & Vinas, I. (2015). Application of modified atmosphere packaging as a safety approach to fresh-cut fruits and vegetables - A review. *Trends in Food Science & Technology*, *46*(1), 13–26. doi:10.1016/j.tifs.2015.07.017

Otles, S., & Yalcin, B. (2008). Intelligent food packaging. *LogForum*, *4*(3), 1–9.

Özdemir, M., & Floros, J. D. (2004). Active food packaging technologies. *Critical Reviews in Food Science and Nutrition*, *44*(3), 185–193. doi:10.1080/10408690490441578 PMID:15239372

Patel, R., Prajapati, J. P., & Balakrishnan, S. (2015). *Recent trends in packaging of dairy and food products*. Paper presented at the meeting National seminar on Indian Dairy Industry - Opportunities and Challenges. Gujarat, India.

Pereira de Abreu, D. A., Cruz, J. M., & Paseiro Losada, P. (2012). Active and intelligent packaging for the food industry. *Food Reviews International*, *28*(2), 146–187. doi:10.1080/87559129.2011.595022

Riva, M., Piergiovanni, L., & Schiraldi, A. (2001). Performances of time–temperature indicators in the study of temperature exposure of packaged fresh food. *Packaging Technology & Science*, *14*(1), 1–9. doi:10.1002/pts.521

Robertson, G. L. (2012). *Food packaging, principles and practice*. London, UK: CRC Press.

Ruiz-Garcia, L., & Lunadai, L. (2011). The role of RFID in agriculture: applications, limitations, and challenges. *Computers and Electronics in Agriculture*, *79*(1), 42–50. doi:10.1016/j.compag.2011.08.010

Sandhya. (2010). Modified atmosphere packaging of fresh produce: Current status and future needs. *Food Science and Technology, 43*, 381-392.

Schnettler, B., Crisostomo, G., Mora, M., Lobos, G., Miranda, H., & Grunert, K. G. (2013). Acceptance of nanotechnology applications and satisfaction with food-related life in southern Chile. *Food Science and Technology (Campinas)*, *34*(1), 157–163. doi:10.1590/S0101-20612014005000001

Siegrist, M., Cousin, M. E., Kastenholz, H., & Wiek, A. (2007). Public acceptance of nanotechnology foods and food packaging: The influence of affect and trust. *Appetite*, *49*(2), 459–466. doi:10.1016/j.appet.2007.03.002 PMID:17442455

Siegrist, M., Stampfli, N., Kastenholz, H., & Keller, C. (2008). Perceived risks and perceived benefits of different nanotechnology foods and nanotechnology food packaging. *Appetite, 51*(2), 283–290. doi:10.1016/j.appet.2008.02.020 PMID:18406006

Sivertsvik, M., Rosnes, J. T., & Bergslien, H. (2002). Modified atmosphere packaging. In T. Ohlsson & N. Bengtsson (Eds.), *Minimal processing technologies in the food industry* (pp. 61–86). Cambridge: CRC Press. doi:10.1533/9781855736795.61

Sivertsvik, M., Rosnes, J. T., & Jeksrud, W. K. (2004). Solubility and absorption rate of carbon dioxide into non-respiring foods. Part 2: Raw fish fillets. *Journal of Food Engineering, 63*(4), 451–458. doi:10.1016/j.jfoodeng.2003.09.004

Smolander, M. (2003). The use of freshness ındicators in packaging. In R. Ahvenainen (Ed.), *Novel food packaging techniques* (pp. 127–143). Cambridge: Woodhead Publishing. doi:10.1533/9781855737020.1.127

Smolander, M. (2008). Freshness indicators and food packaging. In J. Kerry, & P. Butler (Eds.), *Smart packaging technologies for fast moving consumer goods* (pp. 111–127). West Sussex: John Wiley & Sons. doi:10.1002/9780470753699.ch7

Suppakul, P., Miltz, J., Sonneveld, K., & Bigger, S. W. (2003). Active packaging technologies with an emphasis on antimicrobial packaging and its applications. *Journal of Food Science, 68*(2), 408–420. doi:10.1111/j.1365-2621.2003.tb05687.x

Sürengil, G., & Kılınç, B. (2011). Gıda - Ambalaj sektöründe nanoteknolojik uygulamalar ve su ürünleri açısından önemi. *Journal of Fisheries Sciences Com, 5*(4), 317–325.

Tajima, M. (2007). Strategic value of RFID in supply chain management. *Journal of Purchasing and Supply Management, 13*(4), 261–273. doi:10.1016/j.pursup.2007.11.001

Taniwaki, M. H., Hocking, A. D., Pitt, J. I., & Fleet, G. H. (2001). Growth of fungi and mycotoksin production on cheese under modified atmospheres. *International Journal of Food Microbiology, 68*(1-2), 125–133. doi:10.1016/S0168-1605(01)00487-1 PMID:11545212

Taoukis, P., & Tsironi, T. (2016). Smart packaging for monitoring and managing food and beverage shelf life. In P. Subramaniam, & P. Wareing (Eds.), *The stability and shelf life of food* (pp. 141–168). Woodhead Publishing. doi:10.1016/B978-0-08-100435-7.00005-8

Taoukis, P. S., & Labuza, T. P. (2003). Time-temperature indicators. In R. Ahvenainen (Ed.), *Novel food packaging techniques* (pp. 103–126). Cambridge: Woodhead Publishing. doi:10.1533/9781855737020.1.103

Tian, F., Decker, E. A., & Goddard, J. M. (2012). Development of an iron chelating polyethylene film for active packaging applications. *Journal of Agricultural and Food Chemistry*, *60*(8), 2046–2052. doi:10.1021/jf204585f PMID:22288894

Üçüncü, M. (2011). *Gıda ambalajlanma teknolojisi*. İstanbul, Turkey: Ambalaj Sanayiciler Derneği.

Van der Steen, C., Jacxsens, L., Devlieghere, F., & Debevere, J. (2002). Combining high oxygen atmospheres with low oxygen modified atmosphere packaging to improve the keeping quality of strawberries and raspberries. *Postharvest Biology and Technology*, *26*(1), 49–58. doi:10.1016/S0925-5214(02)00005-4

Vanderroost, M., Ragaert, P., Devlieghere, F., & Meulenaer, B. D. (2014). Intelligent food packaging: The next generation. *Trends in Food Science & Technology*, *39*(1), 47–62. doi:10.1016/j.tifs.2014.06.009

Velu, S., Abu Bakar, F., Mahyudin, N. A., Saari, N., & Zaman, M. Z. (2013). Effect of modified atmosphere packaging on microbial flora changes in fishery products. *International Food Research Journal*, *20*(1), 17–26.

Vermeiren, L., Devlieghere, F., Van Beest, M., De Kruijf, N., & Debevere, J. (1999). Developments in the active packaging of foods. *Trends in Food Science & Technology*, *10*(3), 77–86. doi:10.1016/S0924-2244(99)00032-1

Wessling, C., Nielsen, T., Leufvén, A., & Jägerstad, M. (1998). Mobility of α-tocopherol and BHT in LDPE in contact with fatty food simulants. *Food Additives and Contaminants*, *15*(6), 709–715. doi:10.1080/02652039809374701 PMID:10209582

Yam, K. L., Takhistov, P. T., & Miltz, J. (2006). Intelligent packaging: Concepts and applications. *Journal of Food Science*, *70*(1), 1–10. doi:10.1111/j.1365-2621.2005.tb09052.x

Yang, H. J., Lee, J. H., Won, M., & Song, K. B. (2016). Antioxidant activities of distiller dried grains with solubles as protein films containing tea extracts and their application in the packaging of pork meat. *Food Chemistry*, *196*, 174–179. doi:10.1016/j.foodchem.2015.09.020 PMID:26593480

Yüksel, M. E., & Zaim, A. H. (2009). *Yeni nesil teknoloji olarak RFID, RFID sistem yapıları ve bir RFID sistem tasarımı yaklaşımı.* Paper presented at the meeting 5th International Advanced Technologies Symposium. Karabük, Turkey.

Chapter 5 Microbial Inactivation by Ultrasound in the Food Industry

Eda Adal
Gaziantep University, Turkey

ABSTRACT

Pasteurization is the most common processing method for microbial and enzyme inactivation to preserve foods. With this method, foods are exposed to high temperatures and there are disadvantages for many products: thermal treatments cause modifications of sensory attributes (for instance: flavour, colour, nutritional qualities). Now, another method can replace pasteurization: microbial inactivation by ultrasounds. It is a new alternative technology of food processing also called sonication, and it can be used coupled with pressure and/or heat. These techniques inactivate microorganisms in foods. They are effective and energy efficient to kill them, making the techniques promising for the food industry. In this chapter, the method of microbial inactivation by ultrasounds was explained, after that the applications in food industry for instance in milk, orange juice, wastewater, and whole liquid eggs were well-defined, and finally, the advantages, disadvantages, and the limitations of this method were examined.

EXPLANATION OF THE METHOD

How Ultrasound Can Inactivate Microorganisms?

Two types of ultrasound are used in the industry. One of them is low power with high

DOI: 10.4018/978-1-7998-1924-0.ch005

Figure 1. Schematic demonstration of a layer of a material where longitudinal sonic energy is propagated (Suslick et al., 1988)

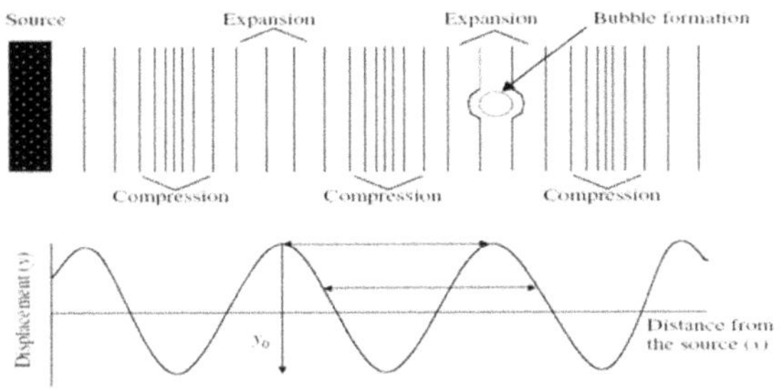

frequency and the other one is high power with low frequency. In the food industry, the high power with a low-frequency application is most commonly used to inactivate microorganisms. This type of ultrasound is called a power ultrasound. The power ultrasounds work with 20 to 100 kHz frequency (Povey, 1989; McClements, 1995).

The inactivation of microorganisms with ultrasounds is achieved by transient cavitation (formation of bubbles in liquid medium) in food processing. The killing of microorganisms is mostly dealing with the thinning of cell membranes, localized heating, and free radical production during cavitation (Leighton et al., 1998). The longitudinal waves occur while ultrasonic waves go through the liquid medium which results in compression and expansion cycles periodically during cavitation. This phenomenon is shown in Figure 1.

As a result of these cycles, gas bubbles are formed in the medium. The surface area of these bubbles is great which causes great diffusion of gas during expansion, in the medium. That is why the size of gas bubbles increases at each cycle. At this point, created ultrasonic energy cannot remain the vapor phase and vapor rapidly passed into the liquid phase (condensation). The bubbles begin to collide with great force which results in the formation of shock waves. The localized high pressure (10^4-10^5 kPa) and high temperature (5500 ºC) are created by these shock waves. The temperature and pressure indicated are generated during very short periods of time at the point was cavitation occurs with an order of temperature variation of 109°C (Manas and Pagan, 2003). This is the mechanical effect of collapsing bubbles in the liquid medium. Also, these bubbles can create the formation of OH^- and H^+ (free radicals) and hydrogen peroxide in the medium known as chemical effects. The bactericidal properties of ultrasound are shown with these effects (Suslick et al., 1988).

There are some important factors that affect cavitation in the ultrasound process. These are hydrostatic pressure, temperature, frequency and amplitude of ultrasonic waves and tensile strength of the liquid. Such as The vapor pressure increase and tensile strength decreases with increasing temperature. As a result of that, the gas bubbles are formed and grown rapidly. However, the intensity of implosion decreases due to the increase of the vapor pressure of liquid and increases with increasing surface tension at the bubble interface (Alliger, 1975). The cavitation increases with greater amplitude because the number of bubbles undergoing cavitation increases with time which supplies the cavitation (Suslick et al., 1988). Also, the hydrostatic pressure effects on the inactivation of microorganism as mentioned above because the intensity of bubbles implosion increases with increasing hydrostatic pressure, the number of bubbles empoisoning decreases with time (Whillock and Harvey, 1997). Also, the type of microorganisms is an important factor for the effectiveness of ultrasound because some microorganisms can resist these effects.

MATERIALS AND METHODS USED IN THIS PROCESS

Ultrasound equipment consists of 3 parts (Figure 2):

1) Generator
2) Transducer
3) Coupler

During ultrasound processing, the transducer converts the electric energy to mechanical energy. This energy transmitted through the liquid or solid material and the result is the creation of sonic energy waves. The displacement of the particles causes compressive and expansion, respectively which mentioned above (Povey, 1989; McClements, 1995; Sala et al., 1995).

WHICH MICROORGANISMS CAN BE INACTIVATED?

Microorganisms that can be inactivated have common points. Firstly, the kind of microorganism is important, secondly, the growth phase, thirdly the growth temperature and finally the sublethal heat treatments is considerable (Alvarez et al., 2000). Inactivation of microorganisms by ultrasound depends on the organism being treated. Most of the studies have been done on *Listeria monocytogenes, different types of Salmonella, Escherichia coli, Staphylococcus aureus, Bacillus subtilis*. Indeed, ultrasounds on *Escherichia coli* in a liquid medium, with the frequency 24kHz show a reduction of the population (McKellar et al., 2003).

Gram-positive microorganisms are more resistant than a Gram-Negative and usually, bacterial spores are more resistant than vegetative cells. Moreover,

Figure 2. Components of Ultrasound equipment (Schläfer et al., 2002)

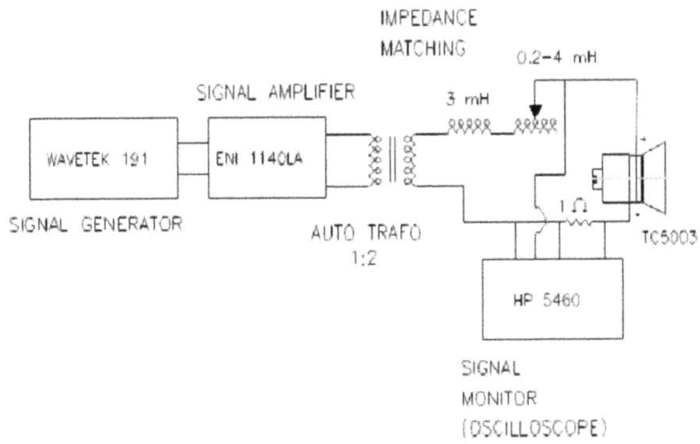

microorganisms that have a rod-shaped form are more sensitive than microorganisms with a coccal form. Scientists do not know exactly the influence of the growth phase on the microbial resistance to ultrasounds but even if they cannot explain that, they observed differences in growth phases. Indeed, cells of *Escherichia coli* at the exponential growth phase were more sensitive to ultrasonic treatment than that at the stationary stage. The effect of the treatment temperature on the success of the treatment depends on the microorganism. In fact, for instance, the resistance *Listeria monocytogenes* and *Yersinia enterocolitis* to ultrasonic treatments do not depend on the growth temperature (Alvarez et al., 2000).

In most cases, heat helps to inactivate microorganisms. Indeed, *Listeria monocytogenes* cells can be inactivated more easily with thermosonic inactivation than with sonication alone (McKellar et al. 2003). Thermosonication treatment at above100°C results in a significant decrease in the population of *Bacillus subtilis. Salmonella Sentfenberg* 775W is one of the *Salmonella* serotypes from foods it is interesting to study because it has high heat resistance. According to some researchers, high-intensity heat treatments are necessary to obtain the required microbial inactivation, which would damage food quality (Alvarez et al., 2006).

Sometimes, microorganism resistance in the treatment increases when microorganisms are exposed to heat: this resistance could be explained by the development of specific proteins called "heat shock proteins" that protect bacteria against heat. Scientifics observed that *Listeria monocytogenes* heat resistance increase after a heat shock, but this shock does not change its resistance to an ultrasound treatment under pressure (Alvarez et al., 2000). Besides, *Listeria monocytogenes* can be inactivated by ultrasounds combined with an increase pressure of 200kPa (McKellar et al., 2003).

Moreover, there are some environmental factors as the water activity which can influence microorganisms' resistance. It depends on the microorganism, indeed, a decrease of the water activity only increases *Salmonella Senftenberg* 775W heat resistance 8-fold at 60° whereas it increased the heat resistance of *Salmonella Enteritidis* 30-fold (Alvarez et al., 2006).

APPLICATIONS OF ULTRASOUND IN THE FOOD INDUSTRY

1) Milk

The use of ultrasounds as a meaning of pasteurization is not used yet in the dairy industry. However, some researches are led in order to find new processes that both permit food safety and enhanced organoleptic qualities.

Concerning milk, the application of ultrasounds aims microorganisms like *Staphylococcus aureus*, *Bacillus subtilis*, *Salmonella typhimurium* which is known as coliform microorganism and some species of *Salmonella, Escherichia coli, Saccharomyces cerevisiae,* and *Listeria monocytogenes.* Dairy products are easily contaminated with *Listeria* because of conditions in the manufacture facilities, the environment, and storage temperatures (Barbosa-Canovas *et al.,* 2009).

Ultrasonic waves at frequencies above 20kHz are generated by mechanical vibrations. The aim is the phenomena of cavitation happens: microscopic bubbles are formed and collapse violently and as a result cell walls break and shears leading to cell death. The use of ultrasonic waves in milk is very useful: fat may be homogenized, gases are removed and the antioxidant activity enhanced (Britz *et al.,* 2009). Sonication perhaps disrupted casein micelles, increasing the effective concentration of casein, which could account for the increased antioxidant activity in the system (Taylor and Richardson, 1980).

Often the sonication treatment is not enough to assure the safety of the milk. That is why thermal treatment should be led at the same time. For instance, an experiment showed that the combination of ultrasound and heat in the inactivation of cells reduces treatment time (from the 30s to 10min), compared to conventional thermal treatment (63°C, 30min for conventional batch pasteurization). However, the most common treatment used today is 72°C for 15s treatment. The combination with ultrasound permits to short the treatment time from 15 to 8s (Barbosa-Canovas *et al.* 2009).

2) Municipal wastewater

Disinfection of wastewater is an important stage in water treatment in order to minimize contamination of the receiving waters. Different technologies have already been tested but they have some drawbacks: chemical oxidation can lead to harmful residues which poisoned waters; UV-C irradiation increases biomass concentration; membranes are rapidly fouled by the slurry. Furthermore, some microorganisms are becoming resistant to existing disinfection techniques. That is why researches are leading to developing new alternative disinfection methods. Sonication is one of these alternatives. Like in milk, microorganisms are inactive thanks to the phenomena of cavitation (Drakopoulou et al., 2009).

Some scientists have studied the impact of ultrasounds on bacteria groups which are mainly presented in wastewater. These bacteria groups are *Faecal coliforms, Pseudomonas spp., Faecal streptococci* and *Clostridium perfringens.* In this study, waves of 24kHz have been used, the wastewater was studied in temperature-controlled batch trials. The scientists have concluded that sonication at the conditions of the study is a method capable of achieving disinfection standards for wastewater reuse (Khanal et al., 2007).

3) Liquid whole egg

Numerous food-borne eruptions included eggs and egg-containing foods infected with bacterial human pathogens. Salmonella inactivation can lead by ultrasonic waves of 20 kHz; the wavelength of 117-μm at nonlethal temperatures and lethal temperatures under pressure of 175 kPa.

Manas et al., 2000 was examined the impact of ultrasound on *Salmonella enteritidis, Salmonella typhimurium,* and *Salmonella senftenberg* under different pressure treatments, at sublethal and lethal temperatures in the citrate-phosphate buffer and in a liquid whole egg. Some of the experiments were shown in Table 1. They observed that increases in pressure cause a rise in the inactivation rate.

The influence of increasing the temperature to 55°C is an improvement in *Salmonella Enteritidis* inactivation in the liquid whole egg. The inactivation depends on the vapor pressure and reduction in the tensile strengths of cavities formed once applying ultrasound (Hust et al., 1995).

4) Orange juice

In fruit juice while not side pulp, ultrasonication only in batch conditions and uncontrolled temperature reduces the entire mesophilic aerobic plate counts. Once the coarse pulp is side to orange juice, the result of ultrasound on numbers of total mesophilic aerobes is irrelevant. Initial yeast and fungous counts aren't tormented by ultrasound altogether experiments (Mason et al., 1996). Ultrasound treatment has virtually no result on the evolution of color rating additionally.

Table 1. Summary of some experiments design to see the effects of ultrasound energy in food preservation (Manas et al., 2000)

Object studied	Media	f (kHz)	P (W)	Remarks	T (C°)	t (mins)
Salmonella typhimurium	egg	20	150-450	Heat and pressure	40	1
Salmonella enteritidis	egg	20	150-450	Heat and pressure	40	1
Salmonella senftenberg	egg	20	150-450	Heat and pressure	40	1

The influence of different ultrasound treatments and conventional heating on the inactivation and potential subsequent growth of micro-organisms in orange juice is investigated by Valero et al. (2007).

They examined changes in limonin content, brown pigment production in non-enzymatic browning reactions and color of orange juices by ultrasound. Ultrasound has no negative effects on all these sensory properties. According to the study combination of ultrasound with other processing, methods will be necessary to inhibit the improvement of food-borne pathogens in orange juice.

ADVANTAGES AND DISADVANTAGES OF THE ULTRASOUND METHOD

Generally, the permeabilization of cell membranes to ions can be achieved by ultrasound and it can decrease the discrimination of the cell membranes considerably. Rises the surface area in contact between the solid and the liquid phase can lead to a reduction in particle size by ultrasonic cavitation. Ultrasound causes mass transfer is increased through the living cells. For gases and liquids, it increases the evaporation rate. For low viscosity liquids, it causes turbulence and kills harmful microbes (Suslick et al., 1988).

Ultrasound provides more yield and rate of extraction in the extraction and isolation of novel potentially bioactive components. Besides, ultrasound can help to increase the effects of enzyme treatment, and lessen the amount of enzyme needed or increase the yield of extractable relevant compounds.

Ultrasound involves minimum undesirable changes of sensory attributes, i.e. texture, flavor, color, smell, and nutritional qualities, i.e. vitamins and proteins. Ultrasound is a competent non-thermal processing alternative. In the ultrasound process, the killing of microorganisms is reached via cavitation which caused by the changes in pressure created by the ultrasonic waves. Furthermore, inactivation

Figure 3. Comparison of energy consumption for microbial inactivation (Knorr et al., 2004)

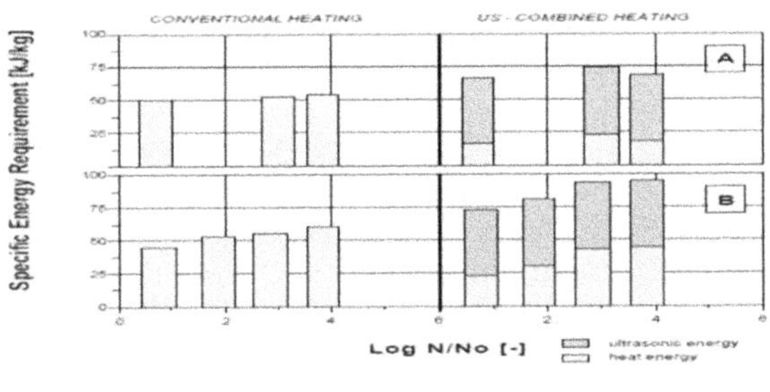

of the Peroxidase enzyme which causes undesirable flavors and browning pigments in most raw and unblanched fruits and vegetables can be reduced considerably by the use of ultrasound. Unlike other non-thermal novel processes, for example, high hydrostatic pressure (HP), compressed Carbon dioxide (cCO_2) and supercritical carbon dioxide (ScCO_2) and high electric field pulses (HELP), ultrasound does not need complex machinery and wide range of technologies (Knorr et al., 2004).

The ultrasound waves can not only be applied safely and environmentally friendly but also efficiently and economically (Figure 3). Ultrasound is generally used for processing, extraction, emulsification, preservation, homogenization of fruit juices and purees (orange, apple, grapefruit) as well as for vegetable sauces and soups, like tomato sauce or asparagus soup (Knorr et al., 2004).

Limitations

The microbial inactivation by ultrasound processing has limits because, for instance, they have to be used with other preservations factors such as high pressure, or high temperature. The combination with other treatments can sterilize foods but the technic cannot be used lonely.

CONCLUSION

Traditional methods like pasteurization and sterilization are most frequently used to protects foods against the deterioration and spoilage in the food industry, but there are some negative effects of these methods on food products. These negative effects cause a loss in nutritional value and sensory characteristics of food. Because of that reason, some new potential techniques are developed.

The ultrasound has a lot of applications in the food industry such as homogenization, microbial inactivation, and enzyme inactivation. The microbial inactivation of ultrasound is used for preservation purposes; this is achieved by cavitation (a form of gas bubbles in liquid medium). The effect of ultrasound depends on several factors. The most important factor is the type of microorganisms because some bacteria which resist the ultrasound cannot effect like spores. Because of that reason, the ultrasound should be combined with other treatments to get more effective inactivation like heat, pressure or both treatments. There are several applications in the food industry like water, milk, orange juice, and eggs. When ultrasound is applied for preservation method, it prevents browning (fruit juice e.g.) and reduces off-flavors in food. If we compare ultrasound treatment and other thermal methods; ultrasound has a lot of advantages. For example, it doesn't change texture, flavor, color, smell, and nutritional qualities, i.e. vitamins and proteins too much.

REFERENCES

Alliger, H. (1975). Ultrasonic distruption. *American Laboratory, 10*, 75–85.

Alvarez, I., Condon, S., Manas, P., & Virto, R. (2006). Inactivation of *Salmonella Sentenberg 775W* by ultrasonic waves under pressure at different water activities. *International Journal of Food Microbiology, 108*(2), 218–225. doi:10.1016/j.ijfoodmicro.2005.11.011 PMID:16488040

Álvarez, I., Pagan, R., Raso, J., Codo, S., & Sala, F. J. (2000). Microbial inactivation by ultrasound. *Food Technology, University of Zaeagoza. Miguel Servet, 177*, 50013.

Barbosa-Canovas, G. V., Bermudez-Aguirre, D., Corradini, M. G., & Mawson, R. (2009). Modeling the inactivation of Listeria innocua in raw whole milk treated under thermo-sonication. *Innovative Food Science & Emerging Technologies, 10*(2), 172–178. doi:10.1016/j.ifset.2008.11.005

Britz, T. J., Cameron, M., & McMaster, L. D. (2009). Impact of ultrasound on dairy spoilage microbes and milk components. *Dairy Science & Technology, 89*(1), 83–98. doi:10.1051/dst/2008037

Drakopoulou, S., Terzakis, S., Fountoulakis, M. S., Mantzavinos, D., & Manios, T. (2009). Ultrasound-induced inactivation of gram-negative and gram-positive bacteria in secondary treated municipal wastewater. *Ultrasonics Sonochemistry, 16*(5), 629–634. doi:10.1016/j.ultsonch.2008.11.011 PMID:19131265

Hust, R. M., Betts, G. D., & Earnshaw, R. G (1995). The antimicrobial effect of power ultrasound. R&D Report, 4.

Khanal, S. K., Grewell, D., Sung, S., & Van Leeuwen, J. (2007). Ultrasound applications in wastewater sludge pretreatment: A review. *Critical Reviews in Environmental Science and Technology*, *37*(4), 277–313. doi:10.1080/10643380600860249

Knorr, D., Zenker, M., Heinz, V., & Lee, D.-U. (2004). Applications and potential of ultrasonics in food processing. *Trends in Food Science & Technology*, *15*(5), 261–266. doi:10.1016/j.tifs.2003.12.001

Leighton, T. G. (1998). The principles of cavitation. *Ultrasound in food processing, 12*.

Manas, P., & Pagan, R. (2003). A review on microbial inactivation by new technologies of food preservation. *Journal of Applied Microbiology*, *98*(6), 1387–1399. doi:10.1111/j.1365-2672.2005.02561.x PMID:15916651

Manas, P., Pagan, R., Raso, J., Sala, F. J., & Condon, S. (2000). Inactivation of Salmonella enteritidis, Salmonella Typhimurium and Salmonella seftenberg by ultrasonic waves under pressure, *Journal of Food Protection*, 63, 451-456.

Mason, T. J., Paniwnyk, L., & Lorimer, J. P. (1996). The uses of ultrasound in food technology. *Ultrasonics Sonochemistry*, *3*(3), S253–S260. doi:10.1016/S1350-4177(96)00034-X

McClements, J. (1995). Advances in the application of ultrasound in food analysis and processing. *Trends in Food Science & Technology*, *6*(9), 293–299. doi:10.1016/S0924-2244(00)89139-6

McKellar, R. C., Mohareb, E., & Piyasena, P. (2003). Inactivation of microbes using ultrasound: A review. *International Journal of Food Microbiology*, *87*(3), 207–216. doi:10.1016/S0168-1605(03)00075-8 PMID:14527793

Povey, J. W. (1989). Ultrasonic in food engineering II. Applications. *Journal of Food Engineering*, *9*(1), 1–20. doi:10.1016/0260-8774(89)90047-2

Sala, F. J., Burgos, J., Condon, S., Lopez, P., & Raso, J. (1995). *Effect of heat and ultrasound on microorganism and enzymes. In G. W. Gound (Ed.), New methods of food preservation* (pp. 176–204). London, UK: Blackie Academic & Professional.

Schläfer, O., Onyeche, T., Bormann, H., Schröder, C., & Sievers, M. (2002). Ultrasound stimulation of micro-organisms for enhanced biodegradation. *Ultrasonics*, *40*(1-8), 25-29.

Suslick, K. S. (1988). *Ultrasound: Its chemical, physical and biological effects*. New York: VCH Publishers.

Taylor, M. J., & Richardson, T. (1980). Antioxidant activity of skim milk: Effect of sonication. *Journal of Dairy Science*, *63*(11), 1938–1942. doi:10.3168/jds.S0022-0302(80)83161-4 PMID:6893712

Valero, M., Recrosio, N., Saura, D., Mun˜oz, N., Martı´, N., & Lizama, V. (2007). Effects of ultrasonic treatments in orange juice processing. *Journal of Food Engineering*, *80*(2), 509–516. doi:10.1016/j.jfoodeng.2006.06.009

Whillock, G. O. H., & Harvey, B. F. (1997). Ultrasonically enhanced corrosion of 304L stainless steel: The effect of temperature and hydrostatic pressure. *Ultrasonics Sonochemistry*, *4*(1), 23–31. doi:10.1016/S1350-4177(96)00014-4 PMID:11233921

Chapter 6
Probiotics as an Alternative Food Therapy:
A Review on the Influence of Microbial Nutraceuticals in Disease Management

Thejaswi Bhandary
Christ University, India

Monisha M
Christ University, India

Ilavarasan Vickraman
https://orcid.org/0000-0002-4327-1375
Christ University, India

Paari K. A.
Christ University, India

ABSTRACT

Antibiotics have been responsible for the evolution of multidrug-resistant microbes. The side-effects of existing drugs and increased treatment costs have led to nutraceuticals gaining popularity. Nutraceuticals have therapeutic applications due to the ability of the probiotics to be viable in encapsulated pills and drinks. Due to their ability to exclude carcinogenic microorganisms by limiting the nutrients available and by competing for receptors nutraceuticals are useful against cancers. Nutraceuticals are useful against diabetes by controlling the genes involved in the insulin-signaling pathway. The future perspective for nutraceuticals includes an increase in production, reduction in manufacturing cost, and enhanced benefits.

DOI: 10.4018/978-1-7998-1924-0.ch006

INTRODUCTION

The word "Nutraceutical" was coined in 1989 by Stephen DeFelice from the two words 'nutrition' and 'pharmaceutical'. Nutraceuticals are "food (or a part of food) that provide medical or health benefits, including the prevention and/or treatment of disease" (Kalra, 2003). This is a term that is used for products obtained from dietary supplements, herbal products and other food sources which are beneficial not only nutritionally, but also medicinally (Nasri et al., 2014). They also fall under the category of "functional foods" since they are foods reserving bioactive compounds that are likely to have beneficial effects on the body beyond the basal nutritional ingredients (i.e. carbohydrates, protein, and fat). The food sources that are classified as nutraceuticals are categorized into dietary fibers, probiotics, and prebiotics (Das et al., 2012). Among the supplements included under these definitions, probiotics and probiotic food sources are of importance (Ku et al., 2016). Currently, the global market for nutraceuticals is estimated to reach a trade worth of 241 billion dollars (Brower, 1998). American Veterinary Medical Association (AVMA) describes nutraceuticals as micronutrients, macronutrients and other nutritional supplements that could be used as an alternative to the therapeutic antibiotics (Claudia, 2011).

Nutraceuticals are broadly classified into traditional and non-traditional nutraceuticals (Newman & Cragg, 2007). Traditional nutraceuticals include naturally available chemical constituents, probiotic microorganisms, and secretory enzymes. Non-traditional nutraceuticals include fortified and recombinant food products. Based on the source of synthesis, nutraceuticals are further grouped as plant-based, animal-based, and microbial-derived products. Recent advances in the field of microbial genetic engineering provide tools to design microbes as cell factories for nutraceutical production. Globally, the usage of nutraceuticals as an alternative to therapeutic drugs has increased over the last 25 years (Singh & Sinha, 2012). Natural bioactive compounds obtained from microbial sources are receiving a huge reception in the health sectors for their medicinal benefits. These products have wide applications as food preservatives, coloring agents and as therapeutic drugs to treat diseases such as cardiovascular disease, osteoporosis, tumor, diabetes, and oxidative stress.

The increased use of antibiotics and synthetic therapeutics has resulted in the evolution of multidrug-resistant microbes (Magiorakos et al., 2012). Several pathogens have been categorized under various degrees of drug-resistance by the World Health Organization based on their resistance to drugs (WHO, 2017). Bacteria such as *Acinetobacter baumannii*, *Pseudomonas aeruginosa* and *Enterobacteriaceae* were categorized as critically resistant. *Enterococcus faecium*, *Staphylococcus aureus*, *Helicobacter pylori*, *Campylobacter* spp., *Salmonellae,* and *Neisseria gonorrhoeae* were grouped as highly resistant. *Streptococcus pneumonia* and *Haemophilus influenza* were placed in the medium category because of their acquired resistance against beta-lactam antibiotics (WHO, 2017). Pan Drug Resistance (PDR) organisms display resistance to all the antibiotics and other useful agents from plant origin (Magiorakos et al., 2012). Adaptation pattern of these PDR organisms is the real challenge in front of the pharmaceutical companies. Though newer antibiotics can serve a temporary solution in controlling the drug-resistant pathogenic organisms, the safety of the antibiotic-associated immune regulations is not guaranteed. Hence, nutraceutical products are now gaining more attention to combat the illness caused by the PDR organisms. Over the years, nutraceuticals are used to treat lifestyle-related diseases such as arthritis, gastroenteritis, bowel syndrome, and some disorders like lactose intolerance and vitamin deficiencies (Chanda, Tiwari, Kumar, & Singh, 2019). Apart from being used for therapeutic applications, they are also utilized in the food industry for bio-coloration, stabilization of food, and as a flavoring and thickening agent (Srinivasan, 2005; Takao Ojima et al., 2018).

CHARACTERISTICS OF NUTRACEUTICALS

Sources

Biosynthesis of nutraceuticals for meeting the demands of the health sector is increasing worldwide. Though plants contain more nutraceutical compounds such as antioxidants, vitamins, polysaccharides, proteins, anti-microbial products, phytochemicals such as polyphenols, betaine, ellagic acid, etc., providing space and time for synthesizing nutraceuticals in plant systems is difficult. Nutraceuticals like omega-3-fatty acids, chondroitin sulfate, glycosaminoglycans, antioxidants, collagen, etc., which are specifically produced for livestock consumption, also face drawbacks during product purification and recovery (Galanakis, 2013; Coppens, Silva, & Pettman, 2006). Microbes can serve as a productive alternative candidate to plant and animal systems in the synthesis of nutraceuticals (Gupta, 2014). Because of the ability of the microbes to synthesize functional secretory products and the presence of a robust technology available to purify the secreted products, microorganisms are preferred over plant and animal systems (Green & Mecsas, 2016). The microbial nutraceutical products are produced in large quantities to meet the increasing demand of society for functional foods. Feasibility of genetically modifying microbial genome using strain improvement techniques has also increased the application and synthesis of microbial secretory products (Kerry et al., 2018).

Biosafety

The organisms employed for the synthesis of nutraceuticals should be non-pathogenic and should not cause any side or adverse effects on human health (Shewale, Sawale, Khedkar, & Singh, 2014). Efficient and strict screening strategies should be employed to produce fermented products from microbial sources after assessing the safety of the strains. Bacteriocins are the most commonly described bacterial products, many of which have not been fully characterized. A well known enterococcal bacteriocin includes enterocins produced by the *Enterococcus faecium*YF5 isolated from the sourdough, tested for the safety assessment in mice revealed no visible damage or illness in mice when fed at the dose of 10^{11} CFU/ day (Tan et al., 2013). The enterocin also showed an antimicrobial character by causing the inhibition of foodborne pathogens such as *Salmonella typhimurium, Shigella* sp., *Escherichia coli* and *Listeria monocytogenes* (Tan et al., 2013). The use of probiotics as supplements in patients with inflammatory bowel syndrome (IBD) and ulcerative colitis (UC) has witnessed a reduction in the severity of the disease (Sanders et al., 2010). However, in some normal healthy humans and immune-compromised persons, probiotic supplements lead to the elevated activation of a non-specific immune response (Sanders et al., 2010). As the risk of involvement of lactobacillus culture and its secretory products in pathogenesis is 1 in 10 million individuals, assessing the safety of probiotic cultures before human applications is mandatory (Bernardeau, Guguen, & Vernoux, 2006).

To show that probiotics are safe for human consumption their effect was studied on patients infected with HIV and also on other immunocompromised individuals. The use of probiotics as adjunctive treatment has not only been effective but also proved to be safe in most of the patients. But, since the effect of every bacteria is different, it is important to be cautious when dealing with individuals with extreme immunosuppression(Happel et al., 2018). Resistance to antibiotics and transfer of resistance to other gut commensals is a vital module for risk assessment of probiotic organisms. Incidence of transfer of tetracycline resistance gene in the commercial probiotic *B. animalis* subsp. *lactis* Bb-12 has been studied (Gueimonde et al., 2010). Although there isn't a single unified international health authority for screening the antibiotic resistance of probiotics, there are projects such as PROSAFE assessing the biosafety of probiotics used for human consumption, ACE-ART for the Assessment and Critical Evaluation of Antibiotic Resistance Transferability in Food Chain and ISO-IDF, the Joint International Organization for Standardization-International Dairy Federation Action Team on Probiotics that are contributing towards this cause (Imperial & Ibana .,2016).

Stability

The nutraceutical secreting micro-organisms utilized for human application should be stable. Any instability in the probiotic strain will reflect on the metabolic activity of the strain. Genetic rearrangements due to mutation during the course of evolution will have an impact on the health contributing factors secreted by the strains (Wind et al., 2010). Genetic rearrangements happen due to the uncontrolled gene transfer and recombination mechanisms undergone by different species or different strains of bacteria occupying the common microbiome. With the advancement in the field of molecular typing and genome sequencing, stability or modifications in the genome can be assessed. Data analysis from the isolates collected from individuals supplied with a probiotic strain gives the clue for the stability of the organism. Exposure of organisms to adverse conditions like oxidative stress, highly acidic/alkaline conditions, and extreme temperature may cause a modification in the metabolic rate of the fed culture which might lead to a loss in the stability of the product. Fecal re-isolates of PRSF-L477 strain did not show any remarkable genomic modifications thus indicating its genetic stability during healthy gastrointestinal tract passage (Wind et al., 2010). In the commercial application of these nutraceutical secreting probiotics, undisturbed stability is a crucial factor for long term storage and functionality. Maintaining the microbial stability and product stability should be carefully monitored in order to obtain similar potential activity of the nutraceuticals in all batches of fermentation. The stability of the nutraceutical products is extended by the process of lyophilization or the dry blending into capsules (Dao et al., 2017)

Viability

Viability of the functional probiotic culture is a critical prerequisite for obtaining the health benefits of the culture. Various studies analyze the microbial and environmental factors that optimize the parameters that alter the cell viability and functionality of the live bio-therapeutic product (LBP). The nutraceuticals and the secreting culture have to be stored either freeze-dried or lyophilized to enhance the viability of the cells (Champagne, Ross, Saarela, Hansen, & Charalampopoulos, 2011). Though viability was not considered as a crucial factor for probiotic action, few reports state that viable culture exerted better probiotic activity compared to non-viable cells. The major difference in viable and non-viable cells is based on the adhesion of probiotics to the host tissues (Nagpal et al., 2012). Colonization of gut probiotics is an important factor for adherence to intestinal tracts to exert beneficial effects to the host organisms. Secretion of antibacterial nutraceutical compounds by the colonized bacteria is necessary for competitive exclusion of the pathogens (Kerry et al.,2018). A study has shown that both live and heat-killed *L. rhamnosus*

HN001 are capable of increasing the phagocytotic activity of leucocytes in mice, but only the live bacteria manage to enhance the antibody response of the gut mucosa against cholera toxin. This demonstrated that while innate immunity is responsive to killed-bacteria, specific gut immunity can be stimulated by live bacteria only (Gill & Rutherfurd, 2001).

Techniques like extrusion, spray drying, emulsion, and encapsulation are used to enhance the viability of the cells. Polysaccharide encapsulation is one common tool used to preserve the viability of probiotic cells (Chávarri, Marañón, & Villarán, 2012). Storage temperature, packaging conditions, and microenvironments influence the viability of probiotic strains. The probiotic strain maintained at 37 °C showed the viability for 18 hours and at 4 °C showed the viability for 15 days (Awaisheh, Al-Dmoor, Omar, Hawari, & Alroyli, 2012) Several other factors like freezing (-30 °C to -80 °C) and thawing, drying, compression, dehydration, etc., also influence the viability of the microbes as products. Alginate gel microgels were used for enhancing the viability of *Pediococcus pentosaceus* Li05. Results of this study showed that microencapsulation provided protection against conditions like heat treatment and gastrointestinal transit (Yao et al.,2018). Cryoprotectants such as sucrose enhanced the viability of the culture up to a period of six months and the probiotic properties of *Leuconostoc mesenteroides* was retained for a longer period of time (Shobharani & Agrawal, 2011). Protecting the viability of the culture will enhance the efficacy of the probiotic culture during storage as well as in the *in-vivo* environment (Chávarri, Marañón, & Villarán, 2012).

Availability

Food and Drug Administration (FDA) has conferred the Generally Regarded as Safe (GRAS) status and the European Food Safety Authority (EFSA), convened the Qualified Presumption of Safety (QPS) to the traditional probiotics and their secretory products to augment their application in human health and disease management (Sanders et al., 2010). As the search for the next generation probiotics is increasing, the safety of the probiotic culture and the secretory products are to be confirmed before applying them for therapeutic purposes. However, with the advancements in microbial research, a new term 'synbiotics' is gaining momentum. It is the fusion of both probiotics and prebiotics resulting in enhanced viability and colonization of the live microbes present in supplements (Tufarelli & Laudadio, 2016). There are also some other bacterial products that have a positive effect on barrier function and signaling pathways even in the absence of live bacteria (Patel & Denning 2013). Such products called postbiotics can be defined as non-living substances produced by probiotics with biological activity in the host. They may be ethanol, diacetyl compounds, bacteriocins, hydrogen peroxide, and acetaldehydes. These postbiotics are antagonistic to the growth of pathogens and can replace antibiotics as they are non-toxic and are resistant to human enzymes (Kerry et al., 2018). The productivity of a probiotic product can be increased by changing its genetic makeup. Vitamin B12 synthesis is reported from bacteria such as *Pseudomonas denitrificans*, *Propionibacterium shermanii*, or *Sinorhizobium meliloti* (Martens, Barg, Warren & Jahn, 2002). Genetic change in the *Pseudomonas denitrificans* and *Propionibacterium shermanii* leads to increased yield of about 20% -30% when compared to the wild strain (Murooka, Piao, Kiatpapan, & Yamashita, 2005). Some of the most commonly utilized microbes in industries include the family of *Trichocomaceae, Lactobacillaceae, Bifidobacteriaceae, Streptomycetaceae and Saccharomycetaceae.* Micro-organisms play an important role in the fermentation of food and the taste, metabolic activity and the effect of fermented food differs from one organism to another. Common micro-organisms used in fermentation are *Aspergillus*, *Rhizopus*, *Mucor*, *Actinomucor*, *Monascus*, *Saccharomyces*, *Neurospora*, *Acetobacter*, *Bacillus*, and *Lactobacillus* (Hesseltine & Wang, 1967). Thus the diversified micro-organisms used in fermentation modify the substrate biochemically and organoleptically for human consumption (Tamang, 1998).

Commercial Production Of Nutraceuticals

Nutraceuticals have been widely used in food technology for preservation, fermentation and other purposes. There are certain probiotics which have been used as starters that enhance the fermentation process and improve the quality of fermented food. These starter microbes along with the other probiotics give a synergic effect for the growth of the microbes in the respective batch and magnifying the effect of the microbes in the fermentation process (Heller, 2001). Mostly used starters are LAB and these are used for their metabolic activities of acid production, ability to improve the fermentation process, detoxification, control of spoilage microbes, and their ability to degrade toxic substances like linamarin and mycotoxins. (Holzapfel, 1997).

Processing of Probiotics

Probiotics can be consumed in various forms. They exist as encapsulated pills containing powdered probiotics or as drinks such as Yakult. They usually contain complex mixtures of more than two microorganisms where the compatibility between the microbes has to be tested prior to use. Generally, the drying methods such as freeze-drying, spray drying or vacuum drying are employed to reduce the cost and to increase the shelf-life. The freeze-dried probiotics have well-preserved morphology of the microencapsulated cells (Chen & Mustapha, 2012). But the frequent freezing and thawing mechanisms reduce the viability of the probiotics

Table 1. List of micro-organisms and their products

Organism	Product	References
Bacillus cereus	Anti-oxidants	(Smith & Alford, 1970)
Aspergillus niger		(Zaika & Smith, 1975)
Aspergillus fumigatus		(Arora & Chandra, 2011)
Aspergillus candidus		(Malpure, Shah & Juvekar,2006)
Streptomyces albulus	Poly amino acids	(Chen et al., 2011)
Kitasatopora sp. MY 5-36		(Chen et al., 2011)
Sterptomyces sp. M-Z18		(Chen et al., 2011)
Bacillus subtilis		(Silva, Cantarelli, & Ayub, 2013)
Lactococcus lactis		(Yu & Osullivan, 2014)
Escherichia coli		(Tseng, Fang, Cho, Chen, & Tsai, 2012)
Xanthophyllomyces dendrorhus	Bio-colorant	(Verwaal et al., 2007)
Cornybacteriuminsidiosum		(Starr, 1958)
Escherichia coli	Prebiotic	(Lee et al., 2012)
Eubacterium rectale		(Scott, Martin, Duncan, & Flint, 2013)
Bifidobacterium longum		(Scott, Martin, Duncan, & Flint, 2013)
Faecalibacterium prausnitzii		(Scott, Martin, Duncan, & Flint, 2013)
Bacillius subtilis	Probiotics	(Hoa et al., 2000)
Bacillus coagulans		(Endres et al., 2009)
Bacillus cereus		(Trapecar et al., 2011)
Sacchromyces boulardii		(Doron, Hibberd, & Gorbach, 2008)
Lactobacillus casei		(Desai, 2008)
Lactobacillus acidophilus		(Gomes & Malcata, 1999;Dave & Shah, 1997)
Lactobacillus fermentum		(Gilliland, 1990)
Bifidobacteriuminfantis		(Härtel et al., 2014)
Bifidobacteriumlactis		(Prescott et al., 2008; Desai, 2008)

(Akın, Akın, & Kırmacı, 2007; Santivarangkna, Kulozik, & Foerst, 2006). Spray drying is the cheaper method when compared to freeze drying. Spray drying involves atomization of the liquid medium containing probiotic organisms at a temperature of around 200°C. The spray dried organism undergoes heat stress (Brennan, 1986). It has been reported that an organism's DNA, RNA, and cellular contents are affected during the spray drying method (Crowe, et al., 1988). Before undergoing the spray drying, mild heat treatment of 52°C for 15 minutes can enhance cell survival as reported by Paéz et al. (2012). Drink such as Yakult contains microorganisms, which have higher cell viability, but with a lower shelf-life. This type of probiotics need storage facilities such as refrigeration. The optimum temperature for storage under refrigeration is 4-5°C (Mortazavian et al., 2008). Proper storage leads to high efficiency and good cell viability.

Traditional Food Sources as Nutraceuticals

Earlier, people consumed wholesome food, which included all the necessary supplements required for growth and good health. For example, fortified food is a rich source of probiotics. The South Indian food porridge (Koozh) containing *Lactobacillus* sp., is abundant in nutritional components (Kumar, et al., 2010). Kefir is a beverage that is rich in probiotics and is used in many clinical aspects. Kefir grains are used in Tibet and China as natural starters for the production of the traditional probiotic drink made by fermenting the milk. It is a source of several bacteria and fungi. The complex microbial community in kefir is composed of Lactic acid bacteria (LAB) and yeasts such as *Saccharomyces, Torula,* and *Kluyveromyces* (Zheng et al., 2013) The organism that constitutes an overwhelming majority in kefir belongs to the genus *Lactobacillus*. Some of the commonly associated species are *Lactobacillus plantarum, Lactobacillus delbrueckii, Lactobacillus casei, Lactobacillus reuteri, Lactobacillus acidophilus, Lactobacillus kefiranofaciens* and *Lactobacillus buchneri*. Strains belonging to *Leuconostoc* and *Lactococcus* are also found (Slattery et al., 2019). Kefir has been associated with the lowering of serum cholesterol levels, improved digestion and gut health, and reduction in hypertension. It is also known to show anti-cancer properties (Slattery et al; 2019). Kefir is known to have a bactericidal effect on gram-negative bacteria (Sarkar, 2007). The bactericidal activity has been reported against *Listeria monocytogenes, Shigella flexneri, Salmonella enteritidis and E. coli* (Santos, San Mauro, Sanchez, Torres, & Marquina, 2003).

The traditional Korean food Doenjang has been consumed as a rich source of protein and as a flavoring agent. It is made out of cooked and crushed soya bean blocks called Meju which is a naturally fermented soybean product that functions as a source of nutrients, flavors, and microorganisms for the Doenjang (Patra et al.,2016). The species of bacteria that are dominantly found in meju such as the *Bacillus spp, Clostridium spp, and Enterococcus spp* are transferred to Doenjang too. The dominant microbes in Doenjang are found to be *B. subtilis* and *B. licheniformis*. Fungi such as

Aspergillus, Mucor, and *Rhizopus* species are also present (Patra et al., 2016; Park et al., 2013). There are several health benefits associated such as its properties of anti-cancer, anti-mutagenicity, antiobesity, antioxidant, anti-inflammatory and anti-melanogenesis actions (Kang et al., 2016). The key constituents of doenjang known to contribute towards its health benefits are the isoflavones and their derivatives (Park et al., 2013). *Doenjang* has a good number of microorganisms like *Leuconostoc mesenteroides, Tetragenococcus halophilus*, and *Enterococcus faecium also bacillus species like Bacillus subtilis* and *B. licheniformis were detected, on further studies few fungi like Mucor plumbeus, Aspergillus oryzae*, and *Debaryomyces hansenii* were also observed (Kim et al.,2009).

NUTRACEUTICALS IN DISEASE

Effect on Cancer

According to WHO, cancer has been the second leading cause of death and about 9.6 million cancer-related deaths have been reported globally in 2018. Majority of the victims belong to low- and middle-income countries such as India. Thus, the need for developing efficient treatment methods is increasing exponentially. Although researchers are constantly involving themselves in proteomics, genomics, nanotechnology and molecular pathology for the discovery of new drugs, the side effects of these have been a major limitation (Kerry et al., 2018). Thus, over the years the focus has shifted towards natural dietary sources conferring anti-cancer effects. Leading this race are the food products probiotics and prebiotics which have evolved as potential agents against cancer(Gayathri & Rashmi, 2016).

Probiotics use different strategies to show their anti-cancer properties. One of them is by causing the inhibition of pathogenic microorganisms (Gayathri & Rashmi,2016). In 2008, 2 million out of the reported 12.7 million cancer cases were attributed to different microorganisms. More than half of the global cancers caused by infectious agents were because of pathogens such as the Epstein barr virus (EBV), Hepatitis B virus (HBV), Human papilloma virus (HPVs), Hepatitis C virus and Helicobacter pylori bacteria (Schottenfeld et al). EBV is associated with Burkitt's lymphoma, hodgkin's and non-hodgkins lymphoma and other carcinomas (Thompson & Kurzrock, 2004). Human papilloma virus is associated with bladder cancer (Cooper, Haffajee & Taylor, 1997). Thus, by eliminating these pathogens probiotics can reduce the cases of infection-associated cancers. There are several reports showing the action of probiotics against different cancer-related pathogens. The probiotic bacteria *Lactobacillus rhamnosus* has been reported to show antiviral activities against Herpes simplex virus -1 by increasing the activity of macrophages

(Khani et al, 2012). Probiotics are believed to show anti-cancer effects due to their ability to exclude the pathogens by limiting the nutrients available and competing for receptors. This way they change the composition of the gut microbiota. They also produce metabolites that protect against cancer, reduce mutagenicity and reduce the genotoxicity of the dietary carcinogens. They are also known to increase the host immune response by producing antibacterial substances such as bacteriocins, hydrogen peroxide and lactic acid (Gayathri & Rashmi,2016). The probiotic strains of *Lactobacillus fermentum* have proven to be highly potent in suppressing colorectal cancer cells in *in-vitro* studies. They are also known to promote the normal growth of colon epithelial cells. The strains such as *Lactobacillus acidophilus* and *Lactobacillus casei* have also shown positive results in *in-vitro* studies against colorectal cancer cell lines by showing antiproliferative activity (Kerry et al., 2018).

The second strategy used by probiotics is by reducing mutagenesis. This can be illustrated using the example of heterocyclic amines (HAA). When the meat is cooked at a high temperature the HAAs are released and these are associated with the development of colon cancer. The bacteria such as *Lactobacilli* and *Bifidobacteria* are known to play an important role in reducing the mutagenesis by directly binding to these mutagenic compounds and their metabolites and thus reducing their absorption in the intestine (Orrhage et al., 1994). Although probiotics have shown promising results against cancer, the research is limited to in-vitro tests (Kerry et al., 2018). Thus, there is a need for increased in vivo experiments and clinical trials in order to explore the full potential of these nutraceuticals.

Effect on Diabetes

Type 2 diabetes mellitus (T2DM) is a group of metabolic diseases characterized by insulin resistance that results in a hyperglycaemic condition. The prolonged hyperglycemia leads to other complications such as dysfunction of eyes, nerves, kidneys, heart, and blood vessels (Chuang Li,2014) The International Diabetes Federation (IDF) of Southeast Asia states that 425 million people worldwide suffer from diabetes and the number is expected to increase to 629 million by 2045 if no action is taken (Kerry et al., 2018) Although several medications are currently available in the market, there isn't an absolute cure for diabetes. Researchers have now started to elucidate the relationship between intestinal microbiota and metabolic diseases like diabetes (Larsen et al., 2010). This is because the existing antidiabetic agents such as biguanides and α-glucosidase inhibitors have adverse side effects such as flatulence, abdominal discomfort, and diarrhea (Chuang Li, 2014). Several reports have postulated that by using probiotics the gut microbiota can be altered to show promising results (Balakumar et al., 2018). Thus, several bacterial strains and dietary fibers have been studied for their effect on glucose metabolism.

Several countries have been using Bifidobacterium species over the years as they help in controlling metabolic disorders such as diabetes. They control genes involved in the insulin-signaling pathway such as IR-β, IRS-1, and Akt which causes an increase in the glucose uptake (Le at al., 2015). *Bifidobacterium lactis* HY 8101 has shown promising results against diabetes because of its control on the expression of important genes such as PPAR-γ (insulin sensitivity-related genes), pp-1 (glycogen synthesis-related enzymes) and GLUT4 (glucose uptake related genes). It also decreases the mRNA expression of GSK-3β (glycogen synthesis related enzymes) and G6PC (gluconeogenesis-related enzymes) (Kim et al., 2014). Adherence to the mucosal barrier and translocation of live bacteria from the intestine to the adipose tissue causes insulin resistance. A report on the treatment with probiotic *Bifidobacterium animalis subsp. lactis* 420 (B420) shows that it provides protection against diabetes by preventing both adherence to the mucosal barrier and translocation of live bacteria from the intestine and thus provides protection (Amar et al, 2011).

Another organism that has been frequently used for its activity against diabetes is *Lactobacillus*. Studies highlighting the effect of probiotics on gene expression suggest that their anti-diabetic effect is because of the control these organisms have over genes involved in glucose uptake and homeostasis such as glucose transporter-4 (GLUT-4) (Miraghajani et al, 2017). A study on *Lactobacillus plantarum* NCU116 has shown their ability to upregulate GLUT-4. Due to this ability, *Lactobacillus plantarum* NCU116 is able to play an important role in glucose uptake and thus control diabetes in rats. Other genes such as PPAR-α and PPAR-γ which regulate insulin sensitivity and glucose homeostasis are also controlled by these bacteria (Chuang Li, 2014). A study that used a combination of the probiotic bacteria *Lactobacillus* and *Bifidobacterium* showed evidence of reduced expression of a chemokine responsible for macrophage infiltration of adipose tissue that causes insulin resistance (Kanda et al., 2006).

The disorders associated with glucose metabolism have been related to an increase in the permeability of the gut barrier that causes the translocation of endotoxins and commensal bacteria that are responsible for causing tissue inflammation and the disorder itself (Stenman et al., 2015). The enhanced endotoxemia and lipopolysaccharides can trigger insulin resistance in mice. Probiotic bacteria can suppress the endotoxemia and exhibit a protective effect. A report on the treatment with *Lactobacillus paracasei* subsp. *paracasei* NTU 101 shows that it improves the intestinal environment and prevents the translocation of bacterial lipopolysaccharides into the systemic circulation (Miraghajani et al., 2017).

Type 1 diabetes is a disease that results from the combined effects of immunological, genetic and environmental factors(Trucco, 2009). It is associated with increased levels of blood sugar and oxidative stress that has shown to cause free radical accumulation resulting in damage of multiple tissues such as B cells of the pancreas and damage to organs such as the kidney and liver. When there is an inflammation of the pancreas, there is also leakage of α-amylase (Carbohydrate digesting enzyme) into the bloodstream increasing its serum levels. Studies show that the use of *L. plantarum* TN627 has shown to reduce the serum α-amylase activity and thus show an improvement in the glycemic index (Bejar et al., 2013). Probiotic bacteria *Lactobacillus kefiranofaciens M* and *Lactobacillus kefir K* have a controlling effect on diabetes type-1 due to their inhibitory action on pro-inflammatory and inflammatory cytokines and increased the production of IL-10 which also inhibits pro-inflammatory cytokines (Wei et al., 2015). Patients with diabetes have lower levels of reactive oxygen species scavengers. Hence, the free radicals generated in these patients may cause lipid peroxidation and the production of malondialdehyde (MDA). Probiotics such as Lactobacillus GG are shown to lower the levels of MDA causing suppression of oxidative stress and increased tolerance towards glucose (Tabuchi et al., 2003). Glucagon-like Peptide 1 (GLP-1) is a hormone that acts as an incretin or a stimulator of insulin secretion. The decrease in GLP-1 secretion has been linked to hyperglycemia. Due to its properties, GLP-1 and its agonists are being evaluated by researchers for their role in therapy. (Holst, 2007) Certain probiotic bacteria show increased levels of incretins like GLP-1produced by the cells of the intestine. They act by increasing the release of insulin by the pancreatic cells. (Miraghajani et al., 2017). These GLP-1 incretins have also been used to study their action against Diabetes type-1. Probiotics such as *Lactobacillus kefiranofaciens M* and *Lactobacillus kefir K* are known to stimulate the production of the GLP-1 (Wei et al., 2015). Some reports also suggest that the action of probiotics is linked to the elevated levels of short-chain fatty acids (SCFAs) such as butyrate in the intestine as SCFAs have shown increased production of GLP-1 in animal models (Yadav et al.,2013). Studies on probiotic yeasts have also shown promising results. The yeast Saccharomyces boulardii has also shown to modify the production of SCFAs such as Butyrate. *Saccharomyces boulardii* also alter the composition of the intestinal environment. It does so by increasing the proportion of *Bacteroidetes* and decreasing *Firmicutes, Tenericutes,* and *Proteobacteria* that have been associated with type 2 diabetes in mice (Everard et al., 2014)

The effects of probiotics on the inflammatory and immune response pathways have also been studied. *Bifidobacterium spp.* increase the production of innate immune response proteins such as protein kinase B (Akt) and extracellular-signal-regulated kinase 2 (ERK2) increase in which causes adipocytes to differentiate into cells that induce insulin sensitivity in diabetic mice that are fed with *Bifidobacterium spp* (Le et al., 2015) Pro-inflammatory adipokines are known to interfere with the pathways of insulin signaling in the peripheral tissues. This causes insulin resistance. A study suggests that the use of *Lactobacillus plantarum No. 14* prevents the development of insulin resistance through the reduction of visceral fat which has an inhibitory action on the adipokines (Okubo et al., 2013). Thus, by shedding light on the different mechanisms by which the probiotic organisms act against diabetes, it is possible to provide a rational reason for the consumption of these organisms by diabetic patients.

Effect on Cardiovascular Disease

Cardiovascular disease is one of the major causes of adult-deaths in developing and developed countries (Ettinger et al., 2014). One of the conditions predominantly seen is Coronary heart disease (CHD). It is a condition that results in reduced efficiency of the coronary arteries supplying blood and oxygen to the heart muscles the main cause for which is plaque accumulation (DiRienzo, 2014; Akbarzadeh & Homayouni, 2012). Cardiovascular diseases are associated with risk factors like diabetes, hypertension, obesity, and increased serum lipid levels (Prabhakaran, 2016). The earliest records of influence of microorganisms on cardiac health date back to the work of Sharper (1963) and Mann (1974) who observed that the cholesterol levels of the people belonging to the Maasai and Samburu tribes of Africa were low despite the consumption of a large number of dairy products. (Pereira & Gibson, 2002) The Tribes also showed a low incidence of cardiovascular diseases. On further investigation, it was found that the milk consumed by them was fermented with a wild Lactobacillus strain and since then several studies have tried to correlate the relationship between probiotics and cholesterol levels (DiRienzo, 2014).

One of the primary risk factors associated with CHD is low-density lipoprotein cholesterol (LDL-C). Other risk factors include serum triacylglycerol (TAG), triglyceride-rich lipoproteins and low levels of high-density lipoprotein cholesterol (HDL-C). Although all the factors contribute to the disease, the chief factor targeted for therapy is the LDL-C as it is associated with atherosclerotic plaques (DiRienzo, 2014). The first line of treatment for those with elevated LDL-C is a controlled diet and exercise. But results using these methods alone have not been satisfactory. Reports suggest that only one-third of the patients receiving treatment achieve their target due to the cost involved and other limitations. Hence, there has been an effort to use nutraceuticals that can lower LDL-C levels. (Jones et al., 2012). In a meta-analytical study of randomized controlled clinical trials, data were pooled from 485 participants and it was found that there was a reduction in LDL-C and total cholesterol due to probiotic consumption. (Guo et al., 2011)

Some of the mechanisms proposed for the mode of action of probiotics are as follows:

- Cellular surfaces and membranes of probiotics bind to the cholesterol (DiRienzo, 2014, Ettinger et al., 2014)
- Cholesterol assimilation in the probiotics (Pereira & Gibson, 2002)

- Bile salt hydrolase (BSH) deconjugation by the microbes causes the excretion of the deconjugated bile salts that result in increased utilization of the body cholesterol for the production of new bile salts. (Jones et al., 2012; Pereira & Gibson, 2002)
- The deconjugation of bile salts in the intestine may also cause reduced absorption of cholesterol by Gram-positive organisms including Lactobacillus and Bifidobacterium (Jones et al., 2012, Ettinger et al., 2014)
- Carbohydrate fermentation leading to the production of short-chain fatty acids that causes a dip in the blood lipid levels and reduced cholesterol production by the liver (Pereira & Gibson, 2002)

In a study, four probiotic strains of *Lactobacillus reuteri* NCIMB 30242, *Enterococcus faecium*, and the combination of *Lactobacillus acidophilus* La5 and *Bifidobacterium lactis Bb12* were used. *L reuteri* NCIMB 30242 showed the most promising result. Two synbiotics i.e. a combination of probiotic and prebiotic were also used. *Lactobacillus acidophilus* CHO-220 with inulin and Lactobacillus acidophilus with fructo-oligosaccharides also decreased LDL-C (DiRienzo, 2014). A probiotic product 'CardiovivaTM' uses *Lactobacillus reuteri* NCIMB 30242 in an encapsulated form showed positive results (Ettinger et al., 2014). The microencapsulation creates a protective barrier against immunoglobulins, enzymes, buffers and pH variations concentrating the bacteria within the microcapsule for efficient deconjugation (Jones et al., 2012).

Another risk factor associated with CVD is obesity (Ettinger et al., 2014). A study was conducted by orally administering probiotics to healthy mice over a period of 3 weeks. It was found that *Lactobacillus rhamnosus* GG or *Lactobacillus sakei* NR28 probiotic bacteria changed the gut microbiome and caused a significant reduction in the weight of the mice receiving treatment. The food consumption was same among the untreated and treated mice indicating that the reduction in the epididymal fat mass was not due to difference in energy consumption, but due to the administration of the probiotics (Ji et al., 2012). A study used a combination of multiple strains of probiotic bacteria administered in the form of a capsule to overweight individuals having a body mass index of above 25. The Capsule contained strains of *Streptococcus thermophilus, Lactobacillus rhamnosus, Lactobacillus acidophilus, Lactobacillus plantarum, Bifidobacterium longum, Bifidobacterium lactis,* and *Bifidobacterium breve*. The twice daily administration resulted in a substantial reduction in weight and levels of serum cholesterol after 8 weeks. These changes correlated with an increase in *Lactobacillus plantarum* populations (Lee et al., 2014).

Probiotics have not only been studied for their effects on factors precursory to heart damage but also for their direct cardioprotective effects (Ettinger et al., 2014). The bacteria *Lactobacillus rhamnosus GG* is known to have a protective effect against apoptosis induced by TNF, IL-1a, and IFN-g in mouse colon cells through the activation of Akt pathway and by the inhibition of the pro-apoptotic p38 mitogen-activated protein kinase pathway. Rat models were used to study the effects of a protein p75 isolated from the bacteria for its action on ischemia/reperfusion (I/R) induced heart cell injury. It was found that the rats pre-treated with p75 30 minutes before surgery showed reduced heart tissue infarction in a dose-dependent manner (Yan et al., 2007). A probiotic drink called "GoodBelly" containing *Lactobacillus plantarum 299* v was administered to rats for 14 days prior to heart surgery showed a cardioprotective effect on the rats and a reduction in ischemia (Lam et al., 2012). With extensive research, it may be possible to replace the medications of CVD with probiotics and reduce the side-effects and cost of medications.

CONCLUSION

Functional foods are a boon to society. Their major advantage is that they are a part of our daily diet and this increases the ease of use by multiple folds. They are also useful because of being non-toxic and having a minimum number of side-effects unlike the antibiotics and other drugs that are currently present in the market. These nutraceuticals not only keep us healthy but can also be used as powerful weapons against deadly diseases such as cancer, obesity, cardiovascular diseases, stress, and depression. Recent years have given rise to industries that produce functional food supplements using live microbes or microbial derived products. There is scope for improvement in the productivity of these nutraceuticals by genetically modifying the organisms. Their properties can be utilized to cure a wider range of diseases, specifically those diseases that do not respond to antibiotics due to multi-drug resistance. The pigments like phycocyanin can act as nutraceuticals that serve as natural dye in the food industries (Kuddus, Singh, Thomas, & Al-Hazimi, 2013). Phycocyanin acts as a free radical scavenger that can prevent the tissue undergoing inflammatory response and apoptosis (Kuddus, Singh, Thomas, & Al-Hazimi, 2013). Nutraceuticals like Omega 3 fatty acids and polyphenols are used to treat artherosclerosis (Massaro et al., 2010). As the trend of personalized nutrition is also evolving, personalized nutraceuticals can also be designed for different individuals.

REFERENCES

Abdrabou, A. M., Osman, E. Y., & Aboubakr, O. A. (2018). Comparative therapeutic efficacy study of Lactobacilli probiotics and citalopram in treatment of acute stress-induced depression in lab murine models. *Human Microbiome Journal, 10*, 33–36. doi:10.1016/j.humic.2018.08.001

Akbarzadeh, F., & Homayouni, A. (2012). Dairy probiotic foods and coronary heart disease: A review on mechanism of action. *Probiotics. InTech*, 121-8.

Akın, M. B., Akın, M. S., & Kırmacı, Z. (2007). Effects of inulin and sugar levels on the viability of yogurt and probiotic bacteria and the physical and sensory characteristics in probiotic ice-cream. *Food Chemistry, 104*(1), 93–99. doi:10.1016/j.foodchem.2006.11.030

Amar, J., Chabo, C., Waget, A., Klopp, P., Vachoux, C., Bermúdez-Humarán, L. G., ... Ouwehand, A. (2011). Intestinal mucosal adherence and translocation of commensal bacteria at the early onset of type 2 diabetes: Molecular mechanisms and probiotic treatment. *EMBO Molecular Medicine, 3*(9), 559–572. doi:10.1002/emmm.201100159 PMID:21735552

Arora, D. S., & Chandra, P. (2011). Antioxidant activity of Aspergillus fumigatus. *ISRN Pharmacology*, 2011. PMID:22084718

Awaisheh, S., Al-Dmoor, H., Omar, S., Hawari, A., & Alroyli, M. (2012). Impact of selected nutraceuticals on viability of probiotic strains in milk during refrigerated storage at 4C for 15 days. *International Journal of Dairy Technology, 65*(2), 268–273. doi:10.1111/j.1471-0307.2011.00817.x

Balakumar, M., Prabhu, D., Sathishkumar, C., Prabu, P., Rokana, N., Kumar, R., ... Mohan, V. (2018). Improvement in glucose tolerance and insulin sensitivity by probiotic strains of Indian gut origin in high-fat diet-fed C57BL/6J mice. *European Journal of Nutrition, 57*(1), 279–295. doi:10.100700394-016-1317-7 PMID:27757592

Bejar, W., Hamden, K., Salah, R. B., & Chouayekh, H. (2013). Lactobacillus plantarum TN627 significantly reduces complications of alloxan-induced diabetes in rats. *Anaerobe, 24*, 4–11. doi:10.1016/j.anaerobe.2013.08.006 PMID:23999246

Bernardeau, M., Guguen, M., & Vernoux, J. P. (2006). Beneficial lactobacilli in food and feed: Long-term use, biodiversity and proposals for specific and realistic safety assessments. *FEMS Microbiology Reviews, 30*(4), 487–513. doi:10.1111/j.1574-6976.2006.00020.x PMID:16774584

Brennan, M., Wanismail, B., Johnson, M. C., & Ray, B. (1986). Cellular damage in dried Lactobacillus acidophilus. *Journal of Food Protection*, *49*(1), 47–53. doi:10.4315/0362-028X-49.1.47 PMID:30959616

Brower, V. (1998). Nutraceuticals: Poised for a healthy slice of the healthcare market? *Nature Biotechnology*, *16*(8), 728–731. doi:10.1038/nbt0898-728 PMID:9702769

Champagne, C. P., Ross, R. P., Saarela, M., Hansen, K. F., & Charalampopoulos, D. (2011). Recommendations for the viability assessment of probiotics as concentrated cultures and in food matrices. *International Journal of Food Microbiology*, *149*(3), 185–193. doi:10.1016/j.ijfoodmicro.2011.07.005 PMID:21803436

Chanda, S., Tiwari, R. K., Kumar, A., & Singh, K. (2019). Nutraceuticals inspiring the current therapy for lifestyle diseases. *Advances in Pharmacological Sciences*, *2019*, 1–5. doi:10.1155/2019/6908716 PMID:30755770

Chávarri, M., Marañón, I., & Villarán, M. C. (2012). Encapsulation technology to protect probiotic bacteria. In Probiotics. IntechOpen. doi:10.5772/50046

Chen, M., & Mustapha, A. (2012). Survival of freeze-dried microcapsules of α-galactosidase producing probiotics in a soy bar matrix. *Food Microbiology*, *30*(1), 68–73. doi:10.1016/j.fm.2011.10.017 PMID:22265285

Chen, X., Ren, X., Dong, N., Li, S., Li, F., Zhao, F., ... Mao, Z. (2011). Culture medium containing glucose and glycerol as a mixed carbon source improves ε-poly-l-lysine production by Streptomyces sp. M-Z18. *Bioprocess and Biosystems Engineering*, *35*(3), 469–475. doi:10.100700449-011-0586-z PMID:21909683

Claudia, A. (2011). Nutraceuticals in pet foods and practice. *World Small Animal Veterinary Association World Congress Proceedings* (pp. 1-3). Knoxville, TN: University of Tennessee College of Veterinary Science.

Cooper, K., Haffajee, Z., & Taylor, L. (1997). Human papillomavirus and schistosomiasis associated bladder cancer. *Molecular Pathology*, *50*(3), 145–148. doi:10.1136/mp.50.3.145 PMID:9292149

Coppens, P., Silva, M. F., & Pettman, S. (2006). European regulations on nutraceuticals, dietary supplements and functional foods: A framework based on safety. *Toxicology*, *221*(1), 59–74. doi:10.1016/j.tox.2005.12.022 PMID:16469424

Crowe, J. H., Crowe, L. M., Carpenter, J. F., Rudolph, A., Wistrom, C. A., Spargo, B., & Anchordoguy, T. (1988). Interactions of sugars with membranes. *Biochimica Et Biophysica Acta (BBA) -. Reviews on Biomembranes*, *947*(2), 367–384.

Dao, H., Lakhani, P., Police, A., Kallakunta, V., Ajjarapu, S. S., Wu, K., ... Murthy, S. N. (2017). Microbial stability of pharmaceutical and cosmetic products. *AAPS PharmSciTech*, *19*(1), 60–78. doi:10.120812249-017-0875-1 PMID:29019083

Das, L., Bhaumik, E., Raychaudhuri, U., & Chakraborty, R. (2012). Role of nutraceuticals in human health. *Journal of Food Science and Technology*, *49*(2), 173–183. doi:10.100713197-011-0269-4 PMID:23572839

Das, L., Bhaumik, E., Raychaudhuri, U., & Chakraborty, R. (2012). Role of nutraceuticals in human health. *Journal of Food Science and Technology*, *49*(2), 173–183. doi:10.100713197-011-0269-4 PMID:23572839

Dashwood, R. H. (2003). Use of transgenic and mutant animal models in the study of heterocyclic amine-induced mutagenesis and carcinogenesis. *Journal of Biochemistry and Molecular Biology*, *36*(1), 35–42. PMID:12542973

Dave, R. I., & Shah, N. P. (1997). Viability of yoghurt and probiotic bacteria in yoghurts made from commercial starter cultures. *International Dairy Journal*, *7*(1), 31–41. doi:10.1016/S0958-6946(96)00046-5

Davì, G., Santilli, F., & Patrono, C. (2010). Nutraceuticals in diabetes and metabolic syndrome. *Cardiovascular Therapeutics*, *28*(4), 216–226. doi:10.1111/j.1755-5922.2010.00179.x PMID:20633024

Desai, A. R. (2008). Strain identification, viability, and probiotics properties of Lactobacillus casei (Doctoral dissertation, Victoria University).

DiRienzo, D. B. (2014). Effect of probiotics on biomarkers of cardiovascular disease: Implications for heart-healthy diets. *Nutrition Reviews*, *72*(1), 18–29. doi:10.1111/nure.12084 PMID:24330093

Doron, S. I., Hibberd, P. L., & Gorbach, S. L. (2008). Probiotics for prevention of antibiotic-associated diarrhea. *Journal of Clinical Gastroenterology*, *42*(Supplement 2), S58–S63. doi:10.1097/MCG.0b013e3181618ab7 PMID:18542041

Endres, J., Clewell, A., Jade, K., Farber, T., Hauswirth, J., & Schauss, A. (2009). Safety assessment of a proprietary preparation of a novel Probiotic, Bacillus coagulans, as a food ingredient. *Food and Chemical Toxicology*, *47*(6), 1231–1238. doi:10.1016/j.fct.2009.02.018 PMID:19248815

Ettinger, G., MacDonald, K., Reid, G., & Burton, J. P. (2014). The influence of the human microbiome and probiotics on cardiovascular health. *Gut Microbes*, *5*(6), 719–728. doi:10.4161/19490976.2014.983775 PMID:25529048

Everard, A., Matamoros, S., Geurts, L., Delzenne, N. M., & Cani, P. D. (2014). Saccharomyces boulardii administration changes gut microbiota and reduces hepatic steatosis, low-grade inflammation, and fat mass in obese and type 2 diabetic db/db mice. *mBio*, *5*(3), e01011–e01014. doi:10.1128/mBio.01011-14 PMID:24917595

Galanakis, C. M. (2013). Emerging technologies for the production of nutraceuticals from agricultural by-products: A viewpoint of opportunities and challenges. *Food and Bioproducts Processing*, *91*(4), 575–579. doi:10.1016/j.fbp.2013.01.004

Gayathri, D., & Rashmi, B. S. (2016). Anti-cancer properties of probiotics: A natural strategy for cancer prevention. *EC Nutrition*, *5*(4), 1191–1202.

Gill, H. S., & Rutherfurd, K. J. (2001). Viability and dose–response studies on the effects of the immunoenhancing lactic acid bacterium Lactobacillus rhamnosus in mice. *British Journal of Nutrition*, *86*(2), 285–289. doi:10.1079/BJN2001402 PMID:11502243

Gilliland, S. E. (1990). Health and nutritional benefits from lactic acid bacteria. *FEMS Microbiology Letters*, *87*(1-2), 175–188. doi:10.1111/j.1574-6968.1990.tb04887.x PMID:2271223

Gomes, A. M., & Malcata, F. (1999). Bifidobacterium spp. and Lactobacillus acidophilus: Biological, biochemical, technological, and therapeutical properties relevant for use as probiotics. *Trends in Food Science & Technology*, *10*(4-5), 139–157. doi:10.1016/S0924-2244(99)00033-3

Green, E. R., & Mecsas, J. (2016). Bacterial secretion systems–an overview. *Microbiology Spectrum*, *4*(1). doi:10.1128/microbiolspec.VMBF-0012-2015 PMID:26999395

Gueimonde, M., Flórez, A. B., van Hoek, A. H., Stuer-Lauridsen, B., Strøman, P., Clara, G., & Margolles, A. (2010). Genetic basis of tetracycline resistance in Bifidobacterium animalis subsp. lactis. *Applied and Environmental Microbiology*, *76*(10), 3364–3369. doi:10.1128/AEM.03096-09 PMID:20348299

Guo, Z., Liu, X. M., Zhang, Q. X., Shen, Z., Tian, F. W., Zhang, H., ... Chen, W. (2011). Influence of consumption of probiotics on the plasma lipid profile: A meta-analysis of randomised controlled trials. *Nutrition, Metabolism, and Cardiovascular Diseases*, *21*(11), 844–850. doi:10.1016/j.numecd.2011.04.008 PMID:21930366

Gupta, C., Prakash, D., & Gupta, S. (2014). Natural useful therapeutic products from microbes. *Journal of Microbiology & Experimentation*, *1*(1), 30-37.

Happel, A. U., Barnabas, S. L., Froissart, R., & Passmore, J. S. (2018). Weighing in on the risks and benefits of probiotic use in HIV-infected and immunocompromised populations. *Beneficial Microbes*, *9*(2), 239–246. doi:10.3920/BM2017.0106 PMID:29345159

Härtel, C., Pagel, J., Rupp, J., Bendiks, M., Guthmann, F., Rieger-Fackeldey, E., ... Göpel, W. (2014). Prophylactic use of Lactobacillus acidophilus/Bifidobacterium infantis probiotics and outcome in very low birth weight infants. *The Journal of Pediatrics*, *165*(2), 285–289.e1. doi:10.1016/j.jpeds.2014.04.029 PMID:24880888

Heller, K. J. (2001). Probiotic bacteria in fermented foods: Product characteristics and starter organisms. *The American Journal of Clinical Nutrition*, *73*(2), 374s–379s. doi:10.1093/ajcn/73.2.374s PMID:11157344

Hesseltine, C. W., & Wang, H. L. (1967). Traditional fermented foods. *Biotechnology and Bioengineering*, *9*(3), 275–288. doi:10.1002/bit.260090302

Hoa, N. T., Baccigalupi, L., Huxham, A., Smertenko, A., Van, P. H., Ammendola, S., ... Cutting, S. M. (2000). Characterization of Bacillus species used for oral bacteriotherapy and bacterioprophylaxis of gastrointestinal disorders. *Applied and Environmental Microbiology*, *66*(12), 5241–5247. doi:10.1128/AEM.66.12.5241-5247.2000 PMID:11097897

Holst, J. J. (2007). The physiology of glucagon-like peptide 1. *Physiological Reviews*, *87*(4), 1409–1439. doi:10.1152/physrev.00034.2006 PMID:17928588

Holzapfel, W. (1997). Use of starter cultures in fermentation on a household scale. *Food Control*, *8*(5-6), 241–258. doi:10.1016/S0956-7135(97)00017-0

Imperial, I. C., & Ibana, J. A. (2016). Addressing the antibiotic resistance problem with probiotics: Reducing the risk of its double-edged sword effect. *Frontiers in Microbiology*, *7*, 1983. doi:10.3389/fmicb.2016.01983 PMID:28018315

Jeong, D. W., Heo, S., Lee, B., Lee, H., Jeong, K., Her, J. Y., ... Lee, J. (2017). Effects of the predominant bacteria from meju and doenjang on the production of volatile compounds during soybean fermentation. *International Journal of Food Microbiology*, *19*(262), 8–13. doi:10.1016/j.ijfoodmicro.2017.09.011 PMID:28950164

Ji, Y., Kim, H., Park, H., Lee, J., Yeo, S., Yang, J., ... Bomba, A. (2012). Modulation of the murine microbiome with a concomitant anti-obesity effect by Lactobacillus rhamnosus GG and Lactobacillus sakei NR28. *Beneficial Microbes*, *3*(1), 13–22. doi:10.3920/BM2011.0046 PMID:22348905

Jones, M. L., Martoni, C. J., Parent, M., & Prakash, S. (2012). Cholesterol-lowering efficacy of a microencapsulated bile salt hydrolase-active Lactobacillus reuteri NCIMB 30242 yoghurt formulation in hypercholesterolaemic adults. *British Journal of Nutrition, 107*(10), 1505–1513. doi:10.1017/S0007114511004703 PMID:22067612

Kalra, E. K. (2003). Nutraceutical--definition and introduction. *AAPS PharmSci, 5*(3), E25. doi:10.1208/ps050325 PMID:14621960

Kanda, H., Tateya, S., Tamori, Y., Kotani, K., Hiasa, K. I., Kitazawa, R., ... Kasuga, M. (2006). MCP-1 contributes to macrophage infiltration into adipose tissue, insulin resistance, and hepatic steatosis in obesity. *The Journal of Clinical Investigation, 116*(6), 1494–1505. doi:10.1172/JCI26498 PMID:16691291

Kang, S. J., Seo, J. Y., Cho, K. M., Lee, C. K., Kim, J. H., & Kim, J. S. (2016). Antioxidant and neuroprotective effects of Doenjang Prepared with Rhizopus, Pichia, and Bacillus. *Preventive Nutrition and Food Science, 21*(3), 221–226. doi:10.3746/pnf.2016.21.3.221 PMID:27752498

Kastelein, J. J., Akdim, F., Stroes, E. S., Zwinderman, A. H., Bots, M. L., Stalenhoef, A. F., ... Groot, E. D. (2008). Simvastatin with or without Ezetimibe in Familial Hypercholesterolemia. *The New England Journal of Medicine, 358*(14), 1431–1443. doi:10.1056/NEJMoa0800742 PMID:18376000

Kerry, R. G., Patra, J. K., Gouda, S., Park, Y., Shin, H., & Das, G. (2018). Benefaction of probiotics for human health: A review. *Journal of Food and Drug Analysis, 26*(3), 927–939. doi:10.1016/j.jfda.2018.01.002 PMID:29976412

Khani, S., Motamedifar, M., Golmoghaddam, H., Hosseini, H. M., & Hashemizadeh, Z. (2012). In vitro study of the effect of a probiotic bacterium Lactobacillus rhamnosus against herpes simplex virus type 1. *The Brazilian Journal of Infectious Diseases, 16*(2), 129–135. doi:10.1016/S1413-8670(12)70293-3 PMID:22552453

Kim, S. H., Huh, C. S., Choi, I. D., Jeong, J. W., Ku, H. K., Ra, J. H., ... Ahn, Y. T. (2014). The anti-diabetic activity of Bifidobacterium lactis HY 8101 in vitro and in vivo. *Journal of Applied Microbiology, 117*(3), 834–845. doi:10.1111/jam.12573 PMID:24925305

Kim, T., Lee, J., Kim, S., Park, M., Chang, H. C., & Kim, H. (2009). Analysis of microbial communities in doenjang, a Korean fermented soybean paste, using nested PCR-denaturing gradient gel electrophoresis. *International Journal of Food Microbiology, 131*(2-3), 265–271. doi:10.1016/j.ijfoodmicro.2009.03.001 PMID:19324443

Ku, S., Park, M., Ji, G., & You, H. (2016). Review on bifidobacterium bifidum bgn4: Functionality and nutraceutical applications as a probiotic microorganism. *International Journal of Molecular Sciences, 17*(9), 1544. doi:10.3390/ijms17091544 PMID:27649150

Ku, S., Park, M. S., Ji, G. E., & You, H. J. (2016). Review on Bifidobacterium bifidum BGN4: Functionality and nutraceutical applications as a probiotic microorganism. *International Journal of Molecular Sciences, 17*(9), 1544. doi:10.3390/ijms17091544 PMID:27649150

Kuddus, M., Singh, P., Thomas, G., & Al-Hazimi, A. (2013). Recent developments in production and biotechnological applications of C-Phycocyanin. *BioMed Research International, 2013*, 1–9. doi:10.1155/2013/742859 PMID:24063013

Kumar, R. S., Varman, D. R., Kanmani, P., Yuvaraj, N., Paari, K. A., Pattukumar, V., & Arul, V. (2010). Isolation, characterization, and identification of a potential probiont from South Indian fermented foods (Kallappam, Koozh, and Mor Kuzhambu) and its use as biopreservative. *Probiotics and Antimicrobial Proteins, 2*(3), 145–151. doi:10.100712602-010-9052-5 PMID:26781237

Lam, V., Su, J., Koprowski, S., Hsu, A., Tweddell, J. S., Rafiee, P., ... Baker, J. E. (2012). Intestinal microbiota determine severity of myocardial infarction in rats. *The FASEB Journal, 26*(4), 1727–1735. doi:10.1096/fj.11-197921 PMID:22247331

Larsen, N., Vogensen, F. K., van den Berg, F. W., Nielsen, D. S., Andreasen, A. S., Pedersen, B. K., ... Jakobsen, M. (2010). Gut microbiota in human adults with type 2 diabetes differs from non-diabetic adults. *PLoS One, 5*(2), e9085. doi:10.1371/journal.pone.0009085 PMID:20140211

Lassen, J., & Yazdankhah, S. (2015). Assessment of probiotics in infant formula and cereal based baby foods containing Bifidobacterium lactis Bb12– Update 2014. *European Journal of Nutrition & Food Safety, 5*(2), 101–103. doi:10.9734/EJNFS/2015/14818

Le, T. K., Hosaka, T., Nguyen, T. T., Kassu, A., Dang, T. O., Tran, H. B., ... Pham, X. D. (2015). Bifidobacterium species lower serum glucose, increase expressions of insulin signaling proteins, and improve adipokine profile in diabetic mice. *Biomedical Research, 36*(1), 63–70. doi:10.2220/biomedres.36.63 PMID:25749152

Le, T. K. C., Hosaka, T., Nguyen, T. T., Kassu, A., Dang, T. O., Tran, H. B., ... Da Pham, X. (2015). Bifidobacterium species lower serum glucose, increase expressions of insulin signaling proteins, and improve adipokine profile in diabetic mice. *Biomedical Research, 36*(1), 63–70. doi:10.2220/biomedres.36.63 PMID:25749152

Lee, B., Kim, J., Kang, Y. M., Lim, J., Kim, Y., Lee, M., ... Je, J. (2010). Antioxidant activity and γ-aminobutyric acid (GABA) content in sea tangle fermented by Lactobacillus brevis BJ20 isolated from traditional fermented foods. *Food Chemistry*, *122*(1), 271–276. doi:10.1016/j.foodchem.2010.02.071

Lee, S. J., Bose, S., Seo, J. G., Chung, W. S., Lim, C. Y., & Kim, H. (2014). The effects of co-administration of probiotics with herbal medicine on obesity, metabolic endotoxemia and dysbiosis: A randomized double-blind controlled clinical trial. *Clinical Nutrition (Edinburgh, Lothian)*, *33*(6), 973–981. doi:10.1016/j.clnu.2013.12.006 PMID:24411490

Lee, W., Pathanibul, P., Quarterman, J., Jo, J., Han, N., Miller, M. J., ... Seo, J. (2012). Whole cell biosynthesis of a functional oligosaccharide, 2′-fucosyllactose, using engineered Escherichia coli. *Microbial Cell Factories*, *11*(1), 48. doi:10.1186/1475-2859-11-48 PMID:22545760

Li, C., Ding, Q., Nie, S.-P., Zhang, Y.-S., Xiong, T., & Xie, M.-Y. (2014). Carrot juice fermented with *Lactobacillus plantarum* NCU116 ameliorates type 2 diabetes in rats. *Journal of Agricultural and Food Chemistry*, *62*(49), 11884–11891. doi:10.1021/jf503681r PMID:25341087

Liu, R. H. (2004). Potential synergy of phytochemicals in cancer prevention: Mechanism of action. *The Journal of Nutrition*, *134*(12), 3479S–3485S. doi:10.1093/jn/134.12.3479S PMID:15570057

Ludwig, D. S. (1999). Dietary fiber, weight gain, and cardiovascular disease risk factors in young adults. *Journal of the American Medical Association*, *282*(16), 1539. doi:10.1001/jama.282.16.1539 PMID:10546693

Magiorakos, A., Srinivasan, A., Carey, R., Carmeli, Y., Falagas, M., Giske, C., ... Monnet, D. (2012). Multidrug-resistant, extensively drug-resistant and pan drug-resistant bacteria: An international expert proposal for interim standard definitions for acquired resistance. *Clinical Microbiology and Infection*, *18*(3), 268–281. doi:10.1111/j.1469-0691.2011.03570.x PMID:21793988

Malpure, P. P., Shah, A. S., & Juvekar, A. R. (2006). Antioxidant and anti-inflammatory activity of extract obtained from *Aspergillus candidus* MTCC 2202 broth filtrate. *Indian Journal of Experimental Biology*, *44*, 468–473. PMID:16784117

Martens, J. H., Barg, H., Warren, M., & Jahn, D. (2002). Microbial production of vitamin B 12. *Applied Microbiology and Biotechnology*, *58*(3), 275–285. doi:10.100700253-001-0902-7 PMID:11935176

Massaro, M., Scoditti, E., & Carluccio, M. A., & DeCaterina, R. (2010). Nutraceuticals and prevention of Atherosclerosis: Focus on ω-3 polyunsaturated fatty acids and mediterranean diet polyphenols. *Cardiovascular Therapeutics*, *28*(4), e13–e19. doi:10.1111/j.1755-5922.2010.00211.x PMID:20633019

Miraghajani, M., Dehsoukhteh, S. S., Rafie, N., Hamedani, S. G., Sabihi, S., & Ghiasvand, R. (2017). Potential mechanisms linking probiotics to diabetes: A narrative review of the literature. *Sao Paulo Medical Journal*, *135*(2), 169–178. doi:10.1590/1516-3180.2016.0311271216 PMID:28538869

Murooka, Y., Piao, Y., Kiatpapan, P., & Yamashita, M. (2005). Production of tetrapyrrole compounds and vitamin B12 using genetically engineering of Propionibacterium freudenreichii. An overview. *Le Lait*, *85*(1-2), 9–22. doi:10.1051/lait:2004035

Nagpal, R., Kumar, A., Kumar, M., Behare, P. V., Jain, S., & Yadav, H. (2012). Probiotics, their health benefits and applications for developing healthier foods: A review. *FEMS Microbiology Letters*, *334*(1), 1–15. doi:10.1111/j.1574-6968.2012.02593.x PMID:22568660

Nair, H. B., Sung, B., Yadav, V. R., Kannappan, R., Chaturvedi, M. M., & Aggarwal, B. B. (2010). Delivery of antiinflammatory nutraceuticals by nanoparticles for the prevention and treatment of cancer. *Biochemical Pharmacology*, *80*(12), 1833–1843. doi:10.1016/j.bcp.2010.07.021 PMID:20654584

Nasri, H., Baradaran, A., Shirzad, H., & Rafieian-Kopaei, M. (2014). New concepts in nutraceuticals as alternative for pharmaceuticals. *International Journal of Preventive Medicine*, *5*(12), 1487. PMID:25709784

Newman, D. J., & Cragg, G. M. (2007). Natural products as sources of new drugs over the last 25 years. *Journal of Natural Products*, *70*(3), 461–477. doi:10.1021/np068054v PMID:17309302

Nishimura, M., Ohkawara, T., Tetsuka, K., Kawasaki, Y., Nakagawa, R., Satoh, H., ... Nishihira, J. (2016). Effects of yogurt containing Lactobacillus plantarum HOKKAIDO on immune function and stress markers. *Journal of Traditional and Complementary Medicine*, *6*(3), 275–280. doi:10.1016/j.jtcme.2015.07.003 PMID:27419093

Ojima, T., Rahman, M. M., Kumagai, Y., Nishiyama, R., Narsico, J., & Inoue, A. (2018). Polysaccharide-degrading enzymes from marine gastropods. *Marine Enzymes and Specialized Metabolism - Part B Methods in Enzymology,* 457-497.

Okubo, T., Takemura, N., Yoshida, A., & Sonoyama, K. (2013). KK/Ta mice administered Lactobacillus plantarum strain no. 14 have lower adiposity and higher insulin sensitivity. *Bioscience of Microbiota, Food and Health, 32*(3), 93–100. doi:10.12938/bmfh.32.93 PMID:24936367

Orrhage, K., Sillerström, E., Gustafsson, J. A., Nord, C. E., & Rafter, J. (1994). Binding of mutagenic heterocyclic amines by intestinal and lactic acid bacteria. *Mutation Research. Fundamental and Molecular Mechanisms of Mutagenesis, 311*(2), 239–248. doi:10.1016/0027-5107(94)90182-1 PMID:7526189

Paéz, R., Lavari, L., Vinderola, G., Audero, G., Cuatrin, A., Zaritzky, N., & Reinheimer, J. (2012). Effect of heat treatment and spray drying on lactobacilli viability and resistance to simulated gastrointestinal digestion. *Food Research International, 48*(2), 748–754. doi:10.1016/j.foodres.2012.06.018

Park, K. Y., Jung, K. O., Rhee, S. H., & Choi, Y. H. (2003). Antimutagenic effects of doenjang (Korean fermented soypaste) and its active compounds. *Mutation Research. Fundamental and Molecular Mechanisms of Mutagenesis, 523*, 43–53. doi:10.1016/S0027-5107(02)00320-2 PMID:12628502

Patel, R. M., & Denning, P. W. (2013). Therapeutic use of prebiotics, probiotics, and postbiotics to prevent necrotizing enterocolitis: What is the current evidence? *Clinics in Perinatology, 40*(1), 11–25. doi:10.1016/j.clp.2012.12.002 PMID:23415261

Patra, J. K., Das, G., Paramithiotis, S., & Shin, H. S. (2016). Kimchi and other widely consumed traditional fermented foods of Korea: A review. *Frontiers in Microbiology, 7*, 1493. doi:10.3389/fmicb.2016.01493 PMID:27733844

Pereira, D. I., & Gibson, G. R. (2002). Effects of consumption of probiotics and prebiotics on serum lipid levels in humans. *Critical Reviews in Biochemistry and Molecular Biology, 37*(4), 259–281. doi:10.1080/10409230290771519 PMID:12236466

Prabhakaran, D., Jeemon, P., & Roy, A. (2016). Cardiovascular diseases in India. *Circulation, 133*(16), 1605–1620. doi:10.1161/CIRCULATIONAHA.114.008729 PMID:27142605

Prabu, S., Suriyaprakash, T. K., Kumar, C., & Kumar, S. (2012). Nutraceuticals and their medicinal importance. *International Journal of Health & Allied Sciences, 1*(2), 47. doi:10.4103/2278-344X.101661

Prescott, S. L., Wickens, K., Westcott, L., Jung, W., Currie, H., Black, P. N., ... Crane, J. (2008). Supplementation with Lactobacillus rhamnosus or Bifidobacterium lactis probiotics in pregnancy increases cord blood interferon-γ and breast milk transforming growth factor-β and immunoglobin A detection. *Clinical and Experimental Allergy*, *38*(10), 1606–1614. doi:10.1111/j.1365-2222.2008.03061.x PMID:18631345

Roh, S. W., Kim, K., Nam, Y., Chang, H., Park, E., & Bae, J. (2009). Investigation of archaeal and bacterial diversity in fermented seafood using barcoded pyrosequencing. *The ISME Journal*, *4*(1), 1–16. doi:10.1038/ismej.2009.83 PMID:19587773

Sanders, M. E., Akkermans, L. M., Haller, D., Hammerman, C., Heimbach, J. T., Hörmannsperger, G., & Huys, G. (2010). Safety assessment of probiotics for human use. *Gut Microbes*, *1*(3), 164–185. doi:10.4161/gmic.1.3.12127 PMID:21327023

Santivarangkna, C., Kulozik, U., & Foerst, P. (2006). Effect of carbohydrates on the survival of Lactobacillus helveticus during vacuum drying. *Letters in Applied Microbiology*, *42*(3), 271–276. doi:10.1111/j.1472-765X.2005.01835.x PMID:16478516

Santos, A., San Mauro, M., Sanchez, A., Torres, J. M., & Marquina, D. (2003). The antimicrobial properties of different strains of Lactobacillus spp. isolated from kefir. *Systematic and Applied Microbiology*, *26*(3), 434–437. doi:10.1078/072320203322497464 PMID:14529186

Sarkar, S. (2007). Potential of kefir as a dietetic beverage – a review. *British Food Journal*, *109*(4), 280–290. doi:10.1108/00070700710736534

Schottenfeld, D., & Beebe-Dimmer, J. (2015). The cancer burden attributable to biologic agents. *Annals of Epidemiology*, *25*(3), 183–187. doi:10.1016/j.annepidem.2014.11.016 PMID:25523895

Scott, K. P., Martin, J. C., Duncan, S. H., & Flint, H. J. (2013). Prebiotic stimulation of human colonic butyrate-producing bacteria and bifidobacteria, in vitro. *FEMS Microbiology Ecology*, *87*(1), 30–40. doi:10.1111/1574-6941.12186 PMID:23909466

Shewale, R. N., Sawale, P. D., Khedkar, C. D., & Singh, A. (2014). Selection criteria for probiotics: A review. *International Journal of Probiotics & Prebiotics*, *9*(1).

Shobharani, P., & Agrawal, R. (2011). Enhancement of cell stability and viability of probiotic Leuconostoc mesenteroides MTCC 5209 on freeze drying. *International Journal of Dairy Technology*, *64*(2), 276–287. doi:10.1111/j.1471-0307.2010.00640.x

Silva, S. B., Cantarelli, V. V., & Ayub, M. A. (2013). Production and optimization of poly-γ-glutamic acid by Bacillus subtilis BL53 isolated from the Amazonian environment. *Bioprocess and Biosystems Engineering*, *37*(3), 469–479. doi:10.100700449-013-1016-1 PMID:23872848

Singh, J., & Sinha, S. (2012, January-March). Classification, regulatory acts, and applications of nutraceuticals for health. *International Journal of Pharma and Bio Sciences*, *2*(1), 177–187.

Slattery, C., Cotter, P. D., & O'Toole, P. W. (2019). Analysis of health benefits conferred by *Lactobacillus* Species from Kefir. *Nutrients*, *11*(6), E1252. doi:10.3390/nu11061252 PMID:31159409

Smith, J. L., & Alford, J. A. (1970). Presence of antioxidant materials in bacteria. *Lipids*, *5*(10), 795–799. doi:10.1007/BF02531970 PMID:4992304

Srinivasan, K. (2005). Role of spices beyond food flavoring: Nutraceuticals with multiple health effects. *Food Reviews International*, *21*(2), 167–188. doi:10.1081/FRI-200051872

Starr, M. P. (1958). The blue pigment of Corynebacterium insidiosum. *Archiv für Mikrobiologie*, *30*(4), 325–334. doi:10.1007/BF00411227 PMID:13595813

Stenman, L. K., Waget, A., Garret, C., Briand, F., Burcelin, R., Sulpice, T., & Lahtinen, S. (2015). Probiotic B420 and prebiotic polydextrose improve efficacy of antidiabetic drugs in mice. *Diabetology & Metabolic Syndrome*, *7*(1), 75. doi:10.118613098-015-0075-7 PMID:26366205

Tabuchi, M., Ozaki, M., Tamura, A., Yamada, N., Ishida, T., Hosoda, M., & Hosono, A. (2003). Antidiabetic effect of Lactobacillus GG in streptozotocin-induced diabetic rats. *Bioscience, Biotechnology, and Biochemistry*, *67*(6), 1421–1424. doi:10.1271/bbb.67.1421 PMID:12843677

Tamang, J. (1998). Role of microorganisms in traditional fermented foods. *Indian Food Industry*, *17*, 162–166.

Tan, Q., Xu, H., Aguilar, Z. P., Peng, S., Dong, S., Wang, B., ... Wei, H. (2013). Safety assessment and probiotic evaluation of enterococcus faeciumyf5 isolated from sourdough. *Journal of Food Science*, *78*(4), M587–M593. doi:10.1111/1750-3841.12079 PMID:23488799

Thompson, M. P., & Kurzrock, R. (2004). Epstein-Barr virus and cancer. *Clinical Cancer Research*, *10*(3), 803–821. doi:10.1158/1078-0432.CCR-0670-3 PMID:14871955

Trapecar, M., Leouffre, T., Faure, M., Jensen, H. E., Granum, P. E., Cencic, A., & Hardy, S. P. (2011). The use of a porcine intestinal cell model system for evaluating the food safety risk of Bacillus cereus probiotics and the implications for assessing enterotoxigenicity. *APMIS*, *119*(12), 877–884. doi:10.1111/j.1600-0463.2011.02797.x PMID:22085364

Traprcaar Yu, L., & Osullivan, D. (2014). Production of galactooligosaccharides using a hyperthermophilic β-galactosidase in permeabilized whole cells of Lactococcus lactis. *Journal of Dairy Science*, *97*(2), 694–703. doi:10.3168/jds.2013-7492 PMID:24359820

Trucco, M. (2009). Gene-environment interaction in type 1 diabetes mellitus. *Endocrinología y Nutrición*, *56*, 56–59. doi:10.1016/S1575-0922(09)73521-1 PMID:20629235

Tseng, W., Fang, T., Cho, C., Chen, P., & Tsai, C. (2012). Assessments of growth conditions on the production of cyanophycin by recombinant Escherichia coli strains expressing cyanophycin synthetase gene. *Biotechnology Progress*, *28*(2), 358–363. doi:10.1002/btpr.1513 PMID:22252992

Tufarelli, V., & Laudadio, V. (2016). An overview on the functional food concept: Prospectives and applied researches in probiotics, prebiotics and synbiotics. *The Journal of Experimental Biology*, *4*, 3S.

Verwaal, R., Wang, J., Meijnen, J., Visser, H., Sandmann, G., Berg, J. A., & Ooyen, A. J. (2007). High-level production of Beta-Carotene in Saccharomyces cerevisiae by successive transformation with Carotenogenic Genes from Xanthophyllomyces dendrorhous. *Applied and Environmental Microbiology*, *73*(13), 4342–4350. doi:10.1128/AEM.02759-06 PMID:17496128

Wang, J., Zhao, X., Tian, Z., Yang, Y., & Yang, Z. (2015). Characterization of an exopolysaccharide produced by *Lactobacillus plantarum* YW11 isolated from Tibet Kefir. *Carbohydrate Polymers*, *125*, 16–25. doi:10.1016/j.carbpol.2015.03.003 PMID:25857955

Wang, Y., Yu, R., & Chou, C. (2004). Viability of lactic acid bacteria and bifidobacteria in fermented soymilk after drying, subsequent rehydration and storage. *International Journal of Food Microbiology*, *93*(2), 209–217. doi:10.1016/j.ijfoodmicro.2003.12.001 PMID:15135959

Wei, S. H., Chen, Y. P., & Chen, M. J. (2015). Selecting probiotics with the abilities of enhancing GLP-1 to mitigate the progression of type 1 diabetes in vitro and in vivo. *Journal of Functional Foods*, *18*, 473–486. doi:10.1016/j.jff.2015.08.016

WHO publishes list of bacteria for which new antibiotics are urgently needed (2017, Feb. 27). Retrieved from https://www.paho.org/hq/index.php?lang=en

Wind, R. D., Tolboom, H., Klare, I., Huys, G., & Knol, J. (2010). Tolerance and safety of the potentially probiotic strain Lactobacillus rhamnosus PRSF-L477: A randomised, double-blind placebo-controlled trial in healthy volunteers. *British Journal of Nutrition, 104*(12), 1806–1816. doi:10.1017/S0007114510002746 PMID:20691131

Yadav, H., Lee, J. H., Lloyd, J., Walter, P., & Rane, S. G. (2013). Beneficial metabolic effects of a probiotic via butyrate-induced GLP-1 hormone secretion. *The Journal of Biological Chemistry, 288*(35), 25088–25097. doi:10.1074/jbc.M113.452516 PMID:23836895

Yan, F., Cao, H., Cover, T. L., Whitehead, R., Washington, M. K., & Polk, D. B. (2007). Soluble proteins produced by probiotic bacteria regulate intestinal epithelial cell survival and growth. *Gastroenterology, 132*(2), 562–575. doi:10.1053/j.gastro.2006.11.022 PMID:17258729

Yao, M., Li, B., Ye, H., Huang, W., Luo, Q., Xiao, H., ... Li, L. (2018). Enhanced viability of probiotics (Pediococcus pentosaceus Li05) by encapsulation in microgels doped with inorganic nanoparticles. *Food Hydrocolloids, 83*, 246–252. doi:10.1016/j.foodhyd.2018.05.024

Zaika, L. L., & Smith, J. L. (1975). Antioxidants and pigments of Aspergillus niger. *Journal of the Science of Food and Agriculture, 26*(9), 1357–1369. doi:10.1002/jsfa.2740260915

Zheng, Y., Lu, Y., Wang, J., Yang, L., Pan, C., & Huang, Y. (2013). Probiotic properties of Lactobacillus strains isolated from Tibetan kefir grains. *PLoS One, 8*(7), e69868. doi:10.1371/journal.pone.0069868 PMID:23894554

KEY TERMS AND DEFINITIONS

Functional Foods: Eatables that have a positive effect on living beings.

Lactic Acid Bacteria: Bacteria which have the ability to produce lactic acid.

Micro-encapsulation: Process of surrounding a useful material or active agent by a polymer.

Pan-Drug Resistance: Organisms which are resistant to almost all current antibiotics available.

Personalized Nutrition: Nutrition designed according to the requirement of a person since the nutrient need of each and every human will be based on their genotype.

Vacuum Drying: A method which uses vacuum to decrease the water content of products that is usually done for the purpose of storage.

Chapter 7
Ultrasound Pretreatment Applications in the Drying of Agricultural Products

Ahmet Polat
University of Bursa Uludag, Turkey

Nazmi Izli
University of Bursa Uludag, Turkey

ABSTRACT

This chapter presents the effect of ultrasound pretreatment on drying kinetics and product temperature. Drying provides a longer shelf life of agricultural products and low-cost transport and storage. The researchers performed pretreatments to the products to accelerate the drying process and to increase the quality for decades. The ultrasound applications considered as one of these are shown as the promising technology in the drying industry. The ultrasound application is used in drying processes in two different ways as ultrasound-assisted drying or as a pretreatment. In ultrasound-assisted drying, mechanical waves directly move to the cell wall of the product. On the other hand, the agricultural product to be dried is immersed in a liquid medium with different solution and ultrasound are applied at specified times by using as a pretreatment. They demonstrate that the ultrasound pretreatment method could be suitable for industrial production.

DOI: 10.4018/978-1-7998-1924-0.ch007

BACKGROUND

Fruits and vegetables contain fiber, minerals, and vitamins, which are important sources of essential dietary nutrients. Due to the high moisture content of fruits and vegetables, it is classified as highly perishable products. The way to keep the product's main nutritional values high is to keep the product fresh. However, there may be difficulties in keeping the product fresh due to the unfavorable to be experienced during the storage and distribution chain of the product (Sagar & Kumar, 2010). In the food industry, pasteurization, sterilization, freezing, storage in controlled atmosphere and drying are used as food preservation methods. Drying, which promises a longer shelf life, is an important food preservation method. Drying is the process of removing the water contained in the product and thereby reducing the water activity. Furthermore, it is a process which increases the dry matter content in which the wholesomeness and physical appearance are preserved without damaging the tissue due to the reduction of the water content in the food. The purposes of the drying of agricultural products can be listed as follows (Castro et al., 2018):

- Providing inhibition of growth of microorganisms
- Preventing deterioration reactions by reducing water activity
- The reduction of transport and storage costs due to product weight and volume decrease.

In the literature, there are more than 200 drying types used for different purposes. Furthermore, the drying properties (air velocity, product retention time, pressure, relative humidity) vary according to the material and the drying method. It is reported that the total energy requirement of the drying operation in the food sector in developed countries consumes 10% to 15%. Therefore, energy is one of the important parameters of drying, which is the traditional method in product preservation. The conventional method is still the most widely used method in the drying industry, although new drying technologies have emerged. One of the most important reasons for this is the ease of installation and use of the dryer (Onwude et al., 2016). The pretreatment process is being worked on to reduce energy costs and to improve product quality. The pretreatment of agricultural products before drying has been investigated, and many methods have been developed. These methods have been shown in Figure 1 by Deng et al. (2019).

The objectives of the pretreatment can be listed as follows:

- To facilitate the discharge of water contained in the product by affecting the cell membrane.
- Prevent energy consumption loss by increasing the drying rate

Figure 1. Pretreatment of food materials

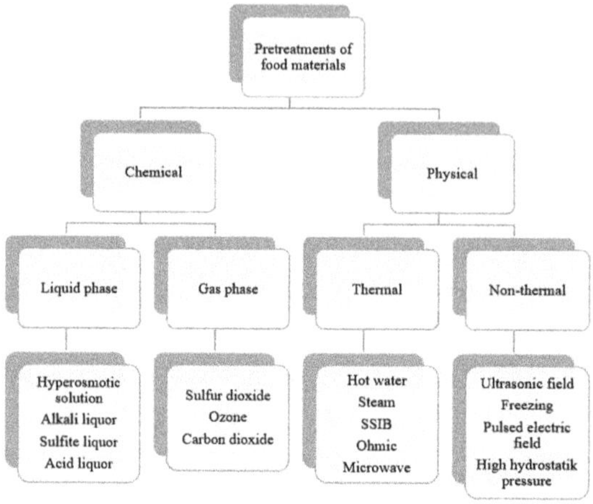

- To enhance retention of the nutrients (fat, protein, carbohydrate, minor and macro minerals) compounds
- Inhibiting enzymes from preventing deterioration during drying and storage

ULTRASOUND APPROACH

The application of ultrasound pretreatment to agricultural products before drying has emerged as an important popular topic for researchers as it greatly reduces the overall processing time of the drying process. This technology has started to be researched since the 1970s. These studies include the use of ultrasound in meat technology and the effect of ultrasound on the diffusion rate (Howkins, 1969; Sajas & Gorbatow 1978). Ultrasound is the sound frequency range above 16 kHz, which is the threshold of the human ear. It is classified in two different ways as the high and low frequency in the food industry. While low energy high frequency (megahertz) ultrasound application is used for food quality monitoring, high energy low frequency (kilohertz) ultrasound application has changed the structure of food materials and has an important role in the drying process (Mothibe et al., 2011).

The ultrasound technology is used in drying processes in two different ways as ultrasound-assisted drying or as a pre-treatment. The effect of ultrasound treatment may vary according to the design of the system. In the ultrasound-assisted drying process, the mechanical waves produced by the machine are carried with air on the product. These waves disrupt the structure of the product and increase the mass transfer by decreasing the external resistance on the product surface (Ricce et al.,

2016). Another method is the use of ultrasound technology as a pretreatment. The application of ultrasound to liquid systems is based on the view that ultrasound can cause medium particle vibration and ultrasonic cavitation. In this method, the ultrasonic waves are transmitted through the molecules induced by molecular compression and sparse wave propagation medium. With this application, occur contraction and loosening of the product. This effect is called the sponge effect. Microchannels are formed in product tissues by the effect of contractions, and these microchannels help to remove the water in the structure. With the accelerated drying process, the quality characteristics of the products are protected by low exposure to temperature. However, cavitation results in pressure above 1000 atm and a temperature of about 5000 K. The result is to increase the operating temperature. This can adversely affect the process. However, it is stated that the temperature can reach up to 70 ° C in the systems (Yang et al., 2018).

The main parts of ultrasound systems are a generator, transducer, and coupler. In ultrasound technology applied in a liquid medium, two different types of devices are used depending on their efficiency and capacities. The acoustic field waves produced by the ultrasound bath are very irregular due to their reflections on the bath walls. Also, the cavitation occurs heterogeneously. This causes the sonication to not be equally effective in all regions. However, in the probed system, acoustic energy is supplied directly to the product. In the area below the impact probe, an intensive sonication zone occurs. Since this region is a limited one, a more effective process can be applied using smaller volumes (Rodríguez et al., 2018).

EXAMPLES IN LITERATURE

Ultrasound is used in the product drying process as both pretreatments and as a drying method, as mentioned in the previous section. This section contains studies where ultrasound is applied as a pretreatment.

Fernandes & Rodrigues (2007) experienced the effect of ultrasound pretreatment before air drying of banana dehydration. In this experiment, Banana samples were immersed in distilled water and submitted to ultrasonic waves for 10, 20, and 30 min. The experiments were applied under ambient water temperature (30 °C) in an ultrasonic bath without mechanical agitation. The ultrasound frequency was 25 kHz, and the intensity was 4870 W/m^2. After the experiment, they found that drying time decreased by 11% due to the increase in water diffusivity. As a result, they have reduced their energy input to produce a more economically beneficial product. They found that the sugar content of the bananas applied ultrasonic treatment was lower than the fresh product, so they commented that the ultrasound pretreatment could be an interesting process to produce dried fruits with low sugar content.

Opalić et al. (2009) investigated changes in quality parameters (sugar content, phenolic compounds, antioxidant activity) of apple (*Malus domestica*, Goldparmâne variety) dried with different duration of ultrasonic pretreatment and air-drying. The ultrasound pretreatment was applied 0, 9, 22.5, 45, and 54 min. Apples were immersed in ultrasonic bath Elmasonic S 120, nominal power 200 W, frequency 37 kHz, and intensity 0.02–0.03 W/cm^3. They observed a decrease in total phenol flavonoid and phenolic and antioxidant activities with prolongation of ultrasound pretreatment under the same drying conditions. They also found that untreated dried apple was determinate as the most sensory acceptable.

Fernandes et al. (2009) studied the influence of ultrasound-assisted osmotic dehydration applied for the different duration on tissue structure of pineapple. The experiments were performed under ambient temperature (30 °C) in an ultrasonic bath (Marconi model Unique USC, Brazil) without mechanical agitation. The ultrasound frequency was 25 kHz, and the intensity was 4870 W/m^2. They used two different liquids (distilled water, 35 and 70°Brix concentration of osmotic solution) for pretreatment. The samples immersed different liquids and subjected ultrasound for 10, 20 and 30 min. They observed that microscopic channels formed by ultrasound application, which allow lower resistance to water and sugar diffusion increase sugar loss and water diffusion ability. Also, with the increase in water diffusion ability, the products dried out in a shorter period, thus reducing the process costs.

Oliveira et al. (2011) determined the influence of the ultrasonic pretreatment before air drying on dehydration of Malay apple (*Syzygium malaccense* L.). This study shows the water loss and sugar gain of the dried product during the pretreatment and the drying process. As previous study (Fernandes et al., 2009), Malay apple samples were immersed in two different liquids (distilled water, 25 and

50 °Brix concentration of osmotic solution) subjected to ultrasonic waves during 10, 20, 30, 45, and 60 min. The experiments were performed under ambient water temperature (25°C) in an ultrasonic bath (Unique model USC 25 kHz, São Paulo, Brazil) without mechanical agitation. The ultrasound frequency was 25 kHz, and the intensity was 60 W or equivalent to 1,785 W/m^2. As a result of the experiments, they determinate that the water effective diffusivity increased by 28.1% after ultrasound application, which caused a reduction of about 27.3% in the total drying time. They also found that Malay apples treated with ultrasound using distilled water lost sugar. They commented that the ultrasound pretreatment stage could be a practical process to produce dried fruits with lower sugar content.

Garcia-Noguera et al. (2010) used the ultrasound-assisted osmotic dehydration method to reduce the diffusivity in water and drying time in strawberries in this study. By applying different pretreatment time (10, 20, 30, and 45 min) and frequency levels (0, 25, and 40 kHz), they observed the effects on drying time; soluble solids gain and water loss. As a result of these pretreatments, microscopic analyses were performed to examine the changes in the product surface. Cell distortion and breakdown were observed by both methods. The strawberry sample pretreated (25 kHz) for 30 min in a 50% w/w osmotic solution of sucrose resulted in the highest drying rate among the pretreatments. In the pretreatment method in which the osmotic method was used alone, the duration of the total processing time increased while the total processing time was decreased in the ultrasound-assisted method, and the effective water diffusivity increased. They stated that ultrasound pre-treatment has a major effect on the drying time of the products.

Fernandes et al. (2008) in another study, they used ultrasound pretreatment aimed to produce dried papaya (*Carica papaya* L.) with low sugar content. For this purpose, in this study, the effect of ultrasound pretreatment on drying of papaya was investigated. Also, changes in effective water diffusion of ultrasound pretreated papaya samples were examined. Papaya samples were immersed in distilled water and submitted to ultrasonic waves for 10, 20, 30, 45, and 90 min. The experiments were carried out under ambient water temperature (30 °C) in an ultrasonic bath (Marconi model Unique USC 25 kHz) without mechanical agitation. The ultrasound frequency was 25 kHz, and the intensity was 4,870 or 100 kW/m^3. When the results were examined, it was found that ultrasound pretreatment increased water effective diffusivity and reduced drying time by 16%. They observed that the sugar content of papaya samples decreased after ultrasound pretreatment. They commented that the ultrasound pretreatment application could be used dried to produce dried food material with lower sugar content.

Fernandes et al. (2008) experienced the influence of ultrasound pretreatment and of osmotic dehydration applied at atmospheric pressure for the different duration on melon tissue structure by using the same brand ultrasonic bath at previous study. The melon samples were submitted to ultrasonic waves during 20 and 30 min and immersed in distilled water. Ultrasound pretreatment provided the formation of microscopic channels in the fruit structure but did not cause the breakdown of the tissue. They observed that microscopic channels appear in the cell structure by using ultrasound pretreatment application. They stated that microscopic channels formed show less resistance to water diffusion and may be responsible for the increase in water diffusion.

Cakmak et al. (2016), in their study, applied electroplasmolysis (EP) and ultrasound pretreatments to mushroom samples and examined the effects of these treatments on drying time and some quality parameters. They performed the ultrasound pre-treatment at 35 kHz for 30 min. They were then dried in a tray dryer at 50 ° C at an air velocity of 1.5 m / s. As a result of the experiments, they found that the control group samples were dried in 3.91 hours, and the samples that were pretreated by ultrasound were dried in 2.66 hours. They observed that the phenolic content and color values of the ultrasound pretreated samples were better preserved compared to the other samples.

Gamboa-Santos et al. (2013) studied the vitamin C content and sensory properties in air-dried carrots subjected to different ultrasound or conventional blanching pretreatments. The samples were sonicated using an ultrasonic system (450 Digital Sonifier, Branson Ultrasonics Coorporation, Danbury, CT, USA). This sonicator was equipped with a temperature sensor and a tip of 13 mm diameter directly attached to a disruptor horn (20 kHz, 400W full power) which was immersed 2 cm in the liquid. Experiments were carried out with generation of heat: US blanching for 10 min at temperatures up to 60 °C, and for 15 min at temperatures up to 70 °C. They found that samples subjected to high temperature and short-term blanching had the highest retention of vitamin C. According to the sensory analysis of carrots, ultrasound pretreated samples represented acceptable quality, but no statistically significant differences were determinate compare to conventionally blanched.

Kek et al. (2013), in their work, used different methods of ultrasound (indirect sonication using an ultrasonic bath system and direct sonication using an ultrasonic probe system) for osmotic dehydration of guava. Three different osmotic concentrations (0, 35, and 70oBrix) at indirect ultrasonic bath and direct ultrasonic probe amplitudes from 0 to 35% was set for pretreatment. Guava samples summited to ultrasonic wave at ultrasound bath (0 to 2.5 kW) for 20–60 min and with an ultrasonic probe for 6–20 min. As a result of calorimetric calculations, direct sonication was found to be more intensive than indirect sonication. Using the general linear model, they determinate that the increase in osmotic solution concentrations, ultrasound input (power and amplitude) and immersion time significantly increased water loss, a solid gain, and total color change of the guava slices. Indirect sonication in osmotic solutions contributed to high water loss and solid gain with acceptable total color change than direct sonication. Applied ultrasound pre-osmotic pretreatment in 70oBrix to guava slices, drying time and total color change was reduced by 33% and 38%, respectively. The results also present that effective diffusivity was increased by 35% with the same application.

CASE STUDY: BEETROOT DRYING

In this study, the effect of ultrasound pre-treatment on the drying of beetroot slices was investigated. The moisture contents of the beetroot slices at all drying conditions were increased from 8.58 to approximately 0.1 g water.g dry matter^{-1}. The beetroot slices were pretreated at room temperature with ultrasound by immersing into a 25-kHz ultrasound bath (Intersonic, Model: MIN18, Istanbul/Turkey) operating at 300 W output power. Beetroot slices were submitted to ultrasonic waves for different

Figure 2. Drying curves for convective dried beetroot slices with pretreatment by (a) Control; (b) US 20; and (c) US 40

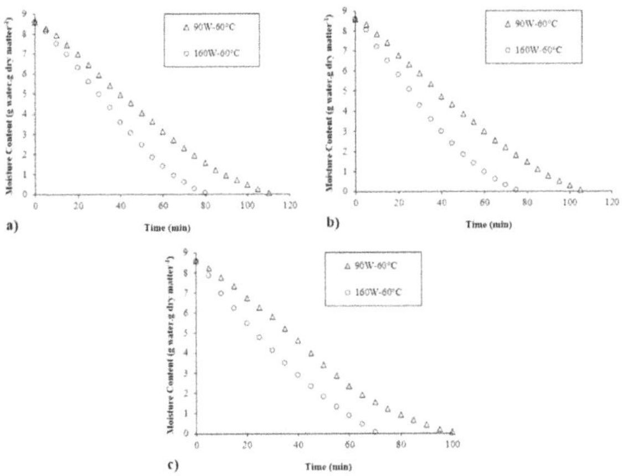

times (0, 20, and 40 min) and then transferred to a modified microwave-convection oven. The experiments were performed in triplicate. The moisture changes with the time of beetroot slices applied different time ultrasound pretreatment and dried at 60 °C combined with 90 and 160W microwave power were shown in Figure 2a-c. It was observed that the drying times of the samples varied according to different microwave power levels and applied ultrasound treatment. The drying times of all drying conditions varied between 110 and 70 minutes. The highest value of drying time was found in untreated samples dried at 90W-60 °C and the shortest period was seen in the sample applied 40 minutes by ultrasound pretreatment and dried at 160W-60 °C. The drying time of the untreated beetroot slices dried at 90W-60 °C and 160W-60 °C lasted 110 and 80 minutes, respectively. The drying time of the product decreased with the increase in microwave power. The same observation was seen by Doymaz et al. (2015) with the drying of green bean slices. Figure 2 shows that the drying time decreased with the applied ultrasound pretreatment. The drying time of the samples dried at 90W-60 °C and ultrasound pretreated for 20 and 40 minutes was lower 4.55% and 9.09% compared to non-pretreated samples, respectively. Nowacka et al. (2012) investigate the usage of ultrasound as a mass transfer enhancing method before drying of apple. They reported that ultrasound pretreatment decreased the drying time by 31%. This finding is similar to our study.

The drying rates of dried carrot samples under different drying conditions were shown in Figure 3. In the early stages of drying, moisture was rapidly moving away, and the drying rate decreased in time. This effect is caused by the high moisture content of the product to allow the absorption of microwave power. A similar effect was experienced by Demiray et al. (2017) with onion and Horuz et al. (2018) with

Figure 3. The drying rate changes of beetroots with pretreatment by (a) Control; (b) US 20; and (c) US 40

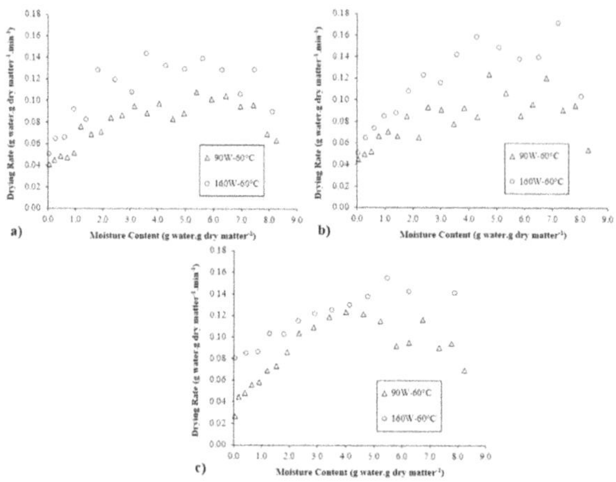

apple. Figure 3 indicates that the effect of ultrasound pretreatment on the rate of drying of beet slices cannot be determined as much as the effect of microwave power levels. The same observation was seen by Horuz et al. (2017).

According to the data obtained as a result of the experiments, moisture content and drying time changes are modeled using non-linear regression analysis. In order to determine the suitability of drying curves, the high value of R^2, the low value of χ^2 and RMSE were taken as criteria. In Table 1, seven different thin layer drying used by Demiray & Tulek (2014), Saxena & Dash (2015), Taskin et al. (2015), Mota et al. (2010), Murthy & Manohar (2014), Arumuganathan et al. (2009) and Midilli et al. (2002) was shown. The data of the dried carrot samples under different conditions as a result of statistical calculations are presented in Table 2, 3 and 4. As seen in Tables, R^2, $\chi 2$, and RMSE values ranged between 0.9232-0.9998, 0.00001373-0.00876589, and 0.0047-0.0933, respectively. From the tables, Midilli et al. model was determinate the most suitable model at these drying conditions. R^2, $\chi^{2,}$ and RMSE values of the most suitable model were changed from 0.9987-0.9998, 0.00001373-0.00009595, and 0.0047-0.0122, respectively. The comparison of the most suitable model with the experimental results in all drying conditions is shown in Figure 4a-c. Figure 5 represented that, Midilli et al. model was found to be in good agreement with the experimental results. This model describes the drying characteristics of the beetroot slices with the microwave-convective combined method.

The effect of microwave and hot air combined drying system on product temperature was investigated. The out temperature of product was measured by an infrared thermometer. The change in the temperature of the product with time is shown in Figure 6. As it is seen from Figure 6, time-dependent evolution of product

Table 1. Selected thin layer drying models used to mathematically model the beetroot drying kinetics

No	Model name	Model	References
1	Henderson and Pabis	$MR = a\exp(-kt)$	Demiray and Tulek (2014)
2	Newton	$MR = \exp(-kt)$	Saxena and Dash (2015)
3	Page	$MR = \exp(-kt^n)$	Taskin et al. (2015)
4	Logarithmic	$MR = a\exp(-kt) + c$	Mota et al. (2010)
5	Two Term	$MR = a\exp(-k_0 t) + b\exp(-k_1 t)$	Murthy and Manohar (2014)
6	Wang and Singh	$MR = 1 + at + bt^2$	Arumuganathan et al.(2009)
7	Midilli et al.	$MR = a\exp(-kt^n) + bt$	Midilli et al. (2002)

Table 2. The statistical results for the drying of beetroot at different conditions for pretreatment by US 0 (control)

No	90W-60°C Control				160W-60°C Control			
	Model coefficients	R^2	RMSE	$\chi^2(10^{-4})$	Model coefficients	R^2	RMSE	$\chi^2(10^{-4})$
1	a=1.126 k=0.01988	0.9504	0.0728	44.0811	a=1.121 k=0.02703	0.9426	0.0805	54.4928
2	k=0.01755	0.9314	0.0856	73.2313	k=0.024	0.9262	0.0913	38.2560
3	k=0.00136 n=1.629	0.9945	0.0243	5.5885	k=0.001848 n=1.682	0.9937	0.0267	82.6623
4	a=2.22 k=0.005907 c=-1.182	0.9971	0.0177	2.3519	a=2.466 k=0.007062 c=-1.431	0.9957	0.0220	6.8294
5	a=95.11 k_0=0.03592 b=-94.04 k_1=0.03625	0.9807	0.0454	16.3139	a=1.212 k_0=0.0293 b=-0.2125 k_1=5.13	0.9512	0.0743	3.6243
6	a=-0.01154 b=0.00002	0.9958	0.0213	4.6232	a=-0.0155 b=0.0000324	0.9946	0.0246	5.7031
7	a=0.9943 k=0.002423 n=1.417 b=-0.001364	0.9996	0.0061	0.2603	a=0.99 k=0.002946 n=1.483 b=-0.001804	0.9992	0.0096	0.6063

Table 3. The statistical results for the drying of beetroot at different conditions for pretreatment by US 20

No	90W-60°C US20				160W-60°C US20			
	Model coefficients	R^2	RMSE	$\chi^2(10^{-4})$	Model coefficients	R^2	RMSE	$\chi^2(10^{-4})$
1	a=1.121 k=0.02046	0.9537	0.0697	42.3146	a=1.105 k=0.03034	0.9575	0.0686	38.9512
2	k=0.01813	0.9357	0.0821	69.4126	k=0.02739	0.9452	0.0780	60.5362
3	k=0.001696 n=1.588	0.9936	0.0259	6.6530	k=0.003677 n=1.549	0.9951	0.0234	5.3821
4	a=2.23 k=0.00605 c=-1.195	0.9983	0.0132	1.2847	a=1.788 k=0.01181 c=-0.7556	0.9972	0.0177	2.2588
5	a=22.56 k_o=0.03793 b=-21.56 k_1=0.03966	0.9876	0.0361	13.3034	a=17.49 k_o=0.05583 b=-16.49 k_1=0.05911	0.9898	0.0337	11.2268
6	a=-0.01196 b=0.0000216	0.9972	0.0172	3.1977	a=-0.01874 b=0.00006917	0.9969	0.0187	3.4319
7	a=1.003 k=0.003668 n=1.311 b=-0.001845	0.9998	0.0051	0.2480	a=0.9967 k=0.006091 n=1.346 b=-0.001733	0.9996	0.0064	0.3437

Table 4. The statistical results for drying of beetroot at different conditions for pretreatment by US 40

No	90W-60 °C US40				160W-60°C US40			
	Model coefficients	R^2	RMSE	$\chi^2(10^{-4})$	Model coefficients	R^2	RMSE	$\chi^2(10^{-4})$
1	a=1.133 k=0.02264	0.9428	0.0806	54.9128	a=1.082 k=0.03078	0.9620	0.0627	33.9531
2	k=0.01993	0.9232	0.0933	87.6589	k=0.02833	0.9546	0.0686	46.4944
3	k=0.001166 n=1.718	0.9953	0.0231	4.8723	k=0.005729 n=1.443	0.9910	0.0306	9.7526
4	a=2.132 k=0.007106 c=-1.086	0.9924	0.0294	6.7231	a=1.861 k=0.01117 c=-0.8508	0.9996	0.0062	0.1480
5	a=27.65 k_o=0.04334 b=-26.66 k_1=0.0451	0.9862	0.0395	15.0278	a=27.16 k_o=0.05427 b=-26.16 k_1=0.05599	0.9861	0.0379	14.6664
6	a=-0.01308 b=0.00002666	0.9908	0.0324	10.0191	a=-0.01956 b=0.00007616	0.9997	0.0054	0.1431
7	a=0.9797 k=0.00137 n=1.631 b=-0.0008061	0.9987	0.0122	0.9595	a=1.001 k=0.01202 n=1.117 b=-0.003507	0.9998	0.0047	0.1373

Figure 4. Comparison of the best model to experimental moisture ratios of beetroot slices with pretreatment by (a) Control; (b) US 20; and (c) US 40

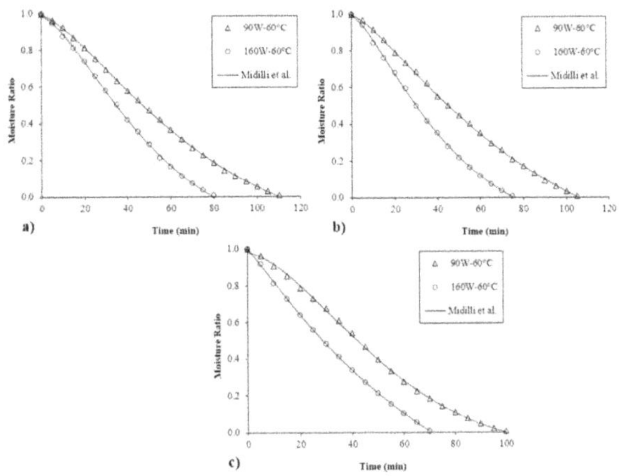

temperature increased rapidly at the initial stage of drying. The initial temperatures of dried agricultural products are lower than the wet bulb temperature. Moisture condensation occurs on the surface of the product, which reveals the latent heat caused by an increase in the initial temperature of the products. As soon as the product temperature reaches the wet bulb temperature, the moisture condensation is replaced by the evaporation of the liquid. Therefore, some of the heat is spent to evaporate the water. As a result, the product temperature slowly increases towards

Figure 5. Comparison of predicted by the best model and experimental moisture ratios at drying conditions for pretreatment by (a) Control; (b) US 20; and (c) US 40

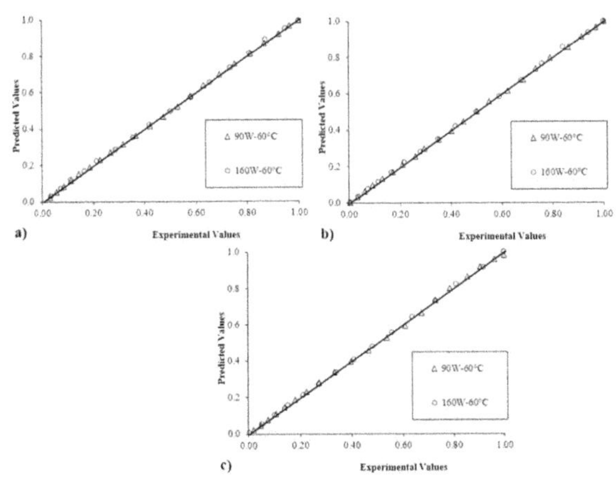

Figure 6. Changes of product temperature at all drying conditions

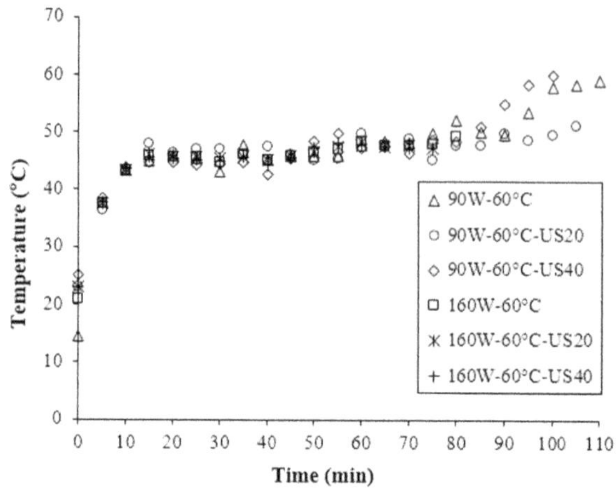

the oven drying temperature (Ju et al., 2016). The temperature of the product has not reached up to 60 °C except 90W-60 C US-40 application. In literature, the researchers examined temperature distribution of product dried with different methods (Kelen et al., 2006; Fiorentini et al., 2016).

FUTURE RESEARCH DIRECTIONS

Drying affects, the nutritional values of dried agricultural products and causes many phytochemicals that are beneficial for health. This results in a loss of quality in the final product. The researchers must focus on to increase drying rate to eliminate these bad effects. As the drying rate increases, energy inputs will decrease, and better quality products have been produced as the product is subjected to less thermal processing. Reducing processing time has been one of the main objectives. It is aimed to preserve the product quality by increasing the product drying rate and reducing the microbial load with the pretreatments to be applied.

Recently, ultrasound technology has been used intensively as a pretreatment to make drying more effective. As a result of laboratory-scale studies, positive results were obtained in terms of energy efficiency, nutritional values , and physical values. The effect of ultrasound pretreatment may not be the same for each product. This effect depends on the structure and shape of the product. Changes in the cell structure of the product by the effect of different ultrasound times and frequency used for different products can be observed using SEM. In addition, the effects of ultrasound pretreatment can be examined by FTIR spectroscopy analysis, biochemical analysis, molecular biological techniques and DNA repair mechanisms. As a result, suitable ultrasound pretreatment method can be determined for different product structure. Also, at high air velocity, acoustic waves are affected and may not show uniform distribution. The studies are carried out on a laboratory scale. Developing the technology to cover more products and shift from lab scale to industrial scale should be a priority step. The focus should be on transferring the energy efficiency and promise of higher quality products to the drying industry by ultrasound technology at the laboratory scale.

CONCLUSION

In this chapter, the place of ultrasound technology in the drying of vegetables and fruits is examined and exemplified by the case study. In the case study, the drying time of the samples dried at 160W-60 °C and ultrasound pretreated for 40 min were 12.5% lower compared to non-pretreated samples. The drying time of the product was reduced with the ultrasound technology and it was less exposed to the thermal effect. In modeling studies, Midilli et al. model was found to be the most suitable model for the drying of beetroot with hot air and microwave combined method in different 7 models. In the first stages of drying, the product temperature increased rapidly, while in the other stages of the drying process, it became a stable form.

Consequently, the ultrasound technology is an encouraging method with its effect of during pretreatment. It can be also a good alternative to reduce drying costs and to obtain better products. It has been observed that ultrasound application has great potential in the drying industry.

REFERENCES

Arumuganathan, T., Manikantan, M. R., Rai, R. D., Anandakumar, S., & Khare, V. (2009). Mathematical modeling of drying kinetics of milky mushroom in a fluidized bed dryer. *International Agrophysics*, *23*(1), 1–7.

Cakmak, R. Ş., Tekeoglu, O., Bozkır, H., Ergun, A. R., & Baysal, T. (2016). Effects of electrical and sonication pretreatments on the drying rate and quality of mushrooms. *Lebensmittel-Wissenschaft + Technologie*, *69*, 197–202. doi:10.1016/j.lwt.2016.01.032

Castro, A. M., Mayorga, E. Y., & Moreno, F. L. (2018). Mathematical modeling of convective drying of fruits: A review. *Journal of Food Engineering*, *223*, 152–167. doi:10.1016/j.jfoodeng.2017.12.012

Demiray, E., Seker, A., & Tulek, Y. (2017). Drying kinetics of onion (Allium cepa L.) slices with convective and microwave drying. *Heat and Mass Transfer*, *53*(5), 1817–1827. doi:10.100700231-016-1943-x

Demiray, E., & Tulek, Y. (2014). Drying characteristics of garlic (Allium sativum L) slices in a convective hot air dryer. *Heat and Mass Transfer*, *50*(6), 779–786. doi:10.100700231-013-1286-9

Deng, L. Z., Mujumdar, A. S., Zhang, Q., Yang, X. H., Wang, J., Zheng, Z. A., ... Xiao, H. W. (2019). Chemical and physical pretreatments of fruits and vegetables: Effects on drying characteristics and quality attributes–a comprehensive review. *Critical Reviews in Food Science and Nutrition*, *59*(9), 1408–1432. doi:10.1080/10408398.2017.1409192 PMID:29261333

Doymaz, I., Kipcak, A. S., & Piskin, S. (2015). Microwave drying of green bean slices: Drying kinetics and physical quality. *Czech Journal of Food Sciences*, *33*(4), 367–376. doi:10.17221/566/2014-CJFS

Fernandes, F. A., Gallão, M. I., & Rodrigues, S. (2008). Effect of osmotic dehydration and ultrasound pre-treatment on cell structure: Melon dehydration. *Lebensmittel-Wissenschaft + Technologie*, *41*(4), 604–610. doi:10.1016/j.lwt.2007.05.007

Fernandes, F. A., Gallão, M. I., & Rodrigues, S. (2009). Effect of osmosis and ultrasound on pineapple cell tissue structure during dehydration. *Journal of Food Engineering*, *90*(2), 186–190. doi:10.1016/j.jfoodeng.2008.06.021

Fernandes, F. A., Oliveira, F. I., & Rodrigues, S. (2008). Use of ultrasound for dehydration of papayas. *Food and Bioprocess Technology*, *1*(4), 339–345. doi:10.100711947-007-0019-9

Fernandes, F. A., & Rodrigues, S. (2007). Ultrasound as pre-treatment for drying of fruits: Dehydration of banana. *Journal of Food Engineering*, *82*(2), 261–267. doi:10.1016/j.jfoodeng.2007.02.032

Fiorentini, C., Demarchi, S. M., Ruiz, N. A. Q., Irigoyen, R. M. T., & Giner, S. A. (2015). Arrhenius activation energy for water diffusion during drying of tomato leathers: The concept of characteristic product temperature. *Biosystems Engineering*, *132*, 39–46. doi:10.1016/j.biosystemseng.2015.02.004

Gamboa-Santos, J., Soria, A. C., Pérez-Mateos, M., Carrasco, J. A., Montilla, A., & Villamiel, M. (2013). Vitamin C content and sensorial properties of dehydrated carrots blanched conventionally or by ultrasound. *Food Chemistry*, *136*(2), 782–788. doi:10.1016/j.foodchem.2012.07.122 PMID:23122127

Garcia-Noguera, J., Oliveira, F. I., Gallão, M. I., Weller, C. L., Rodrigues, S., & Fernandes, F. A. (2010). Ultrasound-assisted osmotic dehydration of strawberries: Effect of pretreatment time and ultrasonic frequency. *Drying Technology*, *28*(2), 294–303. doi:10.1080/07373930903530402

Horuz, E., Bozkurt, H., Karataş, H., & Maskan, M. (2018). Simultaneous application of microwave energy and hot air to whole drying process of apple slices: Drying kinetics, modeling, temperature profile and energy aspect. *Heat and Mass Transfer*, *54*(2), 425–436. doi:10.100700231-017-2152-y

Horuz, E., Jaafar, H. J., & Maskan, M. (2017). Ultrasonication as pretreatment for drying of tomato slices in a hot air–microwave hybrid oven. *Drying Technology*, *35*(7), 849–859. doi:10.1080/07373937.2016.1222538

Howkins, S. D. (1969). Diffusion rates and the effect of ultrasound. *Ultrasonics*, *7*(2), 129–130. doi:10.1016/0041-624X(69)90201-7

Izli, N., Yıldız, G., Ünal, H., Işık, E., & Uylaşer, V. (2014). Effect of different drying methods on drying characteristics, colour, total phenolic content and antioxidant capacity of Goldenberry (P hysalis peruviana L.). *International Journal of Food Science & Technology*, *49*(1), 9–17. doi:10.1111/ijfs.12266

Ju, H. Y., Law, C. L., Fang, X. M., Xiao, H. W., Liu, Y. H., & Gao, Z. J. (2016). Drying kinetics and evolution of the sample's core temperature and moisture distribution of yam slices (Dioscorea alata L.) during convective hot-air drying. *Drying Technology*, *34*(11), 1297–1306. doi:10.1080/07373937.2015.1105814

Kek, S. P., Chin, N. L., & Yusof, Y. A. (2013). Direct and indirect power ultrasound assisted pre-osmotic treatments in convective drying of guava slices. *Food and Bioproducts Processing*, *91*(4), 495–506. doi:10.1016/j.fbp.2013.05.003

Kelen, A., Ress, S., Nagy, T., Pallai, E., & Pintye-Hodi, K. (2006). Mapping of temperature distribution in pharmaceutical microwave vacuum drying. *Powder Technology*, *162*(2), 133–137. doi:10.1016/j.powtec.2005.12.001

Midilli, A., Kucuk, H., & Yapar, Z. (2002). A new model for single-layer drying. *Drying Technology*, *20*(7), 1503–1513. doi:10.1081/DRT-120005864

Mota, C. L., Luciano, C., Dias, A., Barroca, M. J., & Guiné, R. P. F. (2010). Convective drying of onion: Kinetics and nutritional evaluation. *Food and Bioproducts Processing*, *88*(2-3), 115–123. doi:10.1016/j.fbp.2009.09.004

Motevali, A., Minaei, S., Banakar, A., Ghobadian, B., & Darvishi, H. (2016). Energy analyses and drying kinetics of chamomile leaves in microwave-convective dryer. *Journal of the Saudi Society of Agricultural Sciences*, *15*(2), 179–187. doi:10.1016/j.jssas.2014.11.003

Mothibe, K. J., Zhang, M., Nsor-atindana, J., & Wang, Y. C. (2011). Use of ultrasound pretreatment in drying of fruits: Drying rates, quality attributes, and shelf life extension. *Drying Technology*, *29*(14), 1611–1621. doi:10.1080/07373937.2011.602576

Murthy, T. P. K., & Manohar, B. (2014). Hot air-drying characteristics of mango ginger: Prediction of drying kinetics by mathematical modeling and artificial neural network. *Journal of Food Science and Technology*, *51*(12), 3712–3721. doi:10.100713197-013-0941-y PMID:25477637

Nowacka, M., Wiktor, A., Śledź, M., Jurek, N., & Witrowa-Rajchert, D. (2012). Drying of ultrasound pretreated apple and its selected physical properties. *Journal of Food Engineering*, *113*(3), 427–433. doi:10.1016/j.jfoodeng.2012.06.013

Oliveira, F. I., Gallão, M. I., Rodrigues, S., & Fernandes, F. A. N. (2011). Dehydration of Malay apple (Syzygium malaccense L.) using ultrasound as pre-treatment. *Food and Bioprocess Technology*, *4*(4), 610–615. doi:10.100711947-010-0351-3

Onwude, D. I., Hashim, N., Janius, R. B., Nawi, N. M., & Abdan, K. (2016). Modeling the thin-layer drying of fruits and vegetables: A review. *Comprehensive Reviews in Food Science and Food Safety*, *15*(3), 599–618. doi:10.1111/1541-4337.12196

Opalić, M., Domitran, Z., Komes, D., Belščak, A., Horžić, D., & Karlović, D. (2009). The effect of ultrasound pre-treatment and air-drying on the quality of dried apples. *Czech Journal of Food Sciences*, *27*(4), 297. doi:10.17221/606-CJFS

Ricce, C., Rojas, M. L., Miano, A. C., Siche, R., & Augusto, P. E. D. (2016). Ultrasound pre-treatment enhances the carrot drying and rehydration. *Food Research International*, *89*, 701–708. doi:10.1016/j.foodres.2016.09.030 PMID:28460968

Rodríguez, Ó., Eim, V., Rosselló, C., Femenia, A., Cárcel, J. A., & Simal, S. (2018). Application of power ultrasound on the convective drying of fruits and vegetables: Effects on quality. *Journal of the Science of Food and Agriculture*, *98*(5), 1660–1673. doi:10.1002/jsfa.8673 PMID:28906555

Sagar, V. R., & Kumar, P. S. (2010). Recent advances in drying and dehydration of fruits and vegetables: A review. *Journal of Food Science and Technology*, *47*(1), 15–26. doi:10.100713197-010-0010-8 PMID:23572596

Sajas, J. F., & Gorbatow, W. M. (1978). *The use of ultrasonics in meat technology*. Germany: Fleischwirtschaft.

Saxena, J., & Dash, K. K. (2015). Drying kinetics and moisture diffusivity study of ripe Jackfruit. *International Food Research Journal*, *22*(1), 414.

Taskin, O., Izli, G., & Izli, N. Convective drying kinetics and quality parameters of European Cranberrybush. *Journal of Agricultural Sciences*, *24*(3), 349–358.

Yang, F., Zhang, M., Mujumdar, A. S., Zhong, Q., & Wang, Z. (2018). Enhancing drying efficiency and product quality using advanced pretreatments and analytical tools—An overview. *Drying Technology*, *36*(15), 1824–1838. doi:10.1080/07373937.2018.1431658

Chapter 8
Nanoantimicrobials:
An Emerging Technological Approach in Food Preservation

Navneet Kaur
https://orcid.org/0000-0001-8584-5712
Sri Guru Granth Sahib World University, India

ABSTRACT

One of the major issues food technologists deal with is food preservation and safety. Growth of micro-organisms in food poses risk to its quality and safety. Moreover, resistance of food spoilage micro-organisms against various chemical food preservatives has led to an emergence of novel antimicrobial agents with improved action and low rates of microbial resistance. Development in nanotechnology has led to the production of nanoparticles that are not only safe but also effective to resolve the problem of microbial resistance. Nanoantimicrobials have shown improved bioactive performances and controlled toxicity to human beings. They are steadily gaining popularity and the trend will continue in coming years. The chapter gives a comprehensive view of nanoantimicrobials of organic and inorganic origin, various mechanisms adopted by these nanoparticles for the destruction of micro-organisms, factors affecting anti-microbial activities of these particles along with their applications in various fields of food technology.

INTRODUCTION

Invention of novel technologies has dramatically altered the food processing arena over the past few decades. Out of these emerging technologies, "Food Nanotechnology"

DOI: 10.4018/978-1-7998-1924-0.ch008

has become a prominent field with many new opportunities for the food sector (Kumar et al., 2016). The major aim of nanotechnology is identification, assembly, and engineering of living and nonliving entities lesser than 100 nm having distinctive and original functional characteristics (Weiss et al., 2006). As nanotechnology exhibits valuable properties including the potential to improve solubility and bioavailability, and to safeguard bioactive constituents from processing and storage conditions, its applications in different sectors are innumerable such as agriculture, clothing, cosmetics, food, medicine, and public health (Bajpai et al., 2018).

Among other sectors, nanotechnology has served the field of food science meticulously. The various applications of nanotechnology in food include a shelf-life extension of food products, excellent tracking and tracing of contaminants, better storage of food and food related products, safe delivery and release of bioactive compounds, health supplements or antimicrobial agents in food, etc. (Hamad et al., 2017).

Retardation of microbial outbursts is very important in food and hence, researchers are continuously working for the development of new antimicrobial materials and methods of their application. On the basis of activity, antimicrobial substances can be divided into two major groups: bactericidal, which completely eliminate bacteria, and bacteriostatic, which slow bacterial growth. Another classification can be made on the basis of the origin of antimicrobial component, i.e. natural and synthetic antimicrobials. Antimicrobial agents of plant origin such as aldehydes, esters, ketones, terpenes, etc. obtained from the essential oil of herbs or spices come under natural antimicrobials. These compounds are regarded as safe due to their natural origin but not all can be added to food items as they exhibit strong odors that may be imparted to the food product. Moreover, extraction and isolation of these bioactive compounds are both exorbitant and sometimes the essential oil extraction is expensive, laborious, and time consuming (Hosseini et al., 2017). Chemical food preservative agents include acetic acid, benzoic acid, citric acid, ethyl paraben, lactic acid, methyl paraben, nitrites, propyl paraben, sorbic acid, sodium benzoate, sodium diacetate, sodium propionate, and sulfites. However, the adverse effects of these agents on human health have been much discussed and proven.

The recent development in the field of food preservation is nanomaterials processing the antimicrobial properties (Hosseini et al., 2017). Considerable research has been carried out to check the antimicrobial potential of these nanoparticles (Santos et al., 2013).

BACKGROUND OF NANOANTIMICROBIALS

Use of metals as antimicrobial agents dates back thousands of years and Egyptians were the first to utilize astringent properties of copper and its salts. Similarly, the use of copper and silver in food preservation and in the disinfection of water is well documented by Egyptians, Greeks, Indians, Persian kings, Phoenicians, and Romans (Gold et al., 2018). Metal compounds were used as antimicrobial agents in the 20th century which were then taken over by organic antibiotics in the mid-20th century. Then, about 10 years ago, a fascinating fact was discovered that revealed the efficiency of metals as anti-biofouling agents. This was an important discovery as biofilms show antimicrobial resistance. In addition, metals were also found potent against persister cells, which are the temporarily inactive states of regular cells, and show resistance to antibiotics (Turner, 2017). The most popular metal in history is silver, which was used to restrain the proliferation of microorganism-induced disease in humans by incorporating it into daily stuff. The first evidence of the therapeutic properties of silver precedes the Han Dynasty in China around 1500 B.C.E. The Phoenician, Macedonian, and Persian empires used silver vessels and plates for food and drinks (Sim et al., 2018).

CLASSIFICATION OF NANOANTIMICROBIALS

Nanomaterials (NMs) having antimicrobial properties or nanoantibiotics, can be classified into two major groups based on the type of material used: inorganic nanoantimicrobials (metals and metal oxides) such as aluminium oxide, copper/copper oxide, gold, iron oxide, magnesium oxide, nanoclays, silver, superparamagnetic iron oxide, titanium oxide/dioxide, and zinc oxide, and organic nanoantimicrobials including carbon nanotubes (CNTs), nanostructured chitosan (CS), etc. (Hosseini et al., 2017).

INORGANIC/METAL NANOANTIMICROBIALS

Aluminium Oxide Nanoparticles: The activity of aluminium oxide NPs as antibacterial agents remains a topic of debate as their bactericidal action is quite mild and their effect is only exerted at elevated concentrations. However, the mode of action includes spreading and building up of these nanoparticles inside the cells thereby causing the development of pits and perforation, which results in dissociation of membrane components, finally leading to cell death (Beyth et al., 2015).

Copper-Based Nanoparticles: The popularity of copper-based NPs among researchers may be attributed to their economical availability, lesser reactivity to human tissues, and high vulnerability towards microorganisms. Copper and copper complexes are not only used as liquid sterilizer but also exhibit antibacterial, antifungal, antiviral, and antifouling properties. They inhibit several microorganisms such as algae, bacteria, fungi, and viruses (Hosseini et al., 2017). Despite having so many advantages, their use is restricted due to their rapid oxidation in air. The mode of action involves perforation of the cell membrane, suppression of enzymes required for cellular function leading to cell death. The antimicrobial activity depends on various factors including the size of NPs, their concentration, and stability in the growth media (Tayel et al., 2017).

Gold Nanoparticles: These are widely used as antimicrobial agents due to the array of properties exhibited by them such as bioactivity against a wide variety of microorganisms, nontoxicity, easy activation by photothermal treatment, high stability, and easy detection (Tayel et al., 2017). Strong antibacterial activity has been seen in gold NPs having a size less than 2nm (Fernando et al., 2018). Their mode of action includes membrane disorganization due to elevated levels of ATP and a decrease in ribosomal tRNA synthesis. Unlike other NPs, their mechanism of inhibition is independent of the generation of ROS (Reactive oxygen species) (Beyth et al., 2015; Tayel et al., 2017).

Iron Oxide Nanoparticles: Iron oxide nanoparticles are generally regarded as inert and inactive against microorganisms. However, at nanosize the characteristics are altered, resulting in inhibition of microbes. Iron oxide nanoantimicrobials prevent biofouling by displaying anti-adherent properties. They are also potent against both gram-positive and gram-negative bacteria (Beyth et al., 2015).

Magnesium Oxide Nanoparticles: Magnesium oxide as nanoantimicrobial possesses strong activity against bacteria, spores, and viruses. It inhibits both Gram-positive and Gram-negative bacteria. Moreover, it is easily available at a low cost. The mode of action involves generation of ROS and inhibition of vital enzymes necessary for function of bacteria. They are mainly utilized in the food industry as anti-fouling agents or as anti-bacterial coatings to help prevent formation of biofilms by *Escherichia coli* and *Staphyloccocus aureus* (Beyth et al., 2015).

Nanoclays: They may be defined as NPs having layers of silicates which are generally used in the preparation of nanocomposites. Widely used nanoclays include Montmorillonite (MMT), hectorite, and saponite. Layered silicates exhibit two substituted structures—tetrahedral and octahedral. Nanoclays have the ability to disperse UV (Ultravoilet) visible radiation and thus, they are used to provide protection against light in food packaging materials such as dairy products, fruit and vegetable products, and in sports drinks, which contain light-sensitive materials. Moreover, they are also used as active packaging material (Garcia & Lagaron, 2012). Cloisite

30B and Cloisite 20A are organically modified nanoclays. The modifications were done to improve the properties of nanoclays such as antimicrobial activity. Like other inorganic nanoantimicrobials, they also inhibit both Gram-positive and Gram-negative bacteria. The antimicrobial activity of these modified nanoclays may be attributed to the presence of a quaternary ammonium group that after encountering bacterial cell, results in disorganization of membrane leading to cell death. These modified nanoclays have been found to be active against inhibition of *Staphylococcus aureus*, *Listeria monocytogenes*, *Salmonella typhimurium*, and *E. coli* O157: H7 (Hosseini et al., 2017).

Silver Nanoparticles: Silver belongs to the class of noble metal, exhibiting various properties such as low toxicity, high thermal stability, and antibacterial activity (Sportelli et al., 2015). Silver nanoparticles show a wide spectrum of antimicrobial activity against bacteria, fungi, and viruses (Beyth et al., 2015). They show inhibition potential against both Gram-positive and Gram-negative bacteria. The antimicrobial action of silver NPs has been reported against various food spoilage microorganisms such as *Pseudomonas aeruginosa*, *Pseudomonas putida*, *Candida,* and *Penicillium* spp (Shrividya et al., 2016). The antimicrobial activity of these nanoparticles is size dependent, such as the smaller the particles, the greater will be the inhibition potential. The mode of action of silver nanoantimicrobials includes interaction with DNA, the formation of ROS, and disruption of the cell membrane leading to cell lysis (Sportelli et al., 2015).

Superparamagnetic Iron Oxide Nanoparticles (SPION): Superparamagnetic iron oxide nanoparticles are produced utilizing magnetic particles which produce hyperthermia in a particular confined region in an already existing magnetic field. They are quite efficient in destroying biofilms produced by bacteria along with other NPs such as iron and gold (Beyth et al., 2015).

Titanium Oxide/Dioxide Nanoparticles: Titanium oxide exerts antimicrobial potential against a wide range of pathogenic microorganisms such as bacteria (gram positive and gram negative), fungi, parasites, and viruses. It is widely employed in the food, drug, and cosmetic industry as it is approved by the Food and Drug Administration (FDA). However, the antimicrobial potential of titanium oxide NPs depends on size, shape, and structure. It also exhibits photo catalytic activity. Plant extracts such as *Garcinia zeylanica* have been used to produce modified titanium oxide NPs which along with its photolytic activity results in enhanced microbicidal activity. The various mechanisms by which it acts to destroy micro-organisms include photocatalytic properties, ROS generation, disruption of cellular organization, and cell death (Tayel et al., 2017; Fernando et al., 2018).

Zinc Oxide Nanoparticles: Zinc oxide has been given a GRAS status by FDA, i.e. generally recognized as safe compound because it imposes minimum toxicity towards human beings. Like titanium oxide, it also exhibits biocidal and photo catalytic activity. It is popular in the food industry due to its easy availability and low cost. It is used to preserve natural colours and prevent spoilage-causing micro-organisms. The antimicrobial activity of zinc oxide NPs may be attributed to high surface-to-volume ratio and abrasive nature of zinc oxide nanostructures. It shows antibacterial potential against both Gram positive (*Enterococcus faecalis, Staphylococcus aureus, Streptococcus pyogenes*) and Gram-negative bacteria (*Campylobacter jejuni, E. coli O157:H7, Listeria monocytogenes,* and *S. aureus*). The mode of action involves intracellular aggregation of zinc oxide nanoparticles which result in disorganization of cell membrane and disturbing replication of DNA. Moreover, it may result in the generation of ROS which ultimately leads to the killing of bacterial cells (Tayel et al., 2017; Fernando et al., 2018).

ORGANIC NANOANTIMICROBIALS

Carbon Nanotubes/Fullerenes: Carbon nanotubes (CNTs) may be defined as thin tubular arrangements formed by rotating graphene sheets over each other in the nano range. They are of two types—single-walled carbon nano tubes or SWCNTs having single shell (1-5 nm in diameter) and multi-walled carbon nanotubes (MWCNTs) with several layers of graphene and grapheme (100 nm or more in length). They inculcate numerous properties including tube-like structure, electrical, chemical, mechanical, thermal, optical, and antimicrobial action. The antimicrobial activity has been reported against various spoilage microorganisms such as *E.coli* (Gram negative) and *S. epidermidis* (Gram positive). Moreover, it has been revealed that SWCNTs exhibit greater antimicrobial potential in comparison to MWCNTs. Thus, to improve the antimicrobial efficacy of MWCNTs, they were functionalized using amphiphilic poly (propyleneimine) dendrimer (APPI) along with silver nanoparticles. The mechanism of action included greater permeability of cell membrane leading to dysfunction and lastly to cell death (Hosseini et al., 2017).

Chitosan Nanoparticles: Chitosan nanoparticles exhibit antimicrobial activity against bacterial, fungal, and viral species. They were even conjugated with hydroxy-cinnamic acid to improve bactericidal activity. The popularity of chitosan NPs may be attributed to low toxicity, high biocompatibility, strong antimicrobial properties, and mild immunogenicity. The antibacterial activity, however, depends on pH and solvent used. Chitosan in conjunction with metal nanoparticles (such as Zinc) result in reduced antimicrobial action and hence its functionalization with metals should be avoided. The mechanism of action of chitosan NPs involves the interaction of chitosan with bacterial cell membrane causing destabilization of the membrane and finally leading to cell lysis (Beyth et al., 2015).

FACTORS AFFECTING ANTIMICROBIAL POTENTIAL OF NANOANTIMICROBIALS

The antimicrobial activity of nanoparticles depends on several factors which are illustrated in a comprehensive manner:

- **Size:** The antimicrobial activity of NPs is size dependent. The lesser the size of NPs, the greater their capability to easily penetrate the bacterial cell wall. In Gram negative bacteria, smaller NPs can easily penetrate the microbial cell and reach its nuclear content. Moreover, smaller size results in the greater surface area resulting in better bactericidal effect. NPs smaller than 10 nm have been found to possess better penetration properties leading to disintegration of DNA along with vital enzymes and finally to cell death (Hozeinzadeh et al., 2017).
- **Shape:** Shape of NPs has been found to influence antimicrobial activity exhibited by them such as silver NPs of various shapes viz. spherical, rod, and triangular, were tested against *E.coli* and of all the shapes, triangular NPs were found to possess strong bactericidal action. Similarly, better holding capacity and interaction on bacterial surface were observed in the case of cuboctahedral, icosahedral, and decahedral shaped NPs, as these shapes offer efficient binding of sulphur-containing proteins present on bacterial cell wall (Srividya et al., 2016).
- **Concentration:** Higher concentration of NPs results in stronger antimicrobial effect. The effect of antimicrobial NPs was tested on the same population of bacteria by varying concentration and it was observed that at higher or excessive concentration, disorganization of mitochondria was followed by the release of lactate dehydrogenase. Higher concentration also leads to the high surface area thereby increasing antimicrobial potential (Hozeinzadeh et al., 2017). However, the variable concentration of NPs may be required to inhibit different foodborne microbes (Srividya et al., 2016).
- **Surface charge/Zeta Potential:** Zeta potential or electrostatic force on the surface of microbial cells, in addition to particle size and composition of NPs, is an important parameter affecting antimicrobial potential (Hozeinzadeh et al., 2017). Adhesion of NPs to the surface of bacteria is influenced by zeta potential. The bacterial cell membrane is negatively charged and thus, positively charged NPs are easily attracted to the surface of the bacterial cell wall in comparison to neutral or negatively charged NPs. The accumulation leads to changes in membrane permeability, thereby building up these NPs inside the bacterial cell, subsequently leading to cell death (Wang et al., 2017).

- **Roughness:** Out of several factors affecting antimicrobial action of NPs, few studies have taken into account their rough surface. Size and surface area to mass ratio are promoted with an increase in roughness of NPs thereby increasing the contact of bacterial proteins which results in alleviation of bacterial adherence properties (Wang et al., 2017).
- **Presence of active oxygen:** Out of several mechanisms followed for inhibition of microorganisms, the generation of ROS is one of the most important pathways followed. Some NPs such as Calcium oxide and Magnesium oxide exhibit active oxygen on their surface, which in contact with water forms superoxide radicals. This radical being highly reactive and oxidative in nature consequently leads to toxicity of microbial cells (Hoseinzadeh et al., 2017).

MECHANISM OF ACTION OF NANOANTIMICROBIALS

Nanoantimicrobials follow different mechanisms for inhibition of microorganisms and these pathways are discussed below in detail:

The Disintegration of Cell Membrane: The antimicrobial action exerted by NPs is still not clear and different workers have proposed different mechanisms. The most accepted mechanism is the formation of "pits" or "holes" or "pores" in the membrane (Santos et al., 2013). The first step in this mechanism is adherence of NPs to microbial cell walls leading to their aggregation. The adherence may be due to negative charge on the membrane by which positively charged NPs are pulled or due to sulphur-containing proteins which result in increased membrane permeability leading to the release of membrane components such as lipopolysaccharides and other membrane proteins. Leakage of cellular components affects the activity of vital enzymes subsequently resulting in cell lysis or death (Srividya et al., 2016).

Release of Toxic Ions: Several metal ions such as cadmium, silver, and zinc present in NPs possess toxic effects and thus by interacting with bacterial proteins, result in cell death. In the case of silver NPs, silver forms sparingly soluble salts with chlorine ions, i.e. silver chloride, which inhibits cellular respiration thereby attacking the microbial cells by interfering in other cellular and metabolic functions. Silver ions also interfere with the functioning of major enzymes by reacting with their thiol groups. Like silver, cadmium, and zinc ions form bonds with sulphur-containing proteins present in cell membrane, resulting in increased permeability and finally cell death. Zinc oxide NPs cause toxicity by forming hydrated zinc ions in the presence of water which ultimately enter lysosomes (Santos et al., 2013; Srividya et al., 2016).

DNA Damage: Many antimicrobial NPs have also been found to interact with DNA of microbial cells thereby disturbing their DNA metabolism. Silver NPs interact with phosphorous moieties in DNA and nucleic acids, resulting in suppression of DNA replication and cell division. Similarly, gold NPs hinder uncoiling and DNA transcription leading to cell death (Srividya et al., 2016). Copper NPs act by inhibiting DNA replication and also result in degradation of DNA of microbial cells (Hemeg, 2017).

Generation of ROS: Reactive oxygen species such as hydroxyl radicals, hydrogen peroxide, singlet and triplet oxygen, superoxide radicals, etc. interact with microbial cells followed by their death via different pathways. ROS generation results in the disintegration of the membrane and prevents surface adhesion along with the formation of biofilms which are produced by bacteria by interacting with each other. These radicals such as hydrogen peroxide, and singlet and triplet oxygen, also lead to oxidation of DNA, proteins, and lipopolysaccharides of cell membrane. Silver NPs usually form hydroxyl radicals while zinc oxide and iron oxide NPs produce hydrogen peroxide. However, both these radicals are generated by titanium oxide due to its photocatalytic activity (Santos et al., 2013; Srividya et al., 2016; Hemeg, 2017).

Modulation of Cellular Pathways: Silver NPs have been reported to interfere with signal transduction pathway in bacteria by dephosphorylating the protein substrates and thereby hindering the cell growth. Phosphorylation of these protein substrates is essential for the regulation of this pathway (Srividya et al., 2016).

Alteration in pH and Inhibition of Protein Translation: The internalization and accumulation of NPs inside the microbial cells are dependent on surface charge exhibited by NPs. Silver and zinc oxide NPs exert microbicidal effects by producing oxidative stress. Metallic nanoparticles also cause oxidation of lipids in microbial cell membrane, resulting in decline in pH and elevating membrane potential. A similar pathway is followed by aluminium oxide NPs. Silver and gold NPs, after disrupting cell membranes and accumulating inside microbial cells, damage 30S ribosomal subunit resulting in inhibition of protein translation (Hemeg, 2017). The schematic diagram representing different routes of the mechanism of action of inorganic nanoparticles is given in fig 1.

NANOCOMPOSITES

Trapping/Insertion of antimicrobial NPs inside polymers to form hybrids of lipid-polymer to enhance physical properties such as mechanical strength, thermal stability, biocompatibility, and stability along with controlled release, are termed as nanocomposites (Hemeg, 2017). Nanocomposites may be inserted into different matrices-

- Nanofillers such as carbon nanotubes, cellulose microfibrils, nanoclays, and nanooxides.
- Synthetic polymers such as nylon, polyamide, polystyrene, and polyolefins.
- Natural polymers such as carrageenan, cellulose, and chitosan.

However, demand for biodegradable biopolymers of natural and synthetic origin such as polyvinyl alcohol, polyglycolic acid, and polylactide, etc. is increasing due to problems faced in the disposal of non-biodegradable polymers in environment (Pathakoti et al., 2017). Therefore, polymers consisting of metallic nanoantimicrobials are preferred as they exhibit antimicrobial properties along with better mechanical and barrier characteristics and also inhibit the photodegradation of plastics (Fortunati et al., 2017).

APPLICATION OF VARIOUS NANOANTIMICROBIALS IN FOOD PRESERVATION

Active Packaging

Active packages allow direct interaction between antimicrobial nanoparticles and food for better preservation and safety of food products. Various metal nanoparticles such as silver, copper, titanium, and organic NPs such as CNTs have been shown to be effective antimicrobial agents along with other beneficial properties viz. acting as oxygen and UV scavengers. Kodak Company has engineered and marketed

Figure 1. Different routes followed as mechanism of action by various inorganic nanoparticles (Hoseinnejad et al., 2017)

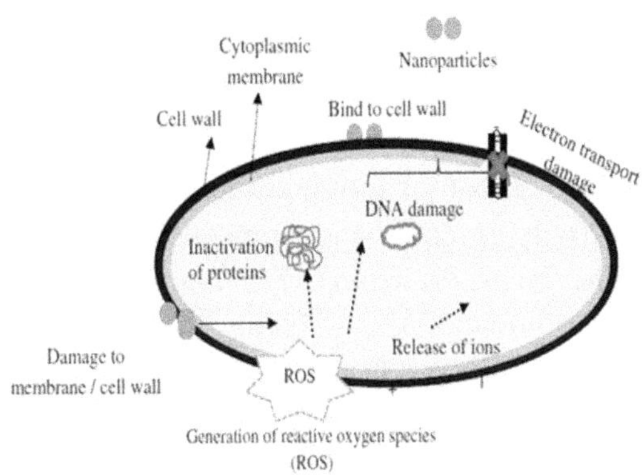

antimicrobial packages having barrier properties against oxygen (Kuswandi, 2016). Some of the nano-antimicrobials employed in active packaging include:

CuNPs: Copper NPs are widely employed in packaging materials to provide hygienic packing conditions, prevention of food borne pathogens, and spoilage causing microorganisms. CuNPs in chitosan films have been found to be effective against inhibiting the growth of bacterial species such as *S. aureus*, *S. thyphimurium,* and fungal species such as *Alternaria solani*, *F. oxysporum,* etc. In addition, the packages were also impermeable to oxygen, moisture, as well as UV light. Similarly, CuNPs embedded in the cellulose matrix inhibited the growth of *Saccharomyces cerevisiae* while the same nanocomposite in pineapple and melon juice displayed antifungal activity by retarding the growth of yeasts and molds. Incorporation of CuNPs in PLA matrix (polylactic acid) was used in the packaging of fresh dairy products such as cheese. The major advantage of using CuNPs is that they do not alter sensory characteristics of the food product for which they are used (Costa et al., 2016; Tayel et al., 2017).

AgNPs: Silver NPs are used in active packaging to increase the shelf life of the product by preventing the growth of spoilage microflora. They have been found to enhance the shelf life of minced meat and various fruits and vegetables (fresh cut carrots and green asparagus). In the case of fruits and vegetables, AgNPs absorb and decompose ethylene (ripening hormone) and hence extend shelf life. They also retard the activity of spoilage microorganisms such as yeast count in case of melons. AgNPs embedded in agar gel were used to store fiordilatte cheese to improve its quality and increase its shelf life. AgNPs enhanced the exchange of ions between agar, brine, and cheese and also inhibited spoilage microorganisms, thereby leading to extended shelf life (Costa et al., 2016).

ZnONPs: Zinc oxide has been given GRAS status by FDA and hence it is extensively utilized in the food industry. ZnONPs coated onto PVC (polyvinyl chloride) films inhibited *E. coli* and *S. aureus*. Nanocomposites of low-density polyethylene (LDPE) containing silver and ZnONPs were used for shelf life extension of up to 28 days in case of orange juice and also retarded the growth of *L. plantarum* significantly. Coating of ZnONPs over glass jars containing liquid albumen effectively reduced the growth of *Salmonella*. Calcium alginate embedded ZnONPs were used in active packaging of Ready-to-eat (RTE) poultry meat and were found to reduce the growth of spoiling microflora i.e. *S. typhimurium* and *S. aureus.* Antifungal activity of ZnO nanocomposites was tested in various fungal species viz. *Aspergillus flavus*, *P. citrinum*, *Pythium* spp., and *Penicillium* spp., and it actively inhibited their growth. Despite being antimicrobial in nature, these NPs also preserve the colour of food products and enhance the barrier properties of the packages (Costa et al., 2016; Hosseini et al., 2017; Tayel et al., 2017).

TiONPs: Titanium oxide being non-toxic and approved by FDA to be utilized in food products is also used in active packaging. It is either used in the form of coating on surface or by embedding them into polymer matrix. TiO_2 coated on oriented polypropylene polymer was used to store fresh cut lettuce where it retarded *E. coli* growth. Similarly, TiO_2 nanocomposites with LDPE film resulted in inactivation of *Pseudomonas* spp. and *P. mucileginosa* along with a decrease in the number of mesophilic bacteria and yeast count. Elastic film coated with TiO_2 NPs showed anti-fungal activity against *P. expansum*. It also absorbed ethylene in LDPE films and increased storage life of strawberries. Shelf life extension was observed in cheese as TiO_2 inhibited lactic acid bacteria and coliforms. Active packaging with titanium oxide and silver NPs retarded the growth of green mold in Chinese bayberries. It also acts as a barrier against oxygen and thus prevents oxidation, rancidity, browning, and spoilage reactions (Costa et al., 2017). Titanium nanotubes embedded in chitosan actively prevented the growth of spoilage microorganisms such as *E.coli*, *S. enterica*, and *S. aureus* (Tayel et al., 2016).

EDIBLE NANOCOATINGS

Edible coatings may be defined as a thin layer that may be applied onto the surface of food either by spraying, immersion, or rubbing to extend the shelf life and at lower cost (Zambarano-Zaragoza et al., 2018). Nanocomposites of organic and inorganic NPs as edible coatings are extensively used nowadays to preserve the quality of fresh foods such as meat, fruits, vegetables, etc.

AgNPs: Asparagus spears were coated using nanocomposites of PVP (polyvinyl pyrollidone) and AgNPs. The edible film inhibited psychrotrophic microorganisms and resulted in the preservation of colour, improvement of texture, and reduced weight loss. Similarly, an antibacterial edible film containing AgNPs and sodium alginate was applied onto the surface of carrots and pears where it inhibited the growth of *E. coli* and *S. aureus*, side by side preserved proteins, and prevented loss of weight thereby extending the shelf life. AgNPs embedded in *Fantasia japonica* leaf extract significantly retarded *E. coli*, *S. aureus,* and *P. italicum*, improving storage life (Zambrano-Zaragoza et al., 2018).

Nanoclays: Nanoclays such as MMT, i.e., Montmorrilonite in conjunction with oregano essential oil, when applied as an edible coating, retarded the growth of physchrophilic microorganisms such as yeast and fungi in fresh cut papaya. It was also used along calcium caseinate as an edible coating in prevention of antifungal activity in strawberries (Zambrano-Zaragoza et al., 2018).

ZnONPs: Edible coating containing 0.2% ZnO_2 and CMC (carboxymethylcellulose) was used to inhibit yeast and mold growth in ready-to-eat pomegranate. Along with preservation action, it also maintained the bioactive components of pomegranate (Zambrano-Zaragoza et al., 2018).

Prevention of Biofouling

Formation of biofilms is an intrinsic defense mechanism shown by microorganisms in which they adhere to either biological or non-biological surface, resulting in fouling. When the film or coating is formed on biological surfaces it is termed as biofouling. Mechanism of biofouling is commonly seen in food industries such as meat, dairy, fruit and vegetable, etc., where the surfaces are continuously in contact with water. Thus, for safety purposes and to prevent foodborne disease, these films should be eliminated. Nanocomposites can either be directly applied to the surfaces or by in situ modification. Several inorganic NPs embedded in polymeric matrices have been used as coatings to prevent biofouling such as copper, silver, titanium, zinc, etc. (Srividya et al., 2016).

ADVANTAGES OF NANO-ANTIMICROBIALS

- Being low cost and easily available, nanoantimicrobials are gaining immense popularity in the food processing sector worldwide. Some of the advantages of nanoantimicrobials are discussed below (Hoseinzadeh et al., 2017):
- Nanoantimicrobials exhibit various properties and the most important is that they can be activated under certain conditions such as light (titanium), heat (gold), magnetic field (SPION), etc. Thus, they can be used under special physico-chemical conditions where there will be no loss in their activity in comparison to chemical or biological antimicrobials.
- Nanoantimicrobials are small in size and exhibit various shapes. They are made up of a wide range of materials and have different mechanisms for elimination of microorganisms. Based on these properties, suitable nanoantimicrobials can be selected against different species of pathogenic microorganisms.
- Many of the nanoantimicrobials used in food industry have been given GRAS (Generally regarded as Safe) status by FAO and if any side-effect is associated with them, that can be easily altered by varying their physico-chemical properties.
- Nanoantimicrobials exhibiting different mechanisms for inhibition of microorganisms can combat high rates of antimicrobial resistance as compared to the standard antibiotics. However, few studies have reported resistance against some of the nanoantimicrobials such as *Cupriavidus metallidurans*

against titanium oxide and aluminium oxide NPs and *Shewanella oneidensis* against copper NPs.

DISADVANTAGES OF NANOANTIMICROBIALS

In spite of nanoantimicrobials having so many advantages, a few limitations encountered while using them are listed below (Hoseinzadeh et al., 2017):

- Due to toxicity exhibited by some of the nanoantimicrobials, they may pose threat to human health and moreover, being stable, they accumulate in the environment and hence may pose ill effects.
- They may pose toxicity against untargeted microorganisms such as gut microflora.

BIOGENIC NANOANTIMICROBIALS

The synthesis of nanoparticles involves two major approaches, top down and bottom down. In the top down approach, lithographic methods such as photolithography, nanomolding, dippen-lithography, and nanofluidics are used. The approach involves breakdown of bulk material into smaller particles using the above methods. However, the major disadvantage of this method is a distorted product and the activity of nanoparticles is dependent on surface properties. In the bottom up approach, nanoparticles are synthesized by aggregation of atoms to form nuclei within nano range and the product obtained is of superior quality (Tyagi, 2016; Qidwai et al., 2018). However, the NPs synthesized by these approaches suffer several limitations such as stability, toxicity, operator and production costs, and recycling issues (Singh et al., 2018).

To overcome these limitations, biogenic NPs came into being and are gaining popularity. These are also called herbal or green NPs and the approach used to produce them is called "Green Synthesis" as no harsh chemicals or methods are used in this technique (Singh et al., 2018). Thus, NPs produced using biological materials such as plants, micro-organisms, animals, etc. are called biogenic NPs. The NPs produced by this approach offer better stability, uniformity in size, and shape. They may either be intracellular or extracellular in nature. Intracellular NPs require cell disruption to separate NPs. Biological organisms synthesize these NPs as part of detoxification process by converting these metal ions to insoluble lumps. Various biological systems used in production of these NPs are given below:

- **Bacteria Induced Production of NPs:** Bacteria usually synthesize metal NPs such as silver, gold, copper, cadmium, etc. and the NPs produced using bacteria are called bacterioform NPs (Nalwa, 2005; Singh et al., 2018). The first report of the production of silver NPs by bacteria came in the year 2000. Bacteria are widely used for the production of metal NPs because they exhibit the property of reducing metal ions. Moreover, they are easy to handle, produce stable NPs, are cost effective and appropriate for large scale production (Qidwai et al., 2018). A few examples of bacterial strains that have been widely studied for the synthesis of bioreduced silver nanoparticles are: *Lactobacillus casei, Aeromonas* sp. SH10, *Arthobacter gangotriensis, Bacillus amyloliquefacians, B. cereus, B. cecembensis, B. indicus, B. subtilis, B. megatarium, Corynebacterium* sp. SH09, *Escherichia coli, Enterobacter cloacae, Geobacter* spp., *Kleibsiella pneumonia, Lactobacillus casei, Phaeocystis antarctica, Pseudomonas proteolytica, Staphyllococcus aureus* and *Shewanella oneidensis*. Similarly, gold nanoparticles are synthesized using *Bacillus megaterium* D01, *Desulfovibrio desulfuricans, Bacillus subtilis* 168, *E. coli* DH5a, *Shewanella alga, Rhodopseudomonas capsulate*, and *Plectonema boryanum* UTEX 485 (Tyagi, 2016; Singh et al., 2018, Qidwai et al., 2018). The disadvantages associated with bacterial nanofactories are a slow rate of synthesis and limited availability of shape and size (Qidwai et al., 2018).
- **Fungi Induced Production of NPs:** Fungi generally synthesize metal NPs such as silver, iron, gold, titanium oxide, zinc oxide, etc. and NPs produced using fungi are called mycogenic NPs. Fungi are widely used for the synthesis of metal and metal oxide NPs because of the presence of enzymes on their cell surface, tolerance to toxicity caused by metals, and ease of handling in laboratory (Nalwa, 2005; Tyagi, 2016; Singh et al., 2018; Qidwai et al., 2018). The enzymes produced by fungi efficiently cause metal reduction and hence the production of NPs. Several advantages are offered by fungi, e.g., enzymes are produced in greater amount which allows their use for large scale production. Moreover, downstream processing is easy as NPs are produced extracellularly. The fungal mycelia are more tolerant to process conditions than bacteria (Tyagi, 2016). The fungal species involved in production of silver NPs are *Aspergillus fumigatus, Aspergillus flavus, Aspergillus niger, Cladoporium cladosporoides, Cariolus versicolor, Fusarium solani, Phaenerochaete chrysosporium, Penicillium brecompactum, Penicillium fellutalum, Rhizopus nigricans,* and *Verticillium.* For synthesis of gold NPs *Fusarium oxysporum, Tricothecium* spp., and *Rhizopus stolonifer* are used, while *Aspergillus tereus* and *Aspergillus flavus* TFR7 are employed for the synthesis of zinc oxide and titanium oxide NPs (Singh et al., 2018).

- **Yeast-induced production of NPs:** Data regarding synthesis of yeast mediated NPs is limited. However, yeast strain MKY3, Candida guilleirmondii, and Saccharomyces cerevisae were investigated for gold and silver NPs (Qidwai et al., 2018).
- **Algae-induced production of NPs:** Similar to yeast, there are only few studies reporting use of algae for the production of NPs. Marine algae *Sargassum wightii* was used to synthesize gold NPs (Tyagi, 2018).
- **Plant-induced production of NPs:** Plants act as biofactories for production of NPs as they can accumulate heavy metals in their plant parts such as leaves, stem, seeds, flowers, shoots, roots, etc. In addition, the approach is relatively simple, efficient, clean, eco-friendly, and accessible (Singh et al., 2018). Biomolecules within plant tissues such as carbohydrates, proteins, and co-enzymes reduce metal ions into NPs. Plant-mediated synthesis of silver NPs was first studied in alfalfa sprouts. The roots of alfalfa absorb silver from the soil and pile up that silver in the shoots in the same oxidation state from where it can be extracted in the form of NPs. Silver and gold NPs are synthesized from various plants including *Aloe barbadensis Miller, Avena sativa, Azadirachta indica, Brassica juncea, Cymbopogon flexuosus, citrus limon, Medicago sativa* and *Osimum sanctum*. While *Brassica juncea, Coriandrum sativum, Calotropis gigantean, Helianthus annus* and *Medicago sativa* are utilized for the synthesis of zinc, nickel, cobalt, and copper NPs (Singh et al., 2018).

CONCLUSION

The involvement of nanotechnology in different sectors of food technology has introduced a number of changes in food packing, preservation, and processing methods. The advent of this new technology has led to significant extension of the shelf life of foods with better management of spoilage microflora associated with food and food products. Although the benefits are uncountable, major concern arises about the toxicity imposed by these nanoparticles. Information related to toxicology of these nanoantimicrobials is scarce and thus, analytical methods for the detection and characterization of these particles in food need to be developed. Research needs to be oriented in the direction of potential risks imposed by the nanoantimicrobials to human health. Data involving clinical trials of direct use of nano-antimicrobials in food is also limited. Legislation should also be imposed to differentiate between nano-based foods and ordinary approaches of manufacturing and storage.

REFERENCES

Bajpai, V. K., Kamle, M., Shukla, S., Mahato, D. K., Chandra, P., Hwang, S. K., ... Han, Y. K. (2018). Prospects of using nanotechnology for food preservation, safety, and security. *Journal of Food and Drug Analysis*, *26*(4), 1201–1214. doi:10.1016/j.jfda.2018.06.011 PMID:30249319

Beyth, N., Houri-Haddad, Y., Domb, A., Khan, W., & Hazan, R. (2015). Alternative antimicrobial approach: Nano-antimicrobial materials. *Evidence-Based Complementary and Alternative Medicine*, *2015*, 1–16. doi:10.1155/2015/246012 PMID:25861355

Costa, C., Conte, A., Alessandro, M., & Nobile, D. (2016). Use of metal nanoparticles. In J. Barros-Velazquez (Ed.), Antimicrobial food packaging (pp. 399-406). Academic Press, Elsevier. doi:10.1016/B978-0-12-800723-5.00031-0

Fernando, S. S. N., Gunasekara, T. D. C. P., & Holton, J. (2018). Antimicrobial nanoparticles: Applications and mechanisms of action. *Sri Lankan Journal of Infectious Diseases*, *8*(1), 2–11. doi:10.4038ljid.v8i1.8167

Fortunati, E., Puglia, D., Armentano, I., Valde, A., Ramos, M., Juarez, N., . . . Kenny, J. M. (2017). Multifunctional antimicrobial nanocomposites for food packaging applications. In A. M. Grumezescu (Ed.), Food preservation- Nanotechnology in the agri food industry (pp. 265-293). Academic Press, Elsevier. doi:10.1016/B978-0-12-804303-5.00008-0

Gold, K., Slay, B., Knackstedt, M., & Gaharwar, A. K. (2018). Antimicrobial activity of metal and metal-oxide based nanoparticles. *Advances in Therapy*, *1*(1700033), 1–15. doi:10.1002/adtp.201700033

Hamad, A. F., Han, J. H., Kim, B. C., & Rather, I. A. (2017). The intertwine of nanotechnology with the food industry. *Saudi Journal of Biological Sciences*, *25*(1), 27–30. doi:10.1016/j.sjbs.2017.09.004 PMID:29379352

Hemeg, H. A. (2017). Nanomaterials for alternative antibacterial therapy. *International Journal of Nanomedicine*, *12*, 8211–8225. doi:10.2147/IJN.S132163 PMID:29184409

Hoseinnejad, M., Jafari, S. M., & Katouzian, I. (2017). Inorganic and metal nanoparticles and their antimicrobial activity in food packaging applications. *Critical Reviews in Microbiology*, *44*(2), 161–181. doi:10.1080/1040841X.2017.1332001 PMID:28578640

Hoseinzadeh, E., Makhdoumi, P., Taha, P., Hossini, H., Stelling, J., Kamal, M. A., & Ashraf, G. M. (2017). A review on nano-antimicrobials: Metal nanoparticles, methods, and mechanisms. *Current Drug Metabolism, 18*(2), 120–128. doi:10.2174/1389200217666161201111146 PMID:27908256

Hosseini, H., Shojaee-Aliabadi, S., Hosseini, S. M., & Mirmoghtadaie, L. (2017). Nanoantimicrobials in Food Industry. In A. E. Oprea, & A. M. Grumezescu (Eds.), *Nanotechnology applications in food-flavour, stability, nutrition, and safety* (pp. 223–238). London, UK: Academic Press; doi:10.1016/B978-0-12-811942-6.00011-X

Kumar, K., Kumar, S., Kumar, S., Kumar, V., & Vyas, P. (2016, October). *Role of nanotechnology in food processing and preservation.* Paper presented at National Conference on Technologies in Sustainable Food Systems, Longowal, Sangrur, Punjab, India.

Nalwa, H. S. (2005). *Encyclopedia of Nanoscience & Nanotechnology, 1.* Valencia, CA: American Scientific Publishers.

Pathakoti, K., Manubolu, M., & Hwang, H. M. (2017). Nanostructures: Current uses and future applications in food science. *Journal of Food and Drug Analysis, 25*(2), 245–253. doi:10.1016/j.jfda.2017.02.004 PMID:28911665

Qidwai, A., Pandey, A., Kumar, R., Shukla, S. K., & Dikshit, A. (2018). Advances in biogenic nanoparticles and the mechanisms of antimicrobial effects. *Indian Journal of Pharmaceutical Sciences, 80*(4), 592–603. doi:10.4172/pharmaceutical-sciences.1000398

Sanchez-Garcia, M. D., & Lagaron, J. M. (2012). Nanocomposites for food and beverage packaging materials. In B. Bhandari, & Y. H. Roos (Eds.), *Food Materials, Science, and Engineering*. Chichester, UK: Blackwell Publishing. doi:10.1002/9781118373903.ch11

Santos, C. L., Albuquerque, A. J. R., Sampaio, F. C., & Keyson, D. (2013). Nanomaterials with antimicrobial properties: Applications in health sciences. In A. Mendez-Vilas (Ed.), *Microbial pathogens and strategies for combating them: Science, Technology, and Education* (pp. 143–154). Formatex.

Sim, W., Barnard, R. T., Blaskovich, M. A. T., & Ziora, Z. M. (2018). Antimicrobial silver in medicinal and consumer applications: A patent review of the past decade (2007–2017). *Antibiotics (Basel, Switzerland), 7*(93), 1–15. doi:10.3390/antibiotics7040093 PMID:30373130

Singh, J., Dutta, T., Kim, K. H., Rawat, M., Samddar, P., & Kumar, P. (2018). Green' synthesis of metals and their oxide nanoparticles: Applications for environmental remediation. *Journal of Nanobiotechnology*, *16*(84), 2–24. doi:10.118612951-018-0408-4 PMID:30373622

Sportelli, M. C., Picca, R. A., & Cioffi, N. (2015). Nano-antimicrobials based on metals. In D. A. Phoenix, F. Harris, & S. R. Dennison, (Eds.), Novel antimicrobial agents and strategies (pp. 181-218). Wiley-VCH Verlag GmbH & Co. KGaA: Weinheim, Germany.

Srividya, N., Ghoora, M. D., & Padmanabh, P. R. (2016). Antimicrobial nanotechnology-research implications and prospects in food safety. In A. Grumezescu (Ed.), *Food preservation-nanotechnology in agri-food industry* (pp. 125–157). Academic Press, Elsevier; doi:10.1016/B978-0-12-804303-5.00014-6

Tayel, A. A. A., Sorour, N. M., El-Baz, A. F., & El-Tras, W. F. (2017). Nanometals appraisal in food preservation and food-related activities. In A. Grumezescu (Ed.), *Food preservation-nanotechnology in agri-food industry* (pp. 487–526). Academic Press, Elsevier; doi:10.1016/B978-0-12-804303-5.00014-6

Turner, J. R. (2017). Metal-based antimicrobial strategies. *Microbial Biotechnology*, *10*(5), 1062–1065. doi:10.1111/1751-7915.12785 PMID:28745454

Tyagi, P. K. (2016). Production of metal nanoparticles from biological resources. *International Journal of Current Microbiology and Applied Sciences*, *5*(3), 548–558. doi:10.20546/ijcmas.2016.503.064

Wang, L., Hu, C., & Shao, L. (2017). The antimicrobial activity of nanoparticles: Present situation and prospects for the future. *International Journal of Nanomedicine*, *12*, 1227–1249. doi:10.2147/IJN.S121956 PMID:28243086

Zambrano-Zaragoza, M. L., Gonzalez-Reza, R., Mendoza-Munoz, N., Miranda-Linares, V., Bernal-Couoh, T. F., Mendoza-Elvira, S., & Quintanar-Guerrero, D. (2018). Nanosystems in edible coatings: A novel strategy for food preservation. *International Journal of Molecular Sciences*, *19*(3), 705. doi:10.3390/ijms19030705 PMID:29494548

KEY TERMS AND DEFINITIONS

Biocidal: Any product or chemical substance which exerts detrimental, destructive, or damaging effect on dangerous organisms is called biocidal.

Biofouling: Biofouling involves building up of benthic creatures such as microorganisms, plants, algae, or animals over aquatic and submerged surfaces. It is a nasty phenomenon.

Hyperthermia: Hyperthermia may be defined as the process in which the body temperature is increased beyond normal.

Nanoantimicrobials/Nanoantibiotics: Nanoantimicrobials or nanoantibiotics are nanomaterials or nanoparticles which exhibit antimicrobial activity or might enhance the activity of encapsulated antimicrobial agent.

Nanomaterials: Nanomaterials may be defined as materials or objects in the range of 1-100 nm in a single dimension.

Nanotechnology: Branch of science and technology dealing with objects at nanometre scale (1-100 nm).

Zeta Potential: Zeta potential may be defined as the charge which is formed between the interfacial surface of two mediums such as between solid and liquid surface due to potential difference between them.

Chapter 9
Recent Trends and Challenges in Sensory and Perception Science

Tugba Aktar
https://orcid.org/0000-0001-8417-868X
Alanya Alaaddin Keykubat University, Turkey

ABSTRACT

Eating is a vital and essential part of life. The routine process is not straightforward but highly sophisticated in that it incorporates dynamic changes on the food structure that triggers different modalities of food sensation. Sensory studies are being expanded with contributions from physiology, psychology, dentistry, neuroscience, and food science. All these experts seek answers to catastrophic and complicated procedures created by the oral process. This chapter provides information about the past and present of sensory science in 5 main sections: Sensation and perception (to provide an insight to psychological and physiological approaches); psychophysical laws (to interpret the missing link between the stimulus and perception); sensory market success and consumer behavior (to highlight the necessity of sensory testing for the industry and consumer); sensory evaluation methods; and recent developments in the field.

INTRODUCTION

Sensory science is one of the neglected fields of study and it has been getting the deserved attention only since mid-90s (Heymann, 2019). Even though being a new field, the development rate is very high due to the increasing demand for food

DOI: 10.4018/978-1-7998-1924-0.ch009

products. Despite being a need for survival, food is a social element for our lives. Therefore, a rapid increase in the field of sensory science is more or less expected. As mentioned by Skinner (1989), sensory evaluation of food is one of the needs of the food companies, especially for new product development, reformulation, optimizing competitor responding capability as well as strengthening the product position in the market. The aim of this chapter is to provide basic background of the sensory field as well as emphasizing the significance of the field. Meanwhile, recent developments about the sensory science and technological applications will be provided in order to image the bigger picture of the food technology and sensory collaboration.

Sensory science got its star era when the food science collides with engineering, physics, human physiology as well as psychology. That multi-science collision initiated with the question of "How do we perceive the senses?". In order to understand the perception mechanism of the senses the knowledge is required about, physiology (to understand the neurons and cognitive receptors that transfer the sense), physics (to get familiar with the oral geometry, food geometry as well as the physical aspect and the design of the food itself), food science (to understand the chemical changes that may occur with the integration of saliva), engineering (to understand the human mechanics application during oral processing such as, jaw movement and shearing between the oral surfaces), and psychology (to know the cognitive perception and cultural observations of the previous experiences with the food sample). Hence, to understand the dynamics of the sensory and oral processing, a team that can co-operate those skills is a "must" for full mapping of the eating experience.

Undoubtedly consumption of food is motivated for survival to compensate the energy used by the body. However, starvation nowadays is a term used not only to express the real meaning but also used for expressing the sociological need of food, especially which involves a crowded group of people such as family or friends. In order to fulfill the need of socialization by food, consumers have a highly increasing demand for food which results in chronic energy intake that leads to obesity in longer terms (Krop et al., 2018). Under the light of this information despite the increasing demand on sensory science itself, emerging fields' related to sensory is also developing. Oral processing, perception pathways or behavioral studies to observe the effect of senses on the perception as well as manipulation of sensory systems has been attracting the researcher's attention. Therefore, this chapter will focus on the sensory evaluation principles of the foods, by highlighting the sensation mechanisms and integrated fields of study. The main objective of this chapter is to provide readers with updated knowledge and mechanisms of underpinning principles of sensory testing.

SENSATION AND PERCEPTION

Eating is a need for survival, especially for the essential nutrients and for the energy requirement of the body. However, when considering the big picture, eating is a complex procedure starting from the first eye contact until the excretion. Now it is clear that before the eating we start to perceive or let's say pre-perceive by the visual cues. Those visual cues can be specified by color (e.g., pink so it has a strawberry flavor), texture properties (e.g., the wobbly texture so it is gel-like) and many more. Sensory is defined as a scientific method to evaluate the perception magnitudes of basic senses which are; auditoria, tactile, gustation, olfaction, and sight. Thanks to our motor skills eating is a routine process for the nutrient intake and also it is the key element for sensory pleasure. Therefore, the eating process is the top element for food scientists not only for nutrition, digestion, and sustainable production of the foods but also for the food industry for economic reasons. Measuring the sensory properties of the food creates the basis of the food acceptance which reflects itself in the company profile. In order to understand and estimate what the consumer will sense during eating, we need a systematic and professional approach as well as contemporary skills and knowledge of the sensory designing, testing and evaluating capabilities. Only after those skills are applied the sensory feedback can be stated and by using those feedbacks companies usually undergo trade and marketing policy setting. Unfortunately, due to difficult test applications, which are effort and money consuming, most companies still apply underdeveloped methods for the sensory testing which usually results with product failure and market loss.

Oral processing (oral food consumption) is a more technical term that is used for specification of the "eating" process. As mentioned earlier, thanks to our automated nervous coordination, eating is an easy task for most individuals. However, it is necessary to investigate the oral processing mechanism not only for healthy individuals but also for the consumers that have eating disorders.

In the mid-1990s sensory studies in the field of food science started to accelerate, mainly with the help of economic improvements throughout the world. Especially understanding the importance of taste on consumer preference, consumer behavior tests were triggered. On the other hand, independent disciplines of food science started to investigate the texture, flavor, taste, aroma, color perception. Most studies were either only on instrumental derived or only human studies that involve sensory science. However, since the beginning of 2000s inter-disciplined approach to perception and sensation of the sensory modalities initiated. Only by then it was possible to observe how the consumer reacts to a machine measured magnitude of a food structural property such as; aroma or texture. Integration of physical (machine measured magnitude) and psychological magnitude (sensory measured magnitude) was not an easy task to do. Psychophysical laws were used in order to observe the magnitude of the sensation which will be further discussed in this chapter.

Physical assessments of the senses are mainly measured by human panelists, which is the starring of the sensation and perception concept. Physical measurements of the sensed properties of the foods can be classified with the measurement methods in Table 1.

Meanwhile, the psychological aspect of the sensory still remains uncovered due to variation between the assessors. However, we still have some concepts that are useful for the literature and sensory test design. Of course, psychological observations and background of the sensory analysis for the food science is not the main focus of this book and the chapter. However, as can be seen in Figure 1, sensory modalities have already addressed and linked with broad psychological observation concepts in the human cortex and nervous system.

PSYCHOPHYSICAL LAWS

Physical and psychological observations were studied individually as mentioned earlier. However, the missing link between the two modalities is the psychophysical laws. Undoubtedly, instrumental assessments are robust and reliable. However, for the sensory analysis psychophysical phenomena is determinative and it has non-linear characteristics with a far more complicated system(Kilcast, 1999). The main challenge of the instrumental and sensory relation is the differences between the human and the instrument as represented Table 2.

In order to link the two different modes of sensation tests, psychophysical laws were developed. Psychophysical laws indeed date back to the 18th century. The literature presents the initiation of the experiments by a French philosopher Pierre Bouguer, who thought about a candlelight intensity perception ratio (Hecht, 1924). After a few decades, another French philosopher François Arago claimed by changing the ratio of the intensities one can change the perceived lightness of the candlelight; regardless of the magnitude of intensities. Following the Arago, almost after half a century philosopher Masson reported a constant ratio of the intensities, regardless of the intensity of the stimulus. This century taking a constant ratio of the stimulus was represented as shown as follows.

$$S = \frac{\Delta I}{I} \tag{1}$$

where;

Table 1. Physical properties of the foods that are sensed during sensory tests

Physical property	Examples of Modality	Definition	Senses Involved in the Sensation	Analytical Measurement
Texture	-Firmness -Viscosity -Cohesiveness	The sensory manifestation of the structure of food and the manner in which this structure reacts to the forces applied during handling and, in particular, during consumption (Szczesniak, 1963).	-Vision -Audition -Tactile	Rheology Texture analysis
Taste	-Sweetness -Saltiness -Sourness -Bitterness -Umami	One of the 5 main senses is the taste and can be defined as the sense detected by the taste cells on the tongue.	-Gustation -Olfaction	Sensory analysis
Odor	Any smell cue reported	Any chemicals that are detected by a small area of yellow/brown receptor cells located in the roof of the nose		-Sensory analysis -Olfactometer
Flavor	Any flavor recall and observation	A sense that is created with the cooperation of taste and flavor together	-Gustation -Olfaction	Sensory analysis (However, aromatic components can be analytically measured by headspace volatile assessments with gas chromatography)
Color	Any color observation	One of the 5 main senses is the vision. Visual cues are reported by the nervous system as colors.	-Vision	Colorimeter

S is sensation magnitude,
ΔI is the change in intensity, and
I is the magnitude of intensity (Fechner, 1860).

Figure 1. Senses that are linked with the possible feedback of liking illustrated with food acceptability model (Aktar, 2015; Cardello, 1996)

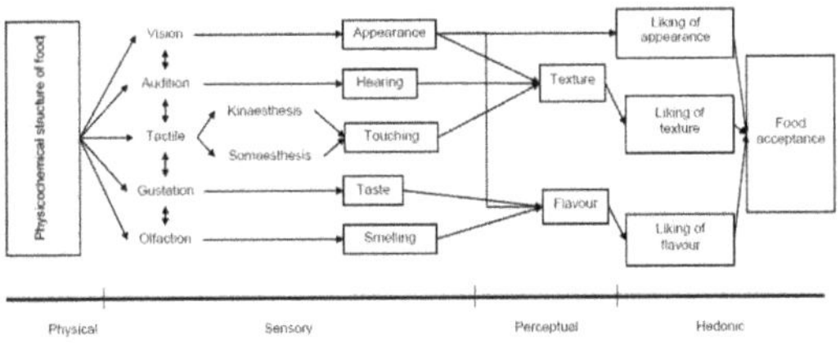

Weber's law is still being used to produce an index of a sensory discrimination capability, by comparing the different modalities regardless of the reference stimulus (Gescheider, 2013).

The 2nd psychophysical concept was introduced at mid- 19th century by a German philosopher Gustav Fechner. He used starlight instead of candlelight to observe astronomical data and introduced a logarithmic relationship for the magnitude of light and its intensity (Fechner, 1860). He invented a psychophysical law named after him as shown below:

$$S = k.logI + C \tag{2}$$

where;

S is Sensation magnitude,

I is the magnitude of intensity, and

C is the constant of integration (Fechner, 1860).

Lastly, 3rd mostly used psychophysical law came out at the mid-1900s by Stevens, which also named after himself. In contradistinction to Fechner, Stevens suggested a power law instead of logarithmic relation for the perception and intensity of the stimuli as follows:

$$\boldsymbol{S = kI^a} \tag{3}$$

where;

S is Sensation magnitude,

I is the magnitude of intensity, and

a is the power exponent (Gescheider, 1997).

These psychophysical laws are widely applied in the sensory tests not only in food science but also for the human physiology related sensory studies. Laws are still valid and have been confirmed by many studies in the literature. Selection of

Table 2. Differences between the instrumental and sensory tests

	Instrument	Sensory
Receptors	Hardware and software is used for the measurement with a previously set algorithm	Human receptors are used and the nervous system detects the term and magnitude of the sensation.
	-Human sensors are more sensitive to the applied force in most cases. -Human sensors have adaptation skills that may mask the real response of the stimuli (Kilcast, 1999) -Instrument software is mostly linear, where human use a logarithmic scale instead	
Temperature	The instrument can run at a certain temperature for a long period of time	Sensory tests can be applied at a set temperature environment. However, during the oral process, the food undergoes structural changes due to the temperature effect within the oral cavity.
	-The oral temperature has been studied and sensory scientists have a consensus over the test temperature (Engelen & de Wijk, 2012; Mela, Langley, & Martin, 1994). Generally accepted test temperature is 37°C (Sund-Levander, Forsberg, & Wahren, 2002) -Noteworthy textural properties are associated with the temperature and they interfere with the temperature differences (Bourne, 1978).	
Physical forces	An instrument only applies set and adjusted the amount of force and in a certain angle or certain direction.	During oral processing, human applies a force that is controlled by the texture of the food. The direction and the shearing of the oral surfaces are complex.
	-Direction, angle and the amount of the forces can be only mimicked in a most realistic way (Le Révérend & Hartmann, 2014; Osborn & Mao, 1993)	
Saliva	Instrumental analysis usually does not encounter the saliva. However, chemically reformulated saliva is also available.	During oral processing, human secrete saliva which facilitates; the digestion by enzymes and dilution effect, alters temperature and lowers the viscosity of the bolus, and creates lubrication on the sample for swallowing.
	- Saliva integration during oral processing is motor controlled ability. The secretion and the concentration highly depends on the food sample (de Almeida, Gregio, Machado, De Lima, & Azevedo, 2008; Humphrey & Williamson, 2001; Neyraud, Palicki, Schwartz, Nicklaus, & Feron, 2012)	
Geometry	Instruments are created with mostly as a solid and concrete geometry. Spindles and probes can be added to behave as a tongue.	The tongue is the dominant component of oral geometry and gives the ability to mix. It creates continues mixing of the food and facilitates the mixing of saliva and bolus.
	- Oral processing predominantly results with catastrophic changes in the bolus. -It also alters the integration of saliva	

the most appropriate law is necessary and is dependent on the sensory modality of interest.

SENSORY, MARKET SUCCESS AND CONSUMER BEHAVIOR

Food is the survival need since the first day of humankind. However, after a certain threshold of evolution, we started to eat for pleasure. It goes without saying that food is a catalyzer for conversation and a stimulator for the mood (Aktar, 2015). Hence, from the perspective of consumer food has an undeniable place. However, the food industry has and should have a more capitalistic point of view. The investors and companies are somewhere in-between business success and general morals. Therefore, they consider producing healthiest possible food with a bigger market share possibility. Healthy food manufacturing is a straightforward procedure which does not need to be discussed. However, a healthy and successful product needs some guidance from an expert, especially for a food product design.

Considering the two determinant factors for a food product design we need sensory tests. Sensory tests show a visionary image of the product when placed in the market and therefore, it helps investors to understand if it will be successful or not.

Sensory test application in the food industry begins back in the mid-1900s from U.S. Army Quartermaster Food and Container Institute (Peryam, 1954). Therefore, the initial move was indeed nonpecuniary and moreover, it was for nutrition and pleasure for the people who cannot reach to various kinds of food. Several governments and some organizations that value consumer satisfaction (such as armed forces of a country) followed what U.S Army Quartermasters introduces and possibly got inspired from the successful approach. In the aggregate, all those success and failures provided some insight and support to the development of sensory science. Sensory science, of course, does not only correspond to food sensory. It involves cosmetic, personal care products, textile, medical products, as well as food products. Experts of the sensory were always on duty to solve the quality related concerns of the consumers. By the efforts of the common consensus, quality standards of the particular product improved and the standardization authorities set standards for specific food groups (such as canning industry).

With the undeniable growth of the economy and population, formulated foods, special care requiring nutrition, functional foods, and recent food processing techniques increased the demand on the sensory adjusted food products. Sensory science conquered the processing lines by realizing the effect of food processing (e.g. homogenization of the milk while pasteurization) which embrace the technological development on the food processing and vice versa.

Meanwhile, sensory techniques started to follow the trends and show development. For instance, sensory fields started to invade new areas such as flavor profile method (Caul, 1957), descriptive analysis (Murray, Delahunty, & Baxter, 2001), temporal dominance sensation (Jager et al., 2014), and taste threshold. These advanced test methods, ease the high complexity of the sensory and provide a better understanding of the several questions.

SENSORY ANALYSES AND TECHNIQUES

Sensory evaluation is defined as a scientific field that evokes, measure, analyze and interpret the responses based in the food characteristics and raw materials as they are perceived by the basic senses (Lawless & Heymann, 2013). By looking at the definition, we can understand that sensory is a cooperative operation of the senses. On the other hand, we have to consider that food is at the center of these analyses and we can only manipulate the basic components in order to meet with the certain regulations and laws of the food type (e.g., max salt content). Moreover, sensory tests are usually applied in order to optimize the instrumental results according to human. A good example of this case is the assessment of textural attribute, let's say crunchiness. We can measure the acoustic performance of a sample under texture analyzer with an acoustic program and a microphone. However, the result will not correspond to a meaningful level, unless the human response and scoring of the crunchiness of different samples were obtained. Therefore, we can say that for most of the cases sensory analysis and sensory responses are used for better instrumental programming and instrumental design.

On the other hand, sensory analyses were underrated for quite long terms, by only investigating taste properties. But later on, more senses were involved in the sensory analysis; such as tactile sensory (Aktar, Chen, Ettelaie, & Holmes, 2015a, 2015b; Aktar, Chen, Ettelaie, Holmes, & Henson, 2017), flavor sensory (Keast, Dalton, & Breslin, 2004; Taylor & Roozen, 1996; Woods, Poliakoff, Lloyd, Dijksterhuis, & Thomas, 2010), auditorial sensory (Albert, Salvador, Schlich, & Fiszman, 2012; Dacremont, 1995; Vickers, 1982), visual sensory (Parker et al., 2004; Weymouth, 1958), astringency sensory (Dinnella, Recchia, Tuorila, & Monteleone, 2011; Thorngate & Noble, 1995) and umami taste (Beauchamp, 2009; Bellisle, 1999; Fuke & Shimizu, 1993). These modalities improve the understanding of the bigger picture and experimental, psychological, physiological, statistical and economic understanding of the sensory mechanism with knowledge of food science and technology.

As mentioned earlier, sensory needs a broad ability to design and analyze the results of the test. Hence, designing and evaluation are as important as the measurement.

Noteworthy, linking the results of the sensory with the market and business factors is significant especially for quality control and assurance. In recent years, there has been a greater requirement of a sophisticated physiological and psychological approach to understanding consumer perception and behavior towards the product. This need for a new approach invented the topic of oral processing which will be covered under this chapter.

Sensory Test Requirements

In order to achieve a contented sensory result, it is necessary to design a systematic test methodology. The well-designed sensory test should include:

1. Objective,
2. Appropriate testing environment,
3. Subjects,
4. Suitable testing method, and
5. Best possible data validation (Kilcast, 1999).

The objective of a test is the heart of the analysis and it should be accurate, clear, cost-effective, and contented enough to solve the problem. It should also highlight the target group and resources. On the other hand, the environment is now certain to affect the decision of the panelists during a sensory test (Genschow, Reutner, & Wänke, 2012; Piqueras-Fiszman, Alcaide, Roura, & Spence, 2012; Piqueras-Fiszman, Giboreau, & Spence, 2013; Piqueras-Fiszman & Spence, 2012). Not only the physical conditions but also other environmental factors such as; odor, air circulation speed, and temperature are effective in the sensory decision of the consumer. Also, choosing the right target group for the sensory test is critical. Sensory specialists usually tend to group the people according to their age, sex, health status, gender, and etc. if there is a special concern about a particular group of population. Moreover, subjects are the "instrument" of the sensory tests, thus, correct grouping and training status of the contributors are very important. Last but not least data validation is the most important part of the sensory analysis. It requires a high understanding of the statistics.

Sensory Evaluation Procedures

As mentioned by Kilcast (1999), sensory procedures are usually classified into two different approaches as; analytical tests and hedonic tests. Sensory classification figure can be seen in detail in Figure 2. According to this proper sensory procedure classification, sensory tests can be categorized according to their motivation.

Analytical tests are considered as laboratory practice tests, where hedonic tests are consumer sensory tests (Bi, 2008). Based on that classification, there are two main analytical sensory methods which are; discrimination (Aktar et al., 2015b, 2017), and descriptive (Hashim, Resurreccion, & McWalters, 1995; Stone, Sidel, Oliver, Woolsey, & Singleton, 2008), where three for the hedonic tests as; preference, acceptability, and relative-to-ideal (Deliza & MacFie, 1996; Drewnowski, Shrager, Lipsky, Stellar, & Greenwood, 1989; Kähkönen, Tuorila, & Hyvönen, 1995).

Discrimination tests basically aim to figure out the capability of detecting certain stimuli. Discrimination tests have four main types which are; paired comparison, duo-trio, triangle, two-out-of-five, and 'A' not 'A'. Sensory scientists usually have one of two main objectives: determine whether a difference exists between samples (recognition threshold) or to measure the lowest level of sensation (absolute threshold).

This kind of sensory analysis uses either comparative or ranking scales (Bi, 2008). For a reliable discrimination test required numbers of panelists were shown in Table 3.

The second type of analytical sensory test is descriptive tests. The descriptive sensory analysis mainly aims to answer the taste profile, perceived characteristics/attributes and to understand whether the changes in the processing/storage/ packaging conditions affect the sensory quality or not. Descriptive sensory tests have six main categories that are; simple descriptive, flavor profile®, texture profile®, quantitative descriptive analysis (The QDA Method) ®, Spectrum descriptive analysis®, and free-choice profiling. More specifically descriptive analysis is defined as a sensory test that provides a quantitative description perceived by a group of qualified subjects. This kind of sensory test is determined as a total sensory description including all modality of basic senses. Therefore, during the assessment word 'product' is used in order to create a total image of the sample (Stone & Sidel, 1992). Numbers of required assessors are presented in Table 4. As represented in the table all descriptive analysis requires highly trained panelists and mostly in a consensus approach (Meilgaard, Civille, & Carr, 2007).

On the other hand, hedonic sensory tests or acceptance tests are a valuable component of sensory testing, especially due to consumer behavior illustration of the product. Specifically, hedonic tests aim to measure the liking or preference properties of the product (Stone & Sidel, 1992). Hedonic tests are mostly required before a considerable amount of capital is invested in a particular product. To be more precise, without proper sensory testing of a product in terms of consumer acceptance, the investments on the equipment, process, production, packaging, storage, distribution, advertising and brand equity will be under a high risk of loss. The hedonic test mainly has three different types as; preference, acceptability, and relative-to-ideal. A minimum number of assessors required for hedonic tests are usually advised of more than 100 (Tomlins, 2000).

Figure 2. Sensory testing procedures (Aktar, 2015; Kilcast, 1999)

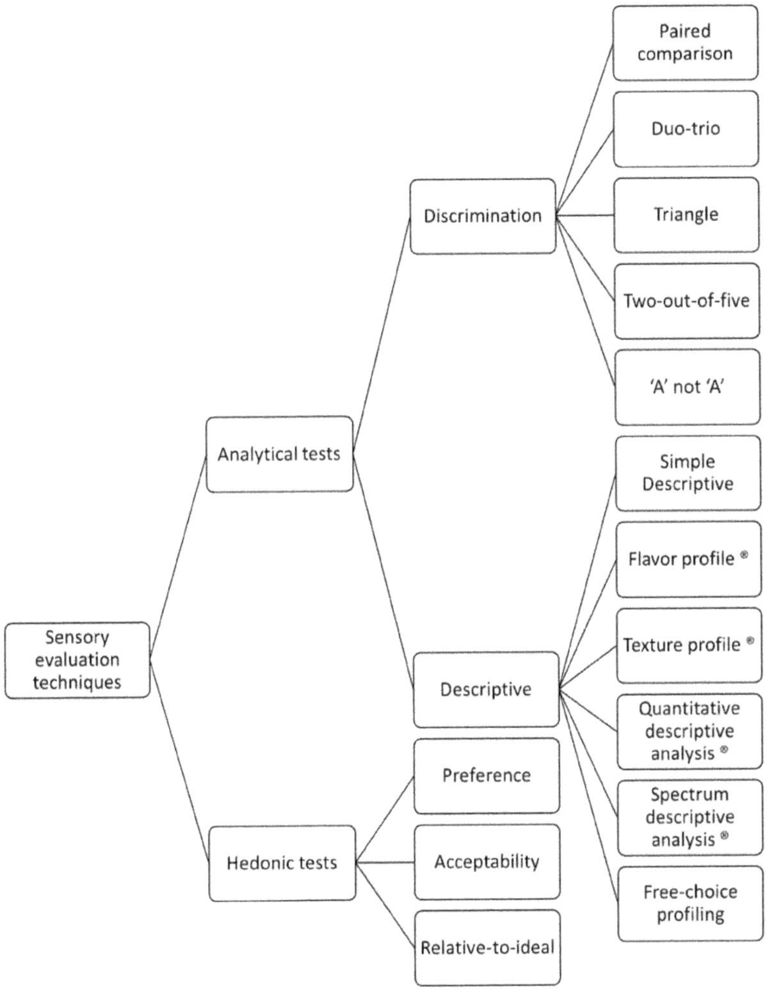

Table 3. Minimum number of assessors required for discrimination test (Tomlins, 2000)

Test Method	Expert assessors	Trained assessors	Untrained assessors	Consumers
Paired comparison	7	20	30	100+
Triangle	5	15	25	
Duo-Trio	-	-	20	
Two-out-of-five	-	10	-	-
'A' not 'A'	-	20	30	-

Table 4. Minimum number of assessors required for Descriptive analysis(Aguirre Cando, 2016; Caul, 1957; Stone et al., 2008)

Test Method	Trained assessors
Flavor profile®	4-6
Texture profile®	6-10
QDA ®	10-12
Spectrum descriptive analysis®	10-12
Free-choice profiling	10-12

RECENT TRENDS IN SENSORY SCIENCE: FOOD ORAL PROCESSING CONCEPT

In the last decade, inter-disciplinary approach to food consumption was introduces following the well-establishment of sensory studies about basic senses (taste, smell, vision, and etc.), and consumer behavior. Increased approach to collaborative way to food and eating triggered the new pathway to the sensory field. Use of physiology to understand the eating operation opened up the new vision and it brought the need for orofacial muscle activity and saliva interaction understanding of the eating process. This new subdivision of sensory especially was needed by food scientists as well as physiology experts, neuroscientists, psychology experts, and oral health experts. Only after then eating process was understood to be underrated and realized to be a highly sophisticated process that creates catastrophic changes on the food at oral surfaces. Therefore, the urge to give a new terminology to this cross-disciplinary approach to sensory sciences was defined as "Food Oral Processing". This new field of sensory not only involves sensory studies but also understanding oral anatomy, oral physiology, oral management, oral processing, and perception.

Oral processing of foods is defined as the first stage of digestion especially due to, amylase interaction with carbohydrates and physical breakdown during chewing. Meanwhile, it is a step of the transportation process from the mouth to the gastrointestinal tract passing from the oral cavity through the oral-pharyngeal-esophagus tract to the stomach (Chen, 2014). Sensory perception is the process that takes place during this oral process and effects the appreciation of food. During oral processing of food, oral geometry acts as well-designed equipment with multifunction. The duty of oral geometry is to act as a food processor for size reduction, mixing, diluting with saliva, shearing, and smearing the food (Chen, 2014). Therefore, the very first step of oral processing is the introduction of the stimuli (food) and followed by the food ingestion. During the food ingestion; food geometry, food rheology, texture, taste, aroma properties, compositional properties, the microstructure of the food, and presentation of the food is determinative. These food-related issues are actively changed by the oral geometry activities which send feedback through our nervous system about perception, appreciation, mouthfeel, and characterization.

Different from those previous single sided sensory studies, food oral processing concept; we have to encounter the dynamic changes that occur on the food structure and continuous mixing of saliva. Based on this recognition, time-intensity analysis (TI) and temporal dominance of sensation analysis (TDS) were the two different sensory tests that introduced (Cliff & Heymann, 1993; Pineau et al., 2009).

In TI sensory evaluation classical scaling method is used with the temporal information about the perception. Assessors continuously monitor their perceptions until the swallowing. Usually, nowadays computerized TI systems are developed which ease the data collection and processing. With the TI sensory test, the researcher obtains a curve that gives information about; certain intensity (initial, maximum, appearance time, etc.), duration (total duration, finish time, extinction time, etc.), rate of increase (maximum rate of adsorption, maximum intensity rate, slope of rising, etc.), rate of decrease (maximum rate of desorption, rate of decay, slope tailing, etc.), and aftertaste (Cliff & Heymann, 1993). TI tests has been well accepted and widely used in the literature specifically on; wine (Guinard, Pangborn, & Lewis, 1986; Noble, 1994; Sokolowsky & Fischer, 2012), cheese (Wendin, Janestad, & Hall, 2003), sweeteners (Harrison & Bernhard, 1984; Larson-Powers & Pangborn, 1978; Swartz, 1980), and meat (Duizer, Gullett, & Findlay, 1993).

On the other hand, TDS was introduced in order to eliminate the time consumption handicap of the TI tests in the year 1999 by the "Centre Européen des Sciences du Goût" in the LIRIS Lab (Pineau et al., 2009). It is a computerized test and the assessors are provided with the full list of attributes before they are asked to assess the most dominant perceived attribute. Assessors are then asked to score that particular attribute dominancy. If they perceive a change on the dominant attribute, they change their perception and so as the score until the ending of the perception.

Both of these trending sensory tests provide more comprehensive results than the single observed sensory tests. However, it is also important that the 1st generation approach to the sensory is still valid and sometimes more informative than TI or TDS.

CONCLUSION

This chapter is organized largely from a food scientist's perspective and the literature available on sensory science and physiology. The link between physiological and psychological parts of the sensation and perception was discussed as it is one of the key elements of sensory science. Whereas, the basic classification of mostly used and needed sensory tests were also enclosed in this chapter, in order to present background data to the readers of this book. Even though the emerging trends in sensory testing were highlighted (such as Time-intensity or Temporal dominance sensation), the challenge still exists to know how to develop a better understanding of human senses during eating and how does the preference and acceptance is affected by the elements of sensory testing. Some excellent work is being undertaken on the food oral processing which highlights new developments and illustrates new protocols.

The majority of the research reviewed in this chapter has focused on the sensory testing practices as well as the food oral processing aspects of the multi-sensory perception. While it has been known for many years that sensory tests are critical for consumer, food manufacturer, and food scientists, there is now growing body of evidence to suggest a cross-disciplinary approach provides a broader and bigger image about a product property that occurs only during eating. Researchers are currently working on ways to require fewer assessors and obtain more reliable results in order to cut the cost needed for the sensory studies. Meanwhile, the expertise needed for lesser assessor is increasing. The hope is that a better understanding of the human sensation/perception mechanisms can be provided by the neuroscientist which will be followed by the better understanding of the change required for a more preferable taste, texture, imager, olfaction, or odor containing food.

ACKNOWLEDGMENT

This research received no specific grant from any funding agency in the public, commercial, or not-for-profit sectors.

REFERENCES

Aguirre Cando, M. E. (2016). Descriptive sensory and texture profile analysis of woody breast in marinated chicken.

Aktar, T. (2015). *Determining the detection threshold for perception of selected textural attributes*. University of Leeds. Retrieved from http://etheses.whiterose.ac.uk/view/creators/Aktar=3ATugba=3A=3A.html

Aktar, T., Chen, J., Ettelaie, R., & Holmes, M. (2015a). Evaluation of the sensory correlation between touch sensitivity and the capacity to discriminate viscosity. *Journal of Sensory Studies*, *30*(2), 98–107. doi:10.1111/joss.12141

Aktar, T., Chen, J., Ettelaie, R., & Holmes, M. (2015b). Tactile sensitivity and capability of soft-solid texture discrimination. *Journal of Texture Studies*, *46*(6), 429–439. doi:10.1111/jtxs.12142

Aktar, T., Chen, J., Ettelaie, R., Holmes, M., & Henson, B. (2017). Human roughness perception and possible factors effecting roughness sensation. *Journal of Texture Studies*, *48*(3), 181–192. doi:10.1111/jtxs.12245 PMID:28573724

Albert, A., Salvador, A., Schlich, P., & Fiszman, S. (2012). Comparison between temporal dominance of sensations (TDS) and key-attribute sensory profiling for evaluating solid food with contrasting textural layers: Fish sticks. *Food Quality and Preference*, *24*(1), 111–118. doi:10.1016/j.foodqual.2011.10.003

Beauchamp, G. K. (2009). Sensory and receptor responses to umami: An overview of pioneering work. *The American Journal of Clinical Nutrition*, *90*(3), 723S–727S. doi:10.3945/ajcn.2009.27462E PMID:19571221

Bellisle, F. (1999). Glutamate and the UMAMI taste: Sensory, metabolic, nutritional and behavioural considerations. A review of the literature published in the last 10 years. *Neuroscience and Biobehavioral Reviews*, *23*(3), 423–438. doi:10.1016/S0149-7634(98)00043-8 PMID:9989429

Bi, J. (2008). *Sensory discrimination tests and measurements: Statistical principles, procedures, and tables*. John Wiley & Sons.

Bourne, M. (1978). Texture profile analysis. *Journal of Food Technology*, *33*, 62–66.

Cardello, A. V. (1996). The role of the human senses in food acceptance. In *Food choice, acceptance and consumption* (pp. 1–82). Springer. doi:10.1007/978-1-4613-1221-5_1

Caul, J. F. (1957). The profile method of flavor analysis. [Elsevier]. *Advances in food research*, *7*, 1–40. doi:10.1016/S0065-2628(08)60245-1

Chen, J. (2014). Food oral processing: Some important underpinning principles of eating and sensory perception. *Food Structure*, *1*(2), 91–105. doi:10.1016/j.foostr.2014.03.001

Cliff, M., & Heymann, H. (1993). Development and use of time-intensity methodology for sensory evaluation: A review. *Food Research International*, *26*(5), 375–385. doi:10.1016/0963-9969(93)90081-S

Dacremont, C. (1995). Spectral composition of eating sounds generated by crispy, crunchy, and crackly foods. *Journal of Texture Studies*, *26*(1), 27–43. doi:10.1111/j.1745-4603.1995.tb00782.x de Almeida, P. D. V., Gregio, A. M., Machado, M. A., De Lima, A. A., & Azevedo, L. R. (2008). Saliva composition and functions: A comprehensive review. *The Journal of Contemporary Dental Practice*, *9*(3), 72–80. doi:10.5005/jcdp-9-3-72 PMID:18335122

Deliza, R., & MacFie, H. J. H. (1996). The generation of sensory expectation by external cues and its effect on sensory perception and hedonic ratings: A review. *Journal of Sensory Studies*, *11*(2), 103–128. doi:10.1111/j.1745-459X.1996.tb00036.x

Dinnella, C., Recchia, A., Tuorila, H., & Monteleone, E. (2011). Individual astringency responsiveness affects the acceptance of phenol-rich foods. *Appetite*, *56*(3), 633–642. doi:10.1016/j.appet.2011.02.017 PMID:21354451

Drewnowski, A., Shrager, E. E., Lipsky, C., Stellar, E., & Greenwood, M. R. C. (1989). Sugar and fat: Sensory and hedonic evaluation of liquid and solid foods. *Physiology & Behavior*, *45*(1), 177–183. doi:10.1016/0031-9384(89)90182-0 PMID:2727131

Duizer, L. M., Gullett, E. A., & Findlay, C. J. (1993). Time-intensity methodology for beef tenderness perception. *Journal of Food Science*, *58*(5), 943–947. doi:10.1111/j.1365-2621.1993.tb06084.x

Engelen, L., & de Wijk, R. A. (2012). Oral processing and texture perception. In J. Chen, & L. Engelen (Eds.), *Food oral processing: Fundamentals of eating and sensory perception* (pp. 159–162). Oxford, UK: Wiley-Blackwell. doi:10.1002/9781444360943.ch8

Fechner, G. T. (1860). Elemente der Psychophysik (Breitkopf und Hèartel, Leipzig, Germany). German.

Fuke, S., & Shimizu, T. (1993). Sensory and preference aspects of umami. *Trends in Food Science & Technology*, *4*(8), 246–251. doi:10.1016/0924-2244(93)90139-2

Genschow, O., Reutner, L., & Wänke, M. (2012). The color red reduces snack food and soft drink intake. *Appetite*, *58*(2), 699–702. doi:10.1016/j.appet.2011.12.023 PMID:22245725

Gescheider, G. (1997). *The classical psychophysical methods. Psychophysics: The Fundamentals* (3rd ed.). Mahwah, NJ: Lawrence Erlbaum Associates.

Gescheider, G. (2013). *Psychophysics: the fundamentals*. Psychology Press. doi:10.4324/9780203774458

Guinard, J.-X., Pangborn, R. M., & Lewis, M. J. (1986). The time-course of astringency in wine upon repeated ingestion. *American Journal of Enology and Viticulture*, *37*(3), 184–189.

Harrison, S. K., & Bernhard, R. A. (1984). Time-intensity sensory characteristics of saccharin, xylitol and galactose and their effect on the sweetness of lactose. *Journal of Food Science*, *49*(3), 780–786. doi:10.1111/j.1365-2621.1984.tb13210.x

Hashim, I. B., Resurreccion, A. V. A., & McWalters, K. H. (1995). Descriptive sensory analysis of irradiated frozen or refrigerated chicken. *Journal of Food Science*, *60*(4), 664–666. doi:10.1111/j.1365-2621.1995.tb06202.x

Hecht, S. (1924). The visual discrimination of intensity and the Weber-Fechner law. *The Journal of General Physiology*, *7*(2), 235–267. doi:10.1085/jgp.7.2.235 PMID:19872133

Heymann, H. (2019). A personal history of sensory science. *Food, Culture, & Society*, 1–21.

Humphrey, S. P., & Williamson, R. T. (2001). A review of saliva: Normal composition, flow, and function. *The Journal of Prosthetic Dentistry*, *85*(2), 162–169. doi:10.1067/mpr.2001.113778 PMID:11208206

Jager, G., Schlich, P., Tijssen, I., Yao, J., Visalli, M., De Graaf, C., & Stieger, M. (2014). Temporal dominance of emotions: Measuring dynamics of food-related emotions during consumption. *Food Quality and Preference*, *37*, 87–99. doi:10.1016/j.foodqual.2014.04.010

Kähkönen, P., Tuorila, H., & Hyvönen, L. (1995). Dairy fat content and serving temperature as determinants of sensory and hedonic characteristics in cheese soup. *Food Quality and Preference*, *6*(2), 127–133. doi:10.1016/0950-3293(95)98555-W

Keast, R., Dalton, P., & Breslin, P. A. S. (2004). *Flavor interactions at the sensory level. Flavor Perception.* Wiley Online Library.

Kilcast, D. (1999). Sensory techniques to study food texture. In A. J. Rosenthal (Ed.), *Food texture: Measurement and perception* (pp. 30–60). Gaithersburg, MD: Aspen Publishers.

Krop, E. M., Hetherington, M. M., Nekitsing, C., Miquel, S., Postelnicu, L., & Sarkar, A. (2018). Influence of oral processing on appetite and food intake–A systematic review and meta-analysis. *Appetite*, *125*, 253–269. doi:10.1016/j.appet.2018.01.018 PMID:29408331

Larson-Powers, N., & Pangborn, R. M. (1978). Paired comparison and time-intensity measurements of the sensory properties of beverages and gelatins containing sucrose or synthetic sweeteners. *Journal of Food Science*, *43*(1), 41–46. doi:10.1111/j.1365-2621.1978.tb09732.x

Lawless, H. T., & Heymann, H. (2013). *Sensory evaluation of food: principles and practices*. Springer Science & Business Media.

Le Révérend, B., & Hartmann, C. (2014). Numerical modeling of human mastication, a simplistic view to design foods adapted to mastication abilities. *Physiology & Behavior*, *124*, 61–64. doi:10.1016/j.physbeh.2013.10.012 PMID:24471180

Meilgaard, M. C., Civille, G. V., & Carr, B. T. (2007). *Sensory evaluation techniques* (3rd ed.). Boca Raton, FL: Taylor & Francis.

Mela, D. J., Langley, K. R., & Martin, A. (1994). No effect of oral or sample temperature on sensory assessment of fat content. *Physiology & Behavior*, *56*(4), 655–658. doi:10.1016/0031-9384(94)90222-4 PMID:7800728

Murray, J. M., Delahunty, C. M., & Baxter, I. A. (2001). Descriptive sensory analysis: Past, present and future. *Food Research International*, *34*(6), 461–471. doi:10.1016/S0963-9969(01)00070-9

Neyraud, E., Palicki, O., Schwartz, C., Nicklaus, S., & Feron, G. (2012). Variability of human saliva composition: Possible relationships with fat perception and liking. *Archives of Oral Biology*, *57*(5), 556–566. doi:10.1016/j.archoralbio.2011.09.016 PMID:22024405

Noble, A. C. (1994). Bitterness in wine. *Physiology & Behavior*, *56*(6), 1251–1255. doi:10.1016/0031-9384(94)90373-5 PMID:7878098

Osborn, J. W., & Mao, J. (1993). A thin bite-force transducer with three-dimensional capabilities reveals a consistent change in bite-force direction during human jaw-muscle endurance tests. *Archives of Oral Biology*, *38*(2), 139–144. doi:10.1016/0003-9969(93)90198-U PMID:8476343

Parker, B. A., Sturm, K., MacIntosh, C. G., Feinle, C., Horowitz, M., & Chapman, I. M. (2004). Relation between food intake and visual analogue scale ratings of appetite and other sensations in healthy older and young subjects. *European Journal of Clinical Nutrition*, *58*(2), 212–218. doi:10.1038j.ejcn.1601768 PMID:14749739

Peryam, D. R. (1954). CHAIRMAN THURSTONE We will next hear the paper by Mr. David R. Peryam. Field Testing of Armed Forces Rations. In *Food Acceptance Testing Methodology: A Symposium Sponsored by the Quartermaster Food and Container Institute for the Armed Forces, Quartermaster Research and Development Command, US Army Quartermaster Corps [at The] Palmer House, Chicago, Oct. 8-9, 1953* (Vol. 1, p. 75). National Academies.

Pineau, N., Schlich, P., Cordelle, S., Mathonnière, C., Issanchou, S., Imbert, A., ... Köster, E. (2009). Temporal dominance of sensations: Construction of the TDS curves and comparison with time–intensity. *Food Quality and Preference*, *20*(6), 450–455. doi:10.1016/j.foodqual.2009.04.005

Piqueras-Fiszman, B., Alcaide, J., Roura, E., & Spence, C. (2012). Is it the plate or is it the food? Assessing the influence of the color (black or white) and shape of the plate on the perception of the food placed on it. *Food Quality and Preference*, *24*(1), 205–208. doi:10.1016/j.foodqual.2011.08.011

Piqueras-Fiszman, B., Giboreau, A., & Spence, C. (2013). Assessing the influence of the color of the plate on the perception of a complex food in a restaurant setting. *Flavour (London)*, *2*(1), 24. doi:10.1186/2044-7248-2-24

Piqueras-Fiszman, B., & Spence, C. (2012). The influence of the color of the cup on consumers' perception of a hot beverage. *Journal of Sensory Studies*, *27*(5), 324–331. doi:10.1111/j.1745-459X.2012.00397.x

Skinner, E. Z. (1989). Commentary, sensory evaluation. In celebration of our beginnings, sensory evaluation of materials and products. Philadelphia, PA: American Society for Testing and Materials.

Sokolowsky, M., & Fischer, U. (2012). Evaluation of bitterness in white wine applying descriptive analysis, time-intensity analysis, and temporal dominance of sensations analysis. *Analytica Chimica Acta*, *732*, 46–52. doi:10.1016/j.aca.2011.12.024 PMID:22688033

Stone, H., & Sidel, J. (1992). Descriptive Analysis. In *Sensory evaluation practices* (2nd ed., pp. 202–242). Academic Press.

Stone, H., Sidel, J., Oliver, S., Woolsey, A., & Singleton, R. C. (2008). Sensory evaluation by quantitative descriptive analysis. *Descriptive Sensory Analysis in Practice*, *28*, 23–34.

Sund-Levander, M., Forsberg, C., & Wahren, L. K. (2002). Normal oral, rectal, tympanic and axillary body temperature in adult men and women: A systematic literature review. *Scandinavian Journal of Caring Sciences*, *16*(2), 122–128. doi:10.1046/j.1471-6712.2002.00069.x PMID:12000664

Swartz, M. (1980). Sensory screening of synthetic sweeteners using time-intensity evaluations. *Journal of Food Science*, *45*(3), 577–581. doi:10.1111/j.1365-2621.1980.tb04105.x

Szczesniak, A. (1963). Classification of textural characteristics. *Journal of Food Science*, *28*(4), 385–389. doi:10.1111/j.1365-2621.1963.tb00215.x

Taylor, A. J., & Roozen, J. P. (1996). Volatile flavor release from foods during eating. *Critical Reviews in Food Science and Nutrition*, *36*(8), 765–784. doi:10.1080/10408399609527749 PMID:8989509

Thorngate, J. H., & Noble, A. C. (1995). Sensory evaluation of bitterness and astringency of 3R (−)-epicatechin and 3S (+)-catechin. *Journal of the Science of Food and Agriculture*, *67*(4), 531–535. doi:10.1002/jsfa.2740670416

Tomlins, K. I. (2000). *Methods for the sensory evaluation of food and drink products*. Chatham Maritime, UK: Natural Resources Institute.

Vickers, Z. M. (1982). Relationships of chewing sounds to judgments of crispness, crunchiness, and hardness. *Journal of Food Science*, *47*(1), 121–124. doi:10.1111/j.1365-2621.1982.tb11041.x

Wendin, K., Janestad, H., & Hall, G. (2003). Modelling and analysis of dynamic sensory data. *Food Quality and Preference*, *14*(8), 663–671. doi:10.1016/S0950-3293(02)00208-2

Weymouth, F. W. (1958). Visual sensory units and the minimal angle of resolution. *American Journal of Ophthalmology*, *46*(1), 102–113. doi:10.1016/0002-9394(58)90042-4 PMID:13545337

Woods, A. T., Poliakoff, E., Lloyd, D. M., Dijksterhuis, G. B., & Thomas, A. (2010). Flavor expectation: The effect of assuming homogeneity on drink perception. *Chemosensory Perception*, *3*(3–4), 174–181. doi:10.100712078-010-9080-2

KEY TERMS AND DEFINITIONS

Analytical Sensory Analyses: One of two main types of sensory analysis techniques that is mainly applied at laboratory applications. It involves discrimination and descriptive analysis.

Descriptive Sensory Analyses: One of the analytical sensory analyses tests, which is applied to observe the full image of the product from the point of the consumer.

Discrimination Sensory Analyses: One of the analytical sensory analyses tests, which aims to understand the perception threshold of the consumer.

Hedonic Sensory Analyses: One of two main types of sensory analysis techniques. It may be named as consumer tests, which is usually applied to observe the consumers acceptance and preference.

Sensory Analysis: It is a field of science that applies experimental design and statistical analysis to measure the human senses, which reflects the acceptance and preference for consumer products.

Temporal Dominance of Sensation Analysis (TDS): An innovative computerized sensory test, that measures the dynamic perception of several attributes at once from the first bite to swallow.

Time-Intensity Analysis (TI): An innovative classical scaling method which involves temporal information about the perception during oral processing.

Chapter 10
Solid-State Fermentation:
A Novel Approach in Food Processing Technology Using Food Industry Wastes

Urvashi Srivastava
Centre of Food Technology, University of Allahabad, India

Zoomi Singh
Centre of Food Technology, University of Allahabad, India

Pinki Saini
Centre of Food Technology, University of Allahabad, India

ABSTRACT

Globally, many byproducts or wastes are produced through diverse food industries. These food industries dispose of their waste in the surroundings; merely a few of them reprocess their waste and use it as functional food ingredients. Fermentation techniques can be adopted as one of the methods to prevail over this waste problem. Among the different fermentation methods, solid state fermentation is reviewed as it is elegantly simple and persists to lift interest of scientists and industries around the world. The last decade has included an unprecedented rise in the significance of solid state fermentation (SSF) for the progress of bioprocesses for nutritional enrichment. This chapter focuses on a general review of the advantages of solid state fermentation over conventional fermentation, bioreactor design for SSF, production of bioactive substances from various food stuffs, bioconversion of agro-industrial wastes, and bio refining strategy. SSF is a remarkable tool to elevate nutritional and functional values of the substrate to a large extent.

DOI: 10.4018/978-1-7998-1924-0.ch010

INTRODUCTION

Environmental problems and public health risk mostly occur due to wrong waste management practices and these wastes generally accumulate on land sites or are consumed by animals (Yazid et al., 2017). The source of these wastes is generally agriculture, household, industries, and certain waste produced by animals or humans (Mussatto et al., 2012). The practices that most people were following for disposal of these wastes were contributing to air and water pollution as people in the past used incinerators to burn waste, producing harmful gases like carbon dioxide and methane (Sánchez et al., 2015) or broken down waste accumulated on land sites by a microorganism which contaminated groundwater (Eco-Cycle, 2011). These problems increase the need for the development of sustainable technology for food processing that will be highly efficient (Fig.1) and able to solve the problem of disposal (Mussatto et al., 2012; Chen and He, 2012).

BACKGROUND

In previous years, solid-state fermentation (SSF) became an emerging technology with sustainable features of bioconversion of waste to functional components. SSF utilizes and reuses the waste, as these wastes are composed of macro and micronutrients like carbohydrates, proteins, and minerals (Mussatto et al., 2012). Under certain standard conditions, the wastes were utilized in the development of microorganisms in differently designed bioreactors and yield products from low-cost residues, with lower energy consumption and hence, were economically feasible. The process of SSF is described in figure 2.

Aspects considered important for SSF process include substrate nature, substrate particle size, selection of bioreactors, selection of microorganisms according to moisture conditions, pH, temperature, water activity, and aeration. SSF technology

Figure 1. The efficiency of Solid-state fermentation
Source: (Yazid et al., 2017)

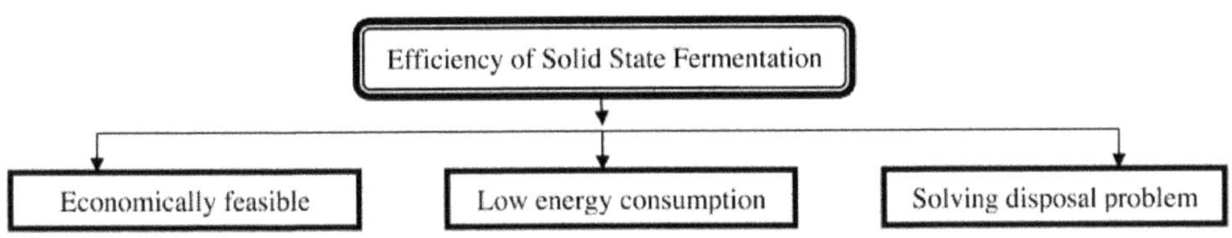

Figure 2. Steps involved in solid state fermentation process

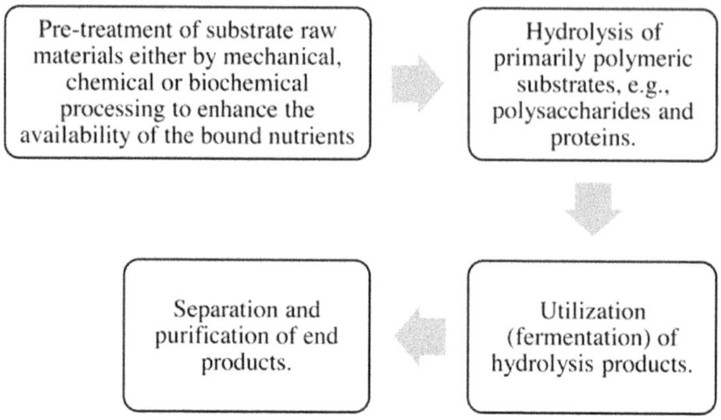

opens up great possibilities of waste management using biochemical engineering and biotechnology field advancement to produce bioactive compounds, chemicals, and functional components (Singhania et al., 2009).

ADVANTAGES AND CHALLENGES OF SSF OVER SUBMERGED LIQUID FERMENTATION (SFL)

Nowadays, the use of SSF is increasing over SFL in the biotechnology field. In the SLF process, liquid media are required for the growth of microorganism. The water content required for fermentation decreases the media concentration and contaminates media due to higher water activity. The liquid waste produced from SLF process causes a dumping problem due to its higher quantity (Manpreet et al., 2005). The overall process of the SLF is expensive as it requires processed substrate and large-scale practice which leads to an increase in the cost for the labors (Kapilan, 2015; Pandey et al., 1996). Whereas the SSF process has many advantages (figure. 3) over SLF, which are as follows:

- It does not require liquid for media preparation, which lowers production of wastewater, reducing contamination and sterility demand (Holker et al., 2004).
- Low-cost solid substrates from agricultural and household waste are available in abundance and economically cheap (Durand, 2003).

Figure 3. Advantages of solid-state fermentation
Source: (Singhania et al., 2009)

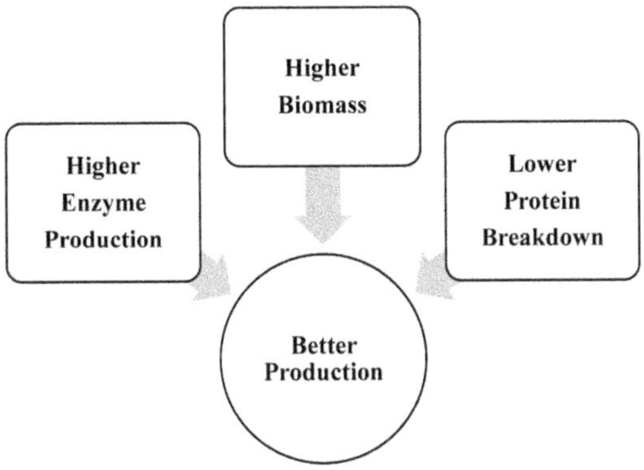

- SSF did not require complex, large-scale bioreactors or machinery because of lesser water requirements (Singhania et al., 2009).
- SSF provides greatest advantage due to lower energy consumption, lower plant and machinery cost, and ultimately reduction in labor cost (Pandey et al., 2001).
- The product yield of SSF is generally high with higher stability, with the higher productivity of fermentation with lower moisture requirements (Sabu et al., 2006; Pandey et al., 2001).

Apart from the above-mentioned advantages, SSF process is also facing certain challenges (figure. 4). Due to accumulated nutrients within the cells, the production of enzymes is delayed, which builds up some toxic substance that hinders the growth of the microorganisms and leads to death. Other big challenges that biotechnological industries have been recently facing are the end product purification, scale-up, and estimation of biomass in SSF, which are of great use for people (Kapilan, 2015).

Figure 4. Challenges of solid-state fermentation
Source: (Singhania et al., 2009)

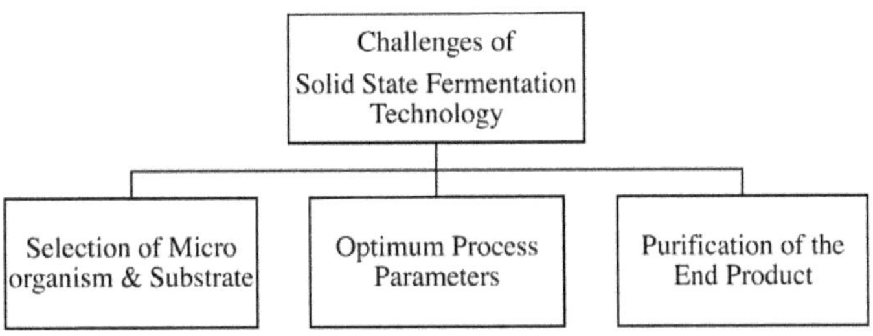

Bioreactors in SSF

Bioreactors are vessels that are mechanically stirred, and microorganisms are grown under a controlled manner on a substrate to produce usable products by certain specific reactions (Musoni et al., 2015). SSF process occurs in bioreactors. Generally, there are two types of bioreactors used in SSF process, i.e., Type A and Type B, where Type B will further divide into three groups based on the SSF process (figure 5). The designs of bioreactors mainly differ by the type aeration and mixing. Bioreactors in SSF are designed, keeping in mind the conditions that will be more favorable for solid-state fermentation.

Type A: Group I

Tray Bioreactor is an example of a Type A where the static bed is used for SSF process, mixing is done occasionally, with no forced aeration (Nigam and Pandey, 2009; Couto and Sanroma´n, 2006; Mitchell et al., 2006).

Type B: Group II

Packed Bioreactors are an example of Group II of Type B where the bed is either static or mixed infrequently with forced-blow air through the bed (Nigam and Pandey, 2009; Couto and Sanroma´n, 2006; Mitchell et al., 2006).

Type B: Group III

These bioreactors consist of cylindrical drums placed horizontally where the bed is mixed continuously and the air has circulated a bed with no forced aeration (Nigam and Pandey, 2009; Couto and Sanroma´n, 2006; Mitchell et al., 2006).

Type B: Group IV

In this type of Bioreactor, the bed is agitated with forced aeration.

The models are further divided into two groups, Group A and Group B. Group A is where the bioreactors are mixed continuously and Group B is where the Bioreactors are mixed intermittently (Nigam and Pandey, 2009; Couto and Sanroma´n, 2006; Mitchell, 2006).

Figure 5. Overview of bioreactors used in SSF
Source: Mitchell et al., 2006

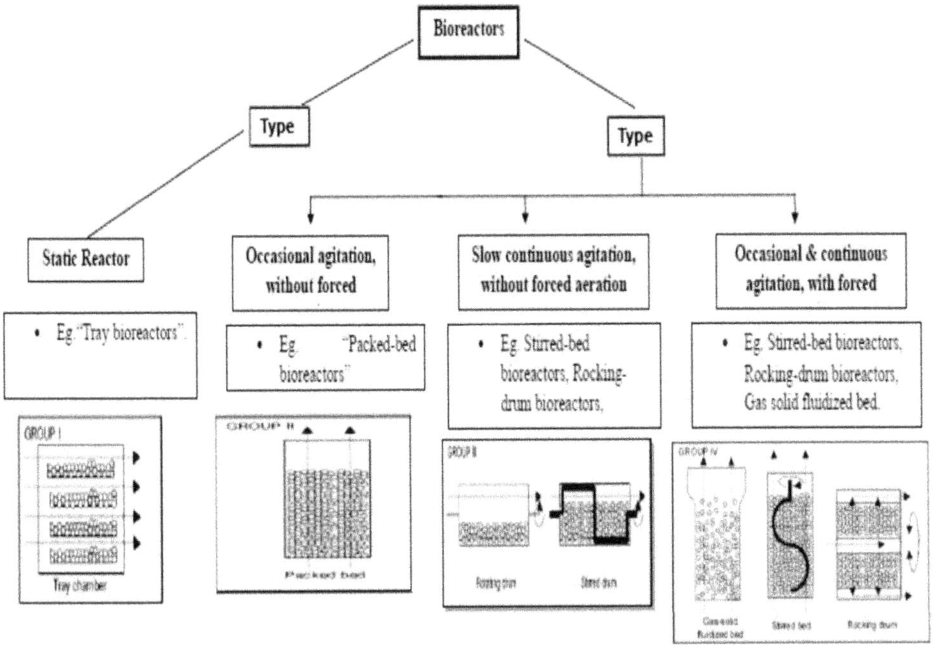

CONTRIBUTION OF SSF IN SUSTAINABLE DEVELOPMENT

The process of bioconversion of wastes like Agricultural, Household, Domestic, Industrial, etc. using SSF shows sustainable characteristics (Figure 6). As it solves many environmental problems by utilizing these wastes and converting it into components that are of therapeutic use and making eco-friendly green materials which are of great importance for future use, by replacing synthetic plastics (Seth et al., 2018).

Bio-Conversion of Agro-Industrial Waste is Using SSF

Many important functional components are produced using SSF by Bio-Conversion of Agro-Industrial wastes, which are as follows:

a) **Bio-Active Compounds:** These are food ingredients that are used as extra-nutritional components. These bioactive compounds are produced from agro-industrial waste by fermentation. SSF shows increasing trends to produce these compounds as they are more efficient, since these compounds are of tremendous value for the environment and humans (Subramaniyam and Vimala, 2012). The details of bio-active compounds produced using certain specific substrates and microorganisms are illustrated in Table 1.

b) **Enzymes:** Currently, an enzyme from microbial sources plays a vital role in human needs. Fermentation serves as a key method to produce various types of enzymes. In studies where the enzyme concentration obtained from SLF and SSF have been compared, it has been observed that in SSF the titles are higher and also, the produced enzymes are stable to wider ranges of temperature and pH. This has been evidenced for enzymes such as the amylase, pectinase, tannase and protease (Aguilar et al., 2008). The details of different studies and enzyme production through SSF process is detailed in Table 2.

c) **Bioethanol Production:** Biomass hydrolysis with well-adapted microorganisms convert cellulose and hemicellulose into sugars and ultimately leads to biofuel production. Therefore, the demand for cellulase production by SSF mistreatment agro-industrial residues is enhancing apace. Ethanol production by SSF using grape and sugar beet pomaces and apple pomace as solid substrates have been recently evaluated and the obtained ethanol production yields were greater than that obtained by Submerged Fermentation. Therefore, considering the importance of the ethanol production in the actual world economy, it is expected to observe an increase in the researches for the development of a suitable process for ethanol production by SSF (Koyani and Rajput, 2015).

d) **Fructooligosaccharides production:** Fructooligosaccharides (FOS) can be used as an artificial or alternative sweetener and are considered a small dietary fiber with low caloric value. FOS also serve as a substrate for microflora in the large intestine, increasing the overall gastrointestinal tract health. Aspergillus japonicus has been considered a potential strain for industrial production of FOS by SSF. In a recent study, some agro-industrial wastes, including corncobs, coffee silverskin, and cork oak were used as support and a nutrient source during the FOS production by Aspergillus japonicas under SSF conditions. Among the wastes, coffee silverskin was the most suitable support for FOS production.

e) **Bioinsecticides:** Several agro-industrial wastes (refused potatoes, coffee husks, and sugarcane bagasse) have been used in SSF to produce spores from Beauveria bassiana to obtain biopesticides for biocontrol of pests of banana, sugarcane, soybean, and coffee. Colletotrichum truncatum is another fungus that has been studied under SSF and it possesses characteristics to be used as mycoherbicide against the difficult weed Sesbania exaltata.

f) **Hypercholesterolemic Agents:** In the liver, the production of cholesterol has been blocked by certain agents known as a hypercholesterolemic agent. High cholesterol levels in blood are believed to be a major risk for atherosclerosis and coronary heart diseases. This pathological condition can be reduced by the use of statins which becomes an emerging scope for researchers to produce them. Statin production can be achieved by the cost-effective method of fermentation. Researchers nowadays are using fungi to produce statins through solid-state fermentation techniques (Subramaniyam and Vimala, 2012).

g) **Antibiotics:** Antibiotics are one of the leading types of bioactive compounds that, with the help of fermentation techniques, can be extracted from microorganisms. The antibiotics like cephalosporin, cyclosporins, streptomycins, tetracycline, and surfactants were produced in huge amounts through fermentation where SSF proves to be suitable for higher production of antibiotics (Subramaniyam and Vimala, 2012).

h) **Antihypertensive peptides:** Production of fermented foods containing antihypertensive compound has become a great achievement in the medical field. Angiotensin-converting enzyme (ACE) is important for regulation of blood pressure and activity of this enzyme can be inhibited by antihypertensive peptides. This creates interest for research to compare different methods of fermentation for the production of these fermented products (Subramaniyam and Vimala, 2012).

Figure 6. Bio-conversion of agro-industrial waste using SSF
Source: Sadh et al., 2018

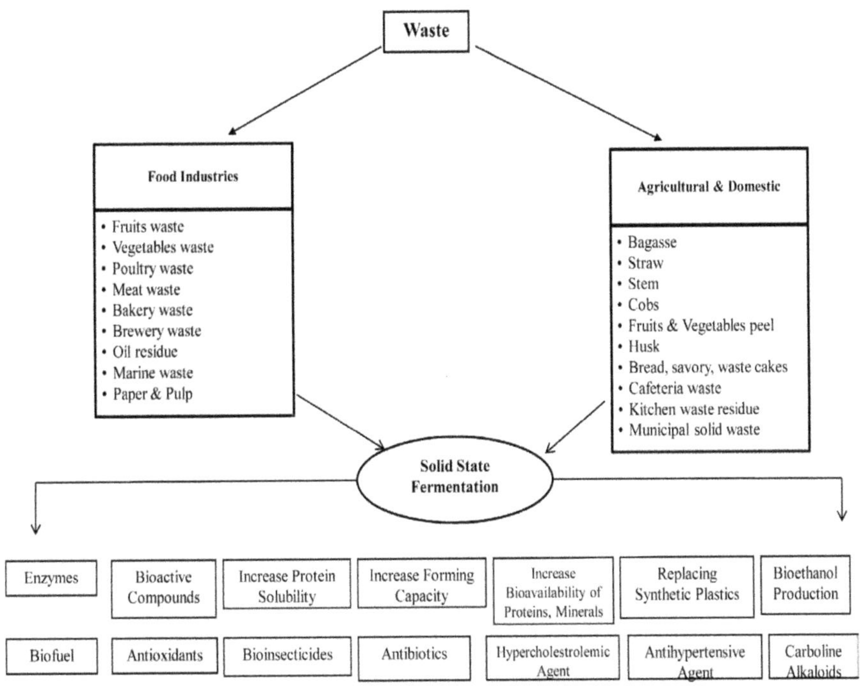

Table 1. Details of bioactive compounds using SSF technology

Author	Substrate (Waste produced)	Microorganism	Bioactive Compound
Hossain et al., **2010**; Dhabekar and Chandak, 2010	Pineapple waste, banana waste, potato starch, palm kernel press cake (PKC)	*Saccharomyces cerevisiae,*	Bioethanol
Swain and Ray, 2008	Cassava fibrous residue	*Bacillus subtilis*	Indole-3-acetic acid
Vidhyalakshmi et al., 2012	Potato peel	*Xanthomonas citri*	Xanthan
Yalemtesfa et al., 2010	Orange peel	*Chaetomium spp. (KC-06) and Aspergillus niger*	Protein
Sadh et al., 2018; Sousa and Correia, 2012	Guava and pineapple waste; Peanut waste (peanut press cake)	*Rhizopus oligosporus: Aspergillus awamori*	Phenolic content
Sadh et al., 2018; Dulf et al., 2017; Ajila et al., 2011	Peanut waste (peanut press cake); apricot pomace; Apple pomace	*Aspergillus awamori; Aspergillus niger and Rhizopus oligosporus; Phanerocheate chrysosporium*	Antioxidants
Vastrad and Neelagund, 2011	Apple pomace, cottonseed meal, soybean powder, and wheat bran	*Streptomyces fradiae*	Neomycin
Vastrad and Neelagund, 2012	Coconut oil cake, groundnut oil cake, ground nutshell, and rice husk	*Amycolatopsis Mediterranean*	Rifamycin
El-Naggar et al., 2009	Rice, wheat bran, quaker, bread, and ground corn	*Streptomyces sp. strain*	Meroparamycin
Klaic et al., 2017	Soybean bran, bagasse, and corn steep liquor	*Phoma sp.*	Bioherbicide
Dhillon et al., 2017	Apple pomace	*Aspergillus niger*	Biosorbents
Dursun and Dalgıç, 2016	Wheat waste	*Yamadazyma guilliermondii*	Astaxanthin
Chandra and Arora, 2016	Wheat straw, Rice straw, Corncob, Peapod, Sugarcane bagasse	*Aspergillus fumigatus, A. terreus, A. wentii, Penicillium citrinum, P. granulatum, P. expansum*	Bioactive phenolic compounds
Ali and Vidhale, 2013; Ali et al., 2016	Apple pomace, brewer's spent grain, citrus waste, sphagnum peat moss; peanut shell	*Aspergillus niger; Aspergillus ornatus and Alternaria alternata*	Citric acid
Das et al., 2015; Das et al., 2016	Apple pomace; pulp and paper solid waste	*Rhizopus oryzae*	Fumaric acid
Rane et al., 2017	Potato peels, orange peels, banana peels, and bagasse	*Bacillus subtilis*	Biosurfactant
Jamal et al., 2016	Tomato waste	*Aspergillus niger*	Lycopene
Razak et al., 2017	Rice bran	*Aspergillus oryzae and Rhizopus oryzae*	Ferulic, p-coumaric, sinapic and syringic

Table 2. Details of enzyme production using SSF technology

Author	Microorganism	Substrate	Enzyme Produced
Mussatto et al., 2012	*Aspergillus niger*	Wheat bran	Glucoamylase
Mussatto et al., 2012	*Bacillus subtilis, Aspergillus ustus, Trichoderma aureoviride, Penicillium citrinum, Trichoderma reesi, Bacillus subtilis, Trichoderma viride*	Banana fruit stalk, Wheat bran, rice straw, Leached beet pulp, Rice husks, Sweet sorghum silage, Banana fruit stalk, Coconut pith.	Cellulases
Vijayaraghavan et al., 2016	*Bacillus halodurans*	Banana peel, black gram husk, paddy straw, rice bran, and wheat bran	Fibrinolytic enzyme
Mussatto et al., 2012	*A. ficuum, A. carbonarius*	Wheat bran	Phytase
Pili et al., 2017	*Aspergillus brasiliensis*	corn steep liquor and orange peel	Pectin lyase
Mussatto et al., 2012	*A. niger*	Palm kernel cake	Tannase
Vintila et al., **2009**; Grover et al., 2012	*Trichoderma viride, Bacillus cereus*	Wheat bran, Rice bran, Corn husks	Cellulase
Mussatto et al., 2012	*Candida rugosa Candida sp. Monascus fulginosus Neurospora sitophila A. niger*	Rice bran, wheat bran, peanut Press cake and coconut oil cake	Lipases
Maldonado and Strasser, 1998	*Aspergillus niger*	Wheat bran, Coffee pulp	Polygalactouronase
Mussatto et al., 2012	*Pleurotus sp. Phanerochaete chrysosporium*	Wheat straw and bagasse	Ligninase
Godoy et al., 2011	*Penicillium simplicissimum*	Castor bean waste	Lipase
Mussatto et al., 2012	*Aspergillus sp. Rhizopus sp. Mucor sp. Bacillus sp. Saccharomyces sp Bacillus subtilis Aeromonas caviae*	Rice husk, coconut cake, tea waste, cassava, cassava bagasse, sugarcane bagasse, banana waste, cornflour	α–amylase and β-amylase
Sharma et al., 2017	*Rheinheimera sp.,*	Peels of citrus fruits	Laccase
Mussatto et al., 2012	*Aspergillus niger A. oryzae Candida sp. Neurospora sitophila P. candidum Mucor sp. Kluyveromyces lactis*	Wheat bran, soybean cake waste	α-galactosidase and β-galactosidase
Chutmanop et al., 2008	*Lactobacillus delbrueckii ssp.*	Rice bran.	Protease production
Mussatto et al., 2012	*Penicillium decumbens,*	Corn silage, Corn straw	Xylanase

Biorefining Strategy

Biorefining strategy is defined by National Renewable Energy Laboratory as a combined process of instrumentation and conversion of biomass to provide chemicals, fuels, and power (Berntson et al., 2012). All kinds of biomass like waste produced by agriculture, industries, lignocellulosic and wood materials, aquatic waste, domestic waste, etc. can be utilized and converted to valuable products (de Jong and Jungmeier, 2015). Thus, the process of biorefining can contribute to sustainable development by renewable waste products and using it for biorefining development. In the future, researchers should use and spread the idea of this process for making valuable functional products for human use.

FUTURE RESEARCH DIRECTION

Reviewed literature suggested that SSF becomes an advantageous process and also mandatory for industrial utilization. However, application of bioengineered microorganisms, biotechnologically modified enzymes, development of bioreactors, and potentialities of synthetic biology increase the possibilities of its practical application in many sectors which are to be encouraged essentially SSF in a whole represents environmental, industrial, and economic feasibility and therefore, would be promoted for their optimum exploitation in an eco-friendly way without any conflicts to the nature.

CONCLUSION

Currently, SSF has become an emerging process and researchers are taking key interest to discover more usage of this technology in various fields by treatment of solid waste. This technique can sort out industrial problems of waste disposal and contribute to an eco-friendly environment.

REFERENCES

Ajila, C. M., Brar, S. K., Verma, M., Tyagi, R. D., & Valéro, J. R. (2011). Solid-state fermentation of apple pomace using *Phanerocheate chrysosporium*–liberation and extraction of phenolic antioxidants. *Food Chemistry*, *126*(3), 1071–1080. doi:10.1016/j.foodchem.2010.11.129

Ali, S. R., Anwar, Z., Irshad, M., Mukhtar, S., & Warraich, N. T. (2016). Bio-synthesis of citric acid from single and co-culture-based fermentation technology using agro-wastes. *J. Radiat. Res. Appl. Sci*, *9*(1), 57–62. doi:10.1016/j.jrras.2015.09.003

Ali, S. S., & Vidhale, N. N. (2013). Protease production by Fusarium oxysporum in solid-state fermentation using rice bran. *African Journal of Microbiological Research*, *1*(3), 45–47. doi:10.12691/ajmr-1-3-2

Vijayaraghavan, P., Vincent, S. G. P., Arasu, M. V., & Al-Dhabi, N. A. (2016). Bioconversion of agro-industrial wastes for the production of fibrinolytic enzyme from *Bacillus halodurans IND18*: Purification and biochemical characterization. *Electronic Journal of Biotechnology*, *20*, 1–8. doi:10.1016/j.ejbt.2016.01.002

Musoni, M., Destain, J., Thonart, P., Bahama, J. B., & Delvigne, F. (2015). Bioreactor design and implementation strategies for the cultivation of filamentous fungi and the production of fungal metabolites: From traditional methods to engineered systems. *Biotechnologie, Agronomie, Société et Environnement*, *19*(4), 430–442.

Chandra, P., & Arora, D. S. (2016). Production of antioxidant bioactive phenolic compounds by solid-state fermentation on agro-residues using various fungi isolated from soil. *Asian J. Biotechnol*, *8*(2), 8–15. doi:10.3923/ajbkr.2016.8.15

Chen, H. Z., & He, Q. (2012). Value-added bioconversion of biomass by solid-state fermentation. *Journal of Chemical Technology and Biotechnology (Oxford, Oxfordshire)*, *87*(12), 1619–1625. doi:10.1002/jctb.3901

Chutmanop, J., Chuichulcherm, S., Chisti, Y., & Srinophakun, P. (2008). Protease production by *Aspergillus oryzae* in solid-state fermentation using agroindustrial substrates. *Journal of Chemical Technology and Biotechnology (Oxford, Oxfordshire)*, *83*(7), 1012–1018. doi:10.1002/jctb.1907

Razak, D. L. A., Rashid, N. Y. A., Jamaluddin, A., Sharifudin, S. A., Kahar, A. A., & Long, K. (2017). Cosmeceutical potentials and bioactive compounds of rice bran fermented with single and mix culture of *Aspergillus oryzae* and *Rhizopus oryzae*. *Journal of the Saudi Society of Agricultural Sciences*, *16*(2), 127–134. doi:10.1016/j.jssas.2015.04.001

Couto, S. R., & Sanroman, M. A. (2006). Application of solid-state fermentation to food industry—A review. *Journal of Food Engineering*, *76*(3), 291–302. doi:10.1016/j.jfoodeng.2005.05.022

Das, R. K., Brar, S. K., & Verma, M. (2015). A fermentative approach towards optimizing directed biosynthesis of fumaric acid by *Rhizopus oryzae 1526* utilizing apple industry waste biomass. *Fungal Biology, 119*(12), 1279–1290. doi:10.1016/j.funbio.2015.10.001 PMID:26615750

Das, R. K., Brar, S. K., & Verma, M. (2016). Potential use of pulp and paper solid waste for the bio-production of fumaric acid through submerged and solid state fermentation. *Journal of Cleaner Production, 112*, 4435–4444. doi:10.1016/j.jclepro.2015.08.108

de Jong, E., & Jungmeier, G. (2015). Biorefinery concepts in comparison to petrochemical refineries. In A. Pandey, R. Hofer, C. Larroche, M. Taherzadeh, & M. Nampoothiri (Eds.), *Industrial biorefineries and white biotechnology* (pp. 3–33). UK: Elsevier. doi:10.1016/B978-0-444-63453-5.00001-X

Dhabekar, A., & Chandak, A. (2010). Utilization of banana peels and beet waste for alcohol production. *Asiatic J. Biotech. Res, 1*, 8–13.

Dulf, F. V., Vodnar, D. C., Dulf, E. H., & Pintea, A. (2017). Phenolic compounds, flavonoids, lipids and antioxidant potential of apricot (*Prunus armeniaca L.*) pomace fermented by two filamentous fungal strains in solid state system. *Chemistry Central Journal, 11*(1), 92. doi:10.118613065-017-0323-z PMID:29086904

Durand, A. (2003). Bioreactor designs for solid state fermentation. *Biochemical Engineering Journal, 13*(2-3), 113–125. doi:10.1016/S1369-703X(02)00124-9

Dursun, D., & Dalgıç, A. C. (2016). Optimization of astaxanthin pigment bioprocessing by four different yeast species using wheat wastes. *Biocatalysis and Agricultural Biotechnology, 7*, 1–6. doi:10.1016/j.bcab.2016.04.006

Eco-Cycle. (2011). Waste of energy-Why incineration is bad for our economy, Environment and Community; Eco-Cycle: Boulder County, CO, (pp. 1–20).

El-Naggar, M. Y., El-Assar, S. A., & Abdul-Gawad, S. M. (2009). Solid-state fermentation for the production of meroparamycin by *Streptomyces sp. strain MAR01*. *Journal of Microbiology and Biotechnology, 19*(5), 468–473. doi:10.4014/jmb.0807.457 PMID:19494694

Sadh, P. K., Kumar, S., Chawla, P., & Duhan, J. S. (2018). Fermentation: A boon for production of bioactive compounds by processing of food industries wastes (by-products). *Molecules (Basel, Switzerland), 23*(2560), 1–33. PMID:30297619

Sharma, A., Gupta, V., Khan, M., Balda, S., Gupta, N., Capalash, N., & Sharma, P. (2017). Flavonoid-rich agro-industrial residues for enhanced bacterial laccase production by submerged and solid-state fermentation. *Biotech*, *7*, 200–212. PMID:28667639

Godoy, M. G., Gutarra, M. L. E., Castro, A. M., Machado, O. L. T., & Freire, D. M. G. (2011). Adding value to a toxic residue from the biodiesel industry: Production of two distinct pool of lipases from *Penicillium simplicissimum* in castor bean waste. *Journal of Industrial Microbiology & Biotechnology*, *38*(8), 945–953. doi:10.100710295-010-0865-8 PMID:20844923

Grover, S., Kathuria, R. S., & Kaur, M. (2012). Energy values and technologies for non woody biomass: As a clean source of energy. *Indian Journal of Science and Technology*, *1*, 10–14.

Holker, U., Hofer, M., & Lenz, J. (2004). Biotechnological advantages of laboratory-scale solid-state fermentation with fungi. *Applied Microbiology and Biotechnology*, *64*, 175–186. doi:10.100700253-003-1504-3 PMID:14963614

Hossain, A. B., & Fazliny, A. R. (2010). Creation of alternative energy by bio-ethanol production from pineapple waste and the usage of its properties for engine. *African Journal of Microbiological Research*, *4*, 813–819.

Manpreet, S., Sawraj, S., Sachin, D., Pankaj, S., & Banerjee, U. C. (2005). Influence of process parameters on the production of metabolites in solid-state fermentation. *Malaysian Journal of Microbiology*, *1*(2), 1–9.

Kapilan, R. (2015). Solid state fermentation for microbial products: A review. *Archives of Applied Science Research*, *7*(8), 21–25.

Klaic, R., Sallet, D., Foletto, E. L., Jacques, R. J. S., Guedes, J. V. C., Kuhn, R. C., & Mazutti, M. A. (2017). Optimization of solid-state fermentation for bioherbicide production by *Phoma sp. Brazilian Journal of Chemical Engineering*, *34*(2), 377–384. doi:10.1590/0104-6632.20170342s20150613

Koyani, R. D., & Rajput, K. S. (2015). Solid state fermentation: comprehensive tool for utilization of lignocellulosic through biotechnology. *Journal of Bioprocessing & Biotechniques*, *5*(10), 258. doi:10.4172/2155-9821.1000258

Maldonado, M. C., & Strasser de Saad, A. M. (1998). Production of pectin estrase and polygalactouronase by *Aspergillus niger* in submerged and solid state systems. *Journal of Industrial Microbiology & Biotechnology*, *20*(1), 34–38. doi:10.1038j.jim.2900470 PMID:9523455

Mitchell, D. A., & Berovic, N. K. M. (Eds.). (2006). Solid-state fermentation bioreactors fundamentals of design and operation. Berlin, Germany: Springer-Verlag. doi:10.1007/3-540-31286-2

Nigam, P. S., & Pandey, A. (2009). Solid-state fermentation technology for bioconversion of biomass and agricultural residues. In P. Singh nee' Nigam, & A. Pandey (Eds.), Biotechnology for agro-industrial residues utilisation (pp. 197-223). Springer Science+Business Media B.V.

Dhillon, G. S., Rosine, G. M. L., Kaur, S., Hegde, K., Brar, S. K., Drogui, P., & Verma, M. (2017). Novel biomaterials from citric acid fermentation as biosorbents for removal of metals from waste chromated copper arsenate wood leachates. *International Biodeterioration & Biodegradation*, *119*, 147–154. doi:10.1016/j.ibiod.2016.09.014

Pandey, A., Soccol, C. R., Leo, J. A. R., & Nigam, P. (2001). *Solid-state fermentation in Biotechnology* (pp. 221–230). New Delhi, India: Asiatech Publishers.

Aguilar, C. N., Gutiérrez-Sánchez, G., Rado-Barragán, P. A., Rodríguez-Herrera, R., Martínez-Hernandez, J. L., & Esquivel, J. C. C. (2008). Perspectives of solid-state fermentation for production of food enzymes. *American Journal of Biochemistry and Biotechnology*, *4*(4), 354–366. doi:10.3844/ajbbsp.2008.354.366

Pili, J., Danielli, A., Nyari, N. L., Zeni, J., Cansian, R. L., Backes, G. T., & Valduga, E. (2017). Biotechnological potential of agro-industrial waste in the synthesis of pectin lyase from *Aspergillus brasiliensis*. *Food Science & Technology International*. doi:10.1177/1082013217733574

Jamal, P., Akbar, I., Yumi, Z., & Irwandi, J. (2016). Process development for maximum lycopene production from selected fruit waste and its antioxidant and antiradical activity. *Journal of Food Processing & Technology*, *7*(4), 576–581. doi:10.4172/2157-7110.1000576

Vintila, T., Dragomirescu, M., Jurcoane, S., Vintila, D., Caprita, R., & Maniu, M. (2009). Production of cellulase by submerged and solid-state cultures and yeasts selection for conversion of lignocellulose to ethanol. *Romanian Biotechnological Letters*, *14*, 4275–4281.

Rane, A. N., Baikar, V. V., Ravi Kumar, V., & Deopurkar, R. L. (2017). Agro-industrial wastes for production of biosurfactant by *Bacillus subtilis ANR 88* and its application in synthesis of silver and gold nanoparticles. *Frontiers in Microbiology*, *8*, 492. doi:10.3389/fmicb.2017.00492 PMID:28392783

Singhania, R. R., Patel, A. K., Soccol, C. R., & Pandey, A. (2009). Recent advances in solid-state fermentation. *Biochemical Engineering Journal, 44*(1), 13–18. doi:10.1016/j.bej.2008.10.019

Sadh, P. K., Chawla, P., & Duhan, J. S. (2018). Fermentation approach on phenolic, antioxidants and functional properties of peanut press cake. *Food Bioscience, 22*, 113–120. doi:10.1016/j.fbio.2018.01.011

Sánchez, A., Artola, A., Font, X., Gea, T., Barrena, R., Gabriel, D., ... Mondini, C. (2015). Greenhouse gas emissions from organic waste composting. *Environmental Chemistry Letters, 13*(3), 223–238. doi:10.100710311-015-0507-5

Singh, R. K., Mishra, S. K., & Kumar, N. (2010). Optimization of culture conditions for amylase production by thermophilic *Bacillus sp.* in submerged fermentation. *Research Journal of Pharmaceutical, Biological and Chemical Sciences, 1*(4), 867–876.

Abu Yazid, N., Barrena, R., Komilis, D., & Sánchez, A. (2017). Solid-state fermentation as a novel paradigm for organicwaste valorization: A review. *Sustainability, 9*(224), 1–28.

Sousa, B. A., & Correia, R. T. P. (2012). Phenolic content, antioxidant activity and antiamylolytic activity of extracts obtained from bioprocessed pineapple and guava wastes. *Brazilian Journal of Chemical Engineering, 29*(1), 25–30. doi:10.1590/S0104-66322012000100003

Subramaniyam, R., & Vimala, R. (2012). Solid state and submerged fermentation for the production of bioactive substances: A comparative study. *International Journal of Science and Nature, 3*(3), 480–486.

Swain, M. R., & Ray, R. C. (2008). Optimization of cultural conditions and their statistical interpretation for production of indole-3-acetic acid by *Bacillus subtilis CM5* using cassava fibrous residue. *Journal of Scientific and Industrial Research (New Delhi, India), 67*, 622–628.

Sabu, A., Augur, C., Swati, C., & Pandey, A. (2006). Tannase production by *Lactobacillus* sp. ASR-S1 under solid-state fermentation, Process. *Bioche, 41*, 575–580.

Mussatto, S. I., Ballesteros, L. F., Martins, S., & Teixeira, J. A. (2012). Use of agro-industrial wastes in solid-state fermentation processes. Industrial Waste (pp. 121–141). In Tech: Rijeka, Croatia.

Vastrad, B. M., & Neelagund, S. E. (2011). Optimization and production of neomycin from different agro industrial wastes in solid state fermentation. *International Journal of Pharmaceutical Sciences and Drug Research*, *3*, 104–111.

Vastrad, B. M., & Neelagund, S. E. (2012). Optimization of process parameters for rifamycin b production under solid state fermentation from *Amycolatopsis mediterranean MTCC14. Int. J. Curr. Pharm. Res*, *4*, 101–108.

Vidhyalakshmi, R., Vallinachiyar, C., & Radhika, R. (2012). Production of xanthan from agro-industrial waste. *J. Adv. Sci. Res*, *3*, 56–59.

Berntsson, T., Sandén, B. A., Olsson, L., & Åsblad, A. (2012). What is biorefinery? In B. Sanden, & K. Pettersson (Eds.), Systems perspective on biorefineries (pp. 16-25). Gothenburg: Chalmers University of Technology.

Yalemtesfa, B., Alemu, T., & Santhanam, A. (2010). Solid substrate fermentation and conversion of orange waste into fungal biomass using *Aspergillus niger KA-06* and *Chaetomium Spp KC-06. African Journal of Microbiological Research*, *4*, 1275–1281.

Compilation of References

Abdrabou, A. M., Osman, E. Y., & Aboubakr, O. A. (2018). Comparative therapeutic efficacy study of Lactobacilli probiotics and citalopram in treatment of acute stress-induced depression in lab murine models. *Human Microbiome Journal*, *10*, 33–36. doi:10.1016/j.humic.2018.08.001

Abu Yazid, N., Barrena, R., Komilis, D., & Sánchez, A. (2017). Solid-state fermentation as a novel paradigm for organicwaste valorization: A review. *Sustainability*, *9*(224), 1–28.

Aguilar, C. N., Gutiérrez-Sánchez, G., Rado-Barragán, P. A., Rodríguez-Herrera, R., Martínez-Hernandez, J. L., & Esquivel, J. C. C. (2008). Perspectives of solid-state fermentation for production of food enzymes. *American Journal of Biochemistry and Biotechnology*, *4*(4), 354–366. doi:10.3844/ajbbsp.2008.354.366

Aguirre Cando, M. E. (2016). Descriptive sensory and texture profile analysis of woody breast in marinated chicken.

Ahmed, J., Ramaswamy, H. S., Kasapis, S., & Boye, I. (2010). *J.* Novel food processing effects on rheological and functional properties. CRC Group.

Ajila, C. M., Brar, S. K., Verma, M., Tyagi, R. D., & Valéro, J. R. (2011). Solid-state fermentation of apple pomace using *Phanerocheate chrysosporium*–liberation and extraction of phenolic antioxidants. *Food Chemistry*, *126*(3), 1071–1080. doi:10.1016/j.foodchem.2010.11.129

Akbarzadeh, F., & Homayouni, A. (2012). Dairy probiotic foods and coronary heart disease: A review on mechanism of action. *Probiotics. InTech*, 121-8.

Akın, M. B., Akın, M. S., & Kırmacı, Z. (2007). Effects of inulin and sugar levels on the viability of yogurt and probiotic bacteria and the physical and sensory characteristics in probiotic ice-cream. *Food Chemistry*, *104*(1), 93–99. doi:10.1016/j.foodchem.2006.11.030

Aktar, T. (2015). *Determining the detection threshold for perception of selected textural attributes*. University of Leeds. Retrieved from http://etheses.whiterose.ac.uk/view/creators/Aktar=3ATugba=3A=3A.html

Aktar, T., Chen, J., Ettelaie, R., & Holmes, M. (2015a). Evaluation of the sensory correlation between touch sensitivity and the capacity to discriminate viscosity. *Journal of Sensory Studies*, *30*(2), 98–107. doi:10.1111/joss.12141

Aktar, T., Chen, J., Ettelaie, R., & Holmes, M. (2015b). Tactile sensitivity and capability of soft-solid texture discrimination. *Journal of Texture Studies*, *46*(6), 429–439. doi:10.1111/jtxs.12142

Aktar, T., Chen, J., Ettelaie, R., Holmes, M., & Henson, B. (2017). Human roughness perception and possible factors effecting roughness sensation. *Journal of Texture Studies*, *48*(3), 181–192. doi:10.1111/jtxs.12245 PMID:28573724

Albert, A., Salvador, A., Schlich, P., & Fiszman, S. (2012). Comparison between temporal dominance of sensations (TDS) and key-attribute sensory profiling for evaluating solid food with contrasting textural layers: Fish sticks. *Food Quality and Preference*, *24*(1), 111–118. doi:10.1016/j.foodqual.2011.10.003

Ali, S. R., Anwar, Z., Irshad, M., Mukhtar, S., & Warraich, N. T. (2016). Bio-synthesis of citric acid from single and co-culture-based fermentation technology using agro-wastes. *J. Radiat. Res. Appl. Sci*, *9*(1), 57–62. doi:10.1016/j.jrras.2015.09.003

Ali, S. S., & Vidhale, N. N. (2013). Protease production by Fusarium oxysporum in solid-state fermentation using rice bran. *African Journal of Microbiological Research*, *1*(3), 45–47. doi:10.12691/ajmr-1-3-2

Alliger, H. (1975). Ultrasonic distruption. *American Laboratory*, *10*, 75–85.

Allouche, Y., Varga, S., Bouden, C., & Oliveira, A. C. (2017). Dynamic simulation of an integrated solar-driven ejector-based air conditioning system with PCM cold storage. *Applied Energy*, *190*, 600–611. doi:10.1016/j.apenergy.2017.01.001

Alpas, H., & Bozoglu, F. (2000, June). The combined effect of high hydrostatic pressure, heat and bacteriocins on inactivation of foodborne pathogens in milk and orange juice. *World Journal of Microbiology & Biotechnology*, *16*(4), 387–392. doi:10.1023/A:1008936607413

Alpas, H., Kalchayanand, N., Bozoglu, F., & Ray, B. (2000). Interactions of high hydrostatic pressure, pressurization temperature and pH on death and injury of pressure-resistant and pressure-sensitive strains of foodborne pathogens. *International Journal of Food Microbiology*, *60*(1), 33–42. doi:10.1016/S0168-1605(00)00324-X PMID:11014520

Alpas, H., Kalchayanand, N., Bozoglu, F., Sikes, A., Dunne, C. P., & Ray, B. (1999). Variation in resistance to hydrostatic pressure among strains of food-borne pathogens. *Applied and Environmental Microbiology*, *65*(9), 4248–4251. PMID:10473446

Alpas, H., Lee, J., Bozoglu, F., & Kaletunç, G. (2003). Evaluation of high hydrostatic pressure sensitivity of Staphylococcus aureus and Escherichia coli O157:H7 by differential scanning calorimetry. *International Journal of Food Microbiology*, *87*(3), 229–237. doi:10.1016/S0168-1605(03)00066-7 PMID:14527795

Altuner, E. M., Alpas, H., Erdem, Y. K., & Bozoglu, F. (2006). Effect of high hydrostatic pressure on physicochemical and biochemical properties of milk. *European Food Research and Technology*, *222*(3–4), 392–396. doi:10.100700217-005-0072-4

Alvarez, I., Condon, S., Manas, P., & Virto, R. (2006). Inactivation of *Salmonella Sentenberg 775W* by ultrasonic waves under pressure at different water activities. *International Journal of Food Microbiology, 108*(2), 218–225. doi:10.1016/j.ijfoodmicro.2005.11.011 PMID:16488040

Álvarez, I., Pagan, R., Raso, J., Codo, S., & Sala, F. J. (2000). Microbial inactivation by ultrasound. *Food Technology, University of Zaeagoza. Miguel Servet, 177*, 50013.

Amar, J., Chabo, C., Waget, A., Klopp, P., Vachoux, C., Bermúdez-Humarán, L. G., ... Ouwehand, A. (2011). Intestinal mucosal adherence and translocation of commensal bacteria at the early onset of type 2 diabetes: Molecular mechanisms and probiotic treatment. *EMBO Molecular Medicine, 3*(9), 559–572. doi:10.1002/emmm.201100159 PMID:21735552

Anonymous. (2009). *EU Guidance to the commission regulation (EC) No 450/2009 of 29 May 2009 on active and intelligent materials and articles intended to come into contact with food. Version 1.0. European Commission Health and Consumers Directorate-General Directorate E-Safety of the Food chain. E6- Innovation and sustainability.* Retrieved from https://ec.europa.eu/food/sites/food/files/safety/docs/cs_fcm_legis_active-intelligent_guidance.pdf

Anonymous. (2019). *Nanotechnology.* European Food Safety Authority. Retrieved from https://www.efsa.europa.eu/en/topics/topic/nanotechnology

Arora, D. S., & Chandra, P. (2011). Antioxidant activity of Aspergillus fumigatus. *ISRN Pharmacology*, 2011. PMID:22084718

Arumuganathan, T., Manikantan, M. R., Rai, R. D., Anandakumar, S., & Khare, V. (2009). Mathematical modeling of drying kinetics of milky mushroom in a fluidized bed dryer. *International Agrophysics, 23*(1), 1–7.

Awad, T. S., Moharram, H. A., Shaltout, O. E., Asker, D., & Youssef, M. M. (2012). Applications of ultrasound in analysis, processing and quality control of food: A review. *Food Research International, 48*(2), 410–427. doi:10.1016/j.foodres.2012.05.004

Awaisheh, S., Al-Dmoor, H., Omar, S., Hawari, A., & Alroyli, M. (2012). Impact of selected nutraceuticals on viability of probiotic strains in milk during refrigerated storage at 4C for 15 days. *International Journal of Dairy Technology, 65*(2), 268–273. doi:10.1111/j.1471-0307.2011.00817.x

Bai, B., Zhao, K., & Li, X. (2019). Application research of nano-storage materials in cold chain logistics of e-commerce fresh agricultural products. *Results in Physics, 13*, 102049. doi:10.1016/j.rinp.2019.01.083

Bajpai, V. K., Kamle, M., Shukla, S., Mahato, D. K., Chandra, P., Hwang, S. K., ... Han, Y. K. (2018). Prospects of using nanotechnology for food preservation, safety, and security. *Journal of Food and Drug Analysis, 26*(4), 1201–1214. doi:10.1016/j.jfda.2018.06.011 PMID:30249319

Balakumar, M., Prabhu, D., Sathishkumar, C., Prabu, P., Rokana, N., Kumar, R., ... Mohan, V. (2018). Improvement in glucose tolerance and insulin sensitivity by probiotic strains of Indian gut origin in high-fat diet-fed C57BL/6J mice. *European Journal of Nutrition, 57*(1), 279–295. doi:10.100700394-016-1317-7 PMID:27757592

Barbosa-Canovas, G. V., Bermudez-Aguirre, D., Corradini, M. G., & Mawson, R. (2009). Modeling the inactivation of Listeria innocua in raw whole milk treated under thermo-sonication. *Innovative Food Science & Emerging Technologies, 10*(2), 172–178. doi:10.1016/j.ifset.2008.11.005

Basu, D. N., & Ganguly, A. (2016). Solar thermal–photovoltaic powered potato cold storage – Conceptual design and performance analyses. *Applied Energy, 165*, 308–317. doi:10.1016/j.apenergy.2015.12.070

Batu, A., Caglar, A., & Kara, H. H. (2008). Afyon kaymagının raf ömrünün uzatılmasında modifiye atmosferde paketleme önerisi. *Gıda Teknolojileri Elektronik Dergisi, 2008*(2), 43-46.

Batu, A., Caglar, A., & Kara, H. H., (2008). Afyon kaymagının raf ömrünün uzatılmasında modifiye atmosferde paketleme önerisi. *Gıda Teknolojileri Elektronik Dergisi, 2008*(2), 43-46.

Batu, A. (1994). Properties of modified atmosphere packaging films and application of fruits vegetables. *Gida, 19*(6), 397–403.

Beauchamp, G. K. (2009). Sensory and receptor responses to umami: An overview of pioneering work. *The American Journal of Clinical Nutrition, 90*(3), 723S–727S. doi:10.3945/ajcn.2009.27462E PMID:19571221

Bejar, W., Hamden, K., Salah, R. B., & Chouayekh, H. (2013). Lactobacillus plantarum TN627 significantly reduces complications of alloxan-induced diabetes in rats. *Anaerobe, 24*, 4–11. doi:10.1016/j.anaerobe.2013.08.006 PMID:23999246

Bellisle, F. (1999). Glutamate and the UMAMI taste: Sensory, metabolic, nutritional and behavioural considerations. A review of the literature published in the last 10 years. *Neuroscience and Biobehavioral Reviews, 23*(3), 423–438. doi:10.1016/S0149-7634(98)00043-8 PMID:9989429

Bernardeau, M., Guguen, M., & Vernoux, J. P. (2006). Beneficial lactobacilli in food and feed: Long-term use, biodiversity and proposals for specific and realistic safety assessments. *FEMS Microbiology Reviews, 30*(4), 487–513. doi:10.1111/j.1574-6976.2006.00020.x PMID:16774584

Berntsson, T., Sandén, B. A., Olsson, L., & Åsblad, A. (2012). What is biorefinery? In B. Sanden, & K. Pettersson (Eds.), Systems perspective on biorefineries (pp. 16-25). Gothenburg: Chalmers University of Technology.

Beyth, N., Houri-Haddad, Y., Domb, A., Khan, W., & Hazan, R. (2015). Alternative antimicrobial approach: Nano-antimicrobial materials. *Evidence-Based Complementary and Alternative Medicine, 2015*, 1–16. doi:10.1155/2015/246012 PMID:25861355

Bhattacharya, S. (2015). *Conventional and advanced food processing technologies*. Wiley Blackwell.

Bi, J. (2008). *Sensory discrimination tests and measurements: Statistical principles, procedures, and tables*. John Wiley & Sons.

Biji, K. B., Ravishankar, C. N., Mohan, C. O., & Gopal, T. S. (2015). Smart packaging systems for food applications: A review. *Journal of Food Science and Technology*, *52*(10), 6125–6135. doi:10.100713197-015-1766-7 PMID:26396360

Bourne, M. (1978). Texture profile analysis. *Journal of Food Technology*, *33*, 62–66.

Brennan, M., Wanismail, B., Johnson, M. C., & Ray, B. (1986). Cellular damage in dried Lactobacillus acidophilus. *Journal of Food Protection*, *49*(1), 47–53. doi:10.4315/0362-028X-49.1.47 PMID:30959616

Brody, A. L., Strupinsky, E. R., & Kline, L. R. (2001). *Active packaging for food applications*. Boca Raton, FL: CRC Press. doi:10.1201/9781420031812

Brower, V. (1998). Nutraceuticals: Poised for a healthy slice of the healthcare market? *Nature Biotechnology*, *16*(8), 728–731. doi:10.1038/nbt0898-728 PMID:9702769

Buzby, J. C. (2010). Nanotechnology for food applications. More questions than answers. *The Journal of Consumer Affairs*, *44*(3), 528–545. doi:10.1111/j.1745-6606.2010.01182.x

Cakmak, R. Ş., Tekeoglu, O., Bozkır, H., Ergun, A. R., & Baysal, T. (2016). Effects of electrical and sonication pretreatments on the drying rate and quality of mushrooms. *Lebensmittel-Wissenschaft + Technologie*, *69*, 197–202. doi:10.1016/j.lwt.2016.01.032

Cameron, M., McMaster, L. D., & Britz, T. J. (2009). Impact of ultrasound on dairy spoilage microbes and milk components. *Dairy Science & Technology*, *89*(1), 83–98. doi:10.1051/dst/2008037

Cardello, A. V. (1996). The role of the human senses in food acceptance. In *Food choice, acceptance and consumption* (pp. 1–82). Springer. doi:10.1007/978-1-4613-1221-5_1

Castro, A. M., Mayorga, E. Y., & Moreno, F. L. (2018). Mathematical modeling of convective drying of fruits: A review. *Journal of Food Engineering*, *223*, 152–167. doi:10.1016/j.jfoodeng.2017.12.012

Caul, J. F. (1957). The profile method of flavor analysis. [Elsevier]. *Advances in food research*, *7*, 1–40. doi:10.1016/S0065-2628(08)60245-1

Çelik, İ., & Tümer, G. (2016). Gıda ambalajlamada son gelişmeler. *Akademik Gıda*, *14*(2), 180–188.

Champagne, C. P., Ross, R. P., Saarela, M., Hansen, K. F., & Charalampopoulos, D. (2011). Recommendations for the viability assessment of probiotics as concentrated cultures and in food matrices. *International Journal of Food Microbiology*, *149*(3), 185–193. doi:10.1016/j.ijfoodmicro.2011.07.005 PMID:21803436

Chanda, S., Tiwari, R. K., Kumar, A., & Singh, K. (2019). Nutraceuticals inspiring the current therapy for lifestyle diseases. *Advances in Pharmacological Sciences*, *2019*, 1–5. doi:10.1155/2019/6908716 PMID:30755770

Chandra, P., & Arora, D. S. (2016). Production of antioxidant bioactive phenolic compounds by solid-state fermentation on agro-residues using various fungi isolated from soil. *Asian J. Biotechnol*, *8*(2), 8–15. doi:10.3923/ajbkr.2016.8.15

Chandrapala, J., Martin, G. J. O., Kentish, S. E., & Ashokkumar, M. (2014). Dissolution and reconstitution of casein micelle containing dairy powders by high shear using ultrasonic and physical methods. *Ultrasonics Sonochemistry*, *21*(5), 1658–1665. doi:10.1016/j.ultsonch.2014.04.006 PMID:24798226

Chau, C. F., Wu, S. H., & Yen, G. C. (2007). The development of regulations for food nanotechnology. *Trends in Food Science & Technology*, *18*(5), 169–280. doi:10.1016/j.tifs.2007.01.007

Chávarri, M., Marañón, I., & Villarán, M. C. (2012). Encapsulation technology to protect probiotic bacteria. In Probiotics. IntechOpen. doi:10.5772/50046

Chemat, F., Zill-e-Huma, & Khan, M. K. (2011). Applications of ultrasound in food technology: Processing, preservation, and extraction. *Ultrasonics Sonochemistry*, *18*(4), 813–835. doi:10.1016/j.ultsonch.2010.11.023 PMID:21216174

Cheng, X., & Zhai, X. (2018). Thermal performance analysis of a cascaded cold storage unit using multiple PCMs. *Energy*, *143*, 448–457. doi:10.1016/j.energy.2017.11.009

Chen, H. Z., & He, Q. (2012). Value-added bioconversion of biomass by solid-state fermentation. *Journal of Chemical Technology and Biotechnology (Oxford, Oxfordshire)*, *87*(12), 1619–1625. doi:10.1002/jctb.3901

Chen, J. (2014). Food oral processing: Some important underpinning principles of eating and sensory perception. *Food Structure*, *1*(2), 91–105. doi:10.1016/j.foostr.2014.03.001

Chen, M., & Mustapha, A. (2012). Survival of freeze-dried microcapsules of α-galactosidase producing probiotics in a soy bar matrix. *Food Microbiology*, *30*(1), 68–73. doi:10.1016/j.fm.2011.10.017 PMID:22265285

Chen, X., Ren, X., Dong, N., Li, S., Li, F., Zhao, F., ... Mao, Z. (2011). Culture medium containing glucose and glycerol as a mixed carbon source improves ε-poly-l-lysine production by Streptomyces sp. M-Z18. *Bioprocess and Biosystems Engineering*, *35*(3), 469–475. doi:10.100700449-011-0586-z PMID:21909683

Chen, X., Worall, M., Omer, S., Su, Y., & Riffat, S. (2014). Experimental investigation on PCM cold storage integrated with ejector cooling system. *Applied Thermal Engineering*, *63*(1), 419–427. doi:10.1016/j.applthermaleng.2013.11.029

Chen, Y., Sheng, L., Gouda, M., & Ma, M. (2019). Impact of ultrasound treatment on the foaming and physicochemical properties of egg white during cold storage. *Lwt*, *113*, 108303. doi:10.1016/j.lwt.2019.108303

Chizoba Ekezie, F. G., Sun, D. W., & Cheng, J. H. (2017). A review on recent advances in cold plasma technology for the food industry: Current applications and future trends. *Trends in Food Science & Technology*, *69*, 46–58. doi:10.1016/j.tifs.2017.08.007

Chutmanop, J., Chuichulcherm, S., Chisti, Y., & Srinophakun, P. (2008). Protease production by *Aspergillus oryzae* in solid-state fermentation using agroindustrial substrates. *Journal of Chemical Technology and Biotechnology (Oxford, Oxfordshire)*, *83*(7), 1012–1018. doi:10.1002/jctb.1907

Claudia, A. (2011). Nutraceuticals in pet foods and practice. *World Small Animal Veterinary Association World Congress Proceedings* (pp. 1-3). Knoxville, TN: University of Tennessee College of Veterinary Science.

Cliff, M., & Heymann, H. (1993). Development and use of time-intensity methodology for sensory evaluation: A review. *Food Research International*, *26*(5), 375–385. doi:10.1016/0963-9969(93)90081-S

Cooper, K., Haffajee, Z., & Taylor, L. (1997). Human papillomavirus and schistosomiasis associated bladder cancer. *Molecular Pathology*, *50*(3), 145–148. doi:10.1136/mp.50.3.145 PMID:9292149

Coppens, P., Silva, M. F., & Pettman, S. (2006). European regulations on nutraceuticals, dietary supplements and functional foods: A framework based on safety. *Toxicology*, *221*(1), 59–74. doi:10.1016/j.tox.2005.12.022 PMID:16469424

Costa, C., Conte, A., Alessandro, M., & Nobile, D. (2016). Use of metal nanoparticles. In J. Barros-Velazquez (Ed.), Antimicrobial food packaging (pp. 399-406). Academic Press, Elsevier. doi:10.1016/B978-0-12-800723-5.00031-0

Coutinho, N. M., Silveira, M. R., Rocha, R. S., Moraes, J., Ferreira, M. V. S., Pimentel, T. C., ... Cruz, A. G. (2018). Cold plasma processing of milk and dairy products. *Trends in Food Science & Technology*, *74*(February), 56–68. doi:10.1016/j.tifs.2018.02.008

Couto, S. R., & Sanroman, M. A. (2006). Application of solid-state fermentation to food industry—A review. *Journal of Food Engineering*, *76*(3), 291–302. doi:10.1016/j.jfoodeng.2005.05.022

Crowe, J. H., Crowe, L. M., Carpenter, J. F., Rudolph, A., Wistrom, C. A., Spargo, B., & Anchordoguy, T. (1988). Interactions of sugars with membranes. *Biochimica Et Biophysica Acta (BBA) -. Reviews on Biomembranes*, *947*(2), 367–384.

Cutter, C. N. (2006). Opportunities for bio-based packaging technologies to improve the quality and safety of fresh and further processed muscle foods. *Meat Science*, *74*(1), 131–142. doi:10.1016/j.meatsci.2006.04.023 PMID:22062722

D'Amico, D. J., Silk, T. M., Wu, J., & Guo, M. (2006). Inactivation of microorganisms in milk and apple cider treated with ultrasound. *Journal of Food Protection*, *69*(3), 556–563. doi:10.4315/0362-028X-69.3.556 PMID:16541685

Dacremont, C. (1995). Spectral composition of eating sounds generated by crispy, crunchy, and crackly foods. *Journal of Texture Studies*, *26*(1), 27–43. doi:10.1111/j.1745-4603.1995.tb00782.x

Dainelli, D., Gontard, N., Spyropoulos, D., Zondervan-van den Beuken, E., & Tobback, P. (2008). Active and intelligent food packaging: Legal aspects and safety concerns. *Trends in Food Science & Technology*, *19*, 103–112. doi:10.1016/j.tifs.2008.09.011

Dao, H., Lakhani, P., Police, A., Kallakunta, V., Ajjarapu, S. S., Wu, K., ... Murthy, S. N. (2017). Microbial stability of pharmaceutical and cosmetic products. *AAPS PharmSciTech*, *19*(1), 60–78. doi:10.120812249-017-0875-1 PMID:29019083

Dashwood, R. H. (2003). Use of transgenic and mutant animal models in the study of heterocyclic amine-induced mutagenesis and carcinogenesis. *Journal of Biochemistry and Molecular Biology*, *36*(1), 35–42. PMID:12542973

Das, L., Bhaumik, E., Raychaudhuri, U., & Chakraborty, R. (2012). Role of nutraceuticals in human health. *Journal of Food Science and Technology*, *49*(2), 173–183. doi:10.100713197-011-0269-4 PMID:23572839

Das, R. K., Brar, S. K., & Verma, M. (2015). A fermentative approach towards optimizing directed biosynthesis of fumaric acid by *Rhizopus oryzae 1526* utilizing apple industry waste biomass. *Fungal Biology*, *119*(12), 1279–1290. doi:10.1016/j.funbio.2015.10.001 PMID:26615750

Das, R. K., Brar, S. K., & Verma, M. (2016). Potential use of pulp and paper solid waste for the bio-production of fumaric acid through submerged and solid state fermentation. *Journal of Cleaner Production*, *112*, 4435–4444. doi:10.1016/j.jclepro.2015.08.108

Dave, R. I., & Shah, N. P. (1997). Viability of yoghurt and probiotic bacteria in yoghurts made from commercial starter cultures. *International Dairy Journal*, *7*(1), 31–41. doi:10.1016/S0958-6946(96)00046-5

Davies, E. S., & Gardner, C. D. (1996). *UK Patent No. GB-2298273*. Oxygen indicating composition, The Victoria University of Manchester.

Davì, G., Santilli, F., & Patrono, C. (2010). Nutraceuticals in diabetes and metabolic syndrome. *Cardiovascular Therapeutics*, *28*(4), 216–226. doi:10.1111/j.1755-5922.2010.00179.x PMID:20633024

de Almeida, P. D. V., Gregio, A. M., Machado, M. A., De Lima, A. A., & Azevedo, L. R. (2008). Saliva composition and functions: A comprehensive review. *The Journal of Contemporary Dental Practice*, *9*(3), 72–80. doi:10.5005/jcdp-9-3-72 PMID:18335122

de Jong, E., & Jungmeier, G. (2015). Biorefinery concepts in comparison to petrochemical refineries. In A. Pandey, R. Hofer, C. Larroche, M. Taherzadeh, & M. Nampoothiri (Eds.), *Industrial biorefineries and white biotechnology* (pp. 3–33). UK: Elsevier. doi:10.1016/B978-0-444-63453-5.00001-X

de Kruijf, N. N., van Beest, M., Rijk, R., Sipilainen-Malm, T., Paseiro, L. P., & De Meulenaer, B. (2002). Active and intelligent packaging: Applications and regulatory aspects. *Food Additives and Contaminants, 19* Suppl, 144-162.

de Kruijf, N. N., van Beest, M., Rijk, R., Sipilainen-Malm, T., Paseiro, L. P., & De Meulenaer, B. (2002). Active and intelligent packaging: Applications and regulatory aspects. *Food Additives and Contaminants, 19,* Suppl, 144-162.

Deliza, R., & MacFie, H. J. H. (1996). The generation of sensory expectation by external cues and its effect on sensory perception and hedonic ratings: A review. *Journal of Sensory Studies, 11*(2), 103–128. doi:10.1111/j.1745-459X.1996.tb00036.x

Demir, M. (1999). *Modified atmosphere packaging.* Retrieved from http://www.apack.com.tr/images/userfiles/146705440090949918.pdf

Demiray, E., Seker, A., & Tulek, Y. (2017). Drying kinetics of onion (Allium cepa L.) slices with convective and microwave drying. *Heat and Mass Transfer, 53*(5), 1817–1827. doi:10.100700231-016-1943-x

Demiray, E., & Tulek, Y. (2014). Drying characteristics of garlic (Allium sativum L) slices in a convective hot air dryer. *Heat and Mass Transfer, 50*(6), 779–786. doi:10.100700231-013-1286-9

Deng, L. Z., Mujumdar, A. S., Zhang, Q., Yang, X. H., Wang, J., Zheng, Z. A., ... Xiao, H. W. (2019). Chemical and physical pretreatments of fruits and vegetables: Effects on drying characteristics and quality attributes–a comprehensive review. *Critical Reviews in Food Science and Nutrition, 59*(9), 1408–1432. doi:10.1080/10408398.2017.1409192 PMID:29261333

Desai, A. R. (2008). Strain identification, viability, and probiotics properties of Lactobacillus casei (Doctoral dissertation, Victoria University).

Devlieghere, F., & Debevere, J. (2000). Influence of dissolved carbon dioxide on the growth of spoilage bacteria. *Lebensmittel-Wissenschaft + Technologie, 33*(8), 531–537. doi:10.1006/fstl.2000.0705

Devlieghere, F., Gil, M. I., & Debevere, J. (2002). Modified atmosphere packaging (MAP). In C. J. K. Henry, & C. Chapman (Eds.), *The nutrition handbook for food processors* (pp. 342–370). England: Woodhead Publishing. doi:10.1533/9781855736658.2.342

Dhabekar, A., & Chandak, A. (2010). Utilization of banana peels and beet waste for alcohol production. *Asiatic J. Biotech. Res, 1,* 8–13.

Dhillon, G. S., Rosine, G. M. L., Kaur, S., Hegde, K., Brar, S. K., Drogui, P., & Verma, M. (2017). Novel biomaterials from citric acid fermentation as biosorbents for removal of metals from waste chromated copper arsenate wood leachates. *International Biodeterioration & Biodegradation, 119,* 147–154. doi:10.1016/j.ibiod.2016.09.014

Dinnella, C., Recchia, A., Tuorila, H., & Monteleone, E. (2011). Individual astringency responsiveness affects the acceptance of phenol-rich foods. *Appetite, 56*(3), 633–642. doi:10.1016/j.appet.2011.02.017 PMID:21354451

DiRienzo, D. B. (2014). Effect of probiotics on biomarkers of cardiovascular disease: Implications for heart-healthy diets. *Nutrition Reviews, 72*(1), 18–29. doi:10.1111/nure.12084 PMID:24330093

Dobrucka, R. (2013). The future of active and intelligent packaging industry. *LogForum*, *9*(2), 103–110.

Dobrucka, R., & Cierpiszewski, R. (2014). Active and intelligent packaging food - Research and development - A review. *Polish Journal of Food and Nutrition Sciences*, *64*(1), 7–15. doi:10.2478/v10222-012-0091-3

Doona, C. J., Kustin, K., & Feeherry, F. (2010). *Case studies in novel food processing technologies*. Woodhead Publishing; doi:10.1533/9780857090713

Doron, S. I., Hibberd, P. L., & Gorbach, S. L. (2008). Probiotics for prevention of antibiotic-associated diarrhea. *Journal of Clinical Gastroenterology*, *42*(Supplement 2), S58–S63. doi:10.1097/MCG.0b013e3181618ab7 PMID:18542041

Doymaz, I., Kipcak, A. S., & Piskin, S. (2015). Microwave drying of green bean slices: Drying kinetics and physical quality. *Czech Journal of Food Sciences*, *33*(4), 367–376. doi:10.17221/566/2014-CJFS

Drakopoulou, S., Terzakis, S., Fountoulakis, M. S., Mantzavinos, D., & Manios, T. (2009). Ultrasound-induced inactivation of gram-negative and gram-positive bacteria in secondary treated municipal wastewater. *Ultrasonics Sonochemistry*, *16*(5), 629–634. doi:10.1016/j.ultsonch.2008.11.011 PMID:19131265

Drewnowski, A., Shrager, E. E., Lipsky, C., Stellar, E., & Greenwood, M. R. C. (1989). Sugar and fat: Sensory and hedonic evaluation of liquid and solid foods. *Physiology & Behavior*, *45*(1), 177–183. doi:10.1016/0031-9384(89)90182-0 PMID:2727131

Duizer, L. M., Gullett, E. A., & Findlay, C. J. (1993). Time-intensity methodology for beef tenderness perception. *Journal of Food Science*, *58*(5), 943–947. doi:10.1111/j.1365-2621.1993.tb06084.x

Dulf, F. V., Vodnar, D. C., Dulf, E. H., & Pintea, A. (2017). Phenolic compounds, flavonoids, lipids and antioxidant potential of apricot (*Prunus armeniaca L.*) pomace fermented by two filamentous fungal strains in solid state system. *Chemistry Central Journal*, *11*(1), 92. doi:10.118613065-017-0323-z PMID:29086904

Duncan, T. V. (2011). Applications of nanotechnology in food packaging and food safety: Barrier materials, antimicrobials, and sensors. *Journal of Colloid and Interface Science*, *363*(1), 1–24. doi:10.1016/j.jcis.2011.07.017 PMID:21824625

Duqué, B., Haddad, N., Rossero, A., Membré, J.-M., & Guillou, S. (2019). Influence of cell history on the subsequent inactivation of Campylobacter jejuni during cold storage under modified atmosphere. *Food Microbiology*, *84*, 103263. doi:10.1016/j.fm.2019.103263 PMID:31421767

Durance, T. (2002). *Handbook of food preservation* (Vol. 35). Food Research International; doi:10.1016/S0963-9969(00)00143-5

Durand, A. (2003). Bioreactor designs for solid state fermentation. *Biochemical Engineering Journal*, *13*(2-3), 113–125. doi:10.1016/S1369-703X(02)00124-9

Dursun, D., & Dalgıç, A. C. (2016). Optimization of astaxanthin pigment bioprocessing by four different yeast species using wheat wastes. *Biocatalysis and Agricultural Biotechnology*, *7*, 1–6. doi:10.1016/j.bcab.2016.04.006

Eco-Cycle. (2011). Waste of energy-Why incineration is bad for our economy, Environment and Community; Eco-Cycle: Boulder County, CO, (pp. 1–20).

El-Naggar, M. Y., El-Assar, S. A., & Abdul-Gawad, S. M. (2009). Solid-state fermentation for the production of meroparamycin by *Streptomyces sp. strain MAR01. Journal of Microbiology and Biotechnology*, *19*(5), 468–473. doi:10.4014/jmb.0807.457 PMID:19494694

Endres, J., Clewell, A., Jade, K., Farber, T., Hauswirth, J., & Schauss, A. (2009). Safety assessment of a proprietary preparation of a novel Probiotic, Bacillus coagulans, as a food ingredient. *Food and Chemical Toxicology*, *47*(6), 1231–1238. doi:10.1016/j.fct.2009.02.018 PMID:19248815

Engelen, L., & de Wijk, R. A. (2012). Oral processing and texture perception. In J. Chen, & L. Engelen (Eds.), *Food oral processing: Fundamentals of eating and sensory perception* (pp. 159–162). Oxford, UK: Wiley-Blackwell. doi:10.1002/9781444360943.ch8

Ergun, M. (2016). Taze meyve ve sebzeler için aktif, zeki veya akıllı paketleme teknolojileri. *Alatarım*, *15*(2), 51–60.

Ettinger, G., MacDonald, K., Reid, G., & Burton, J. P. (2014). The influence of the human microbiome and probiotics on cardiovascular health. *Gut Microbes*, *5*(6), 719–728. doi:10.4161/19490976.2014.983775 PMID:25529048

Everard, A., Matamoros, S., Geurts, L., Delzenne, N. M., & Cani, P. D. (2014). Saccharomyces boulardii administration changes gut microbiota and reduces hepatic steatosis, low-grade inflammation, and fat mass in obese and type 2 diabetic db/db mice. *mBio*, *5*(3), e01011–e01014. doi:10.1128/mBio.01011-14 PMID:24917595

Evrendilek, G. A., Zhang, Q. H., & Richter, E. R. (2004). Application of pulsed electric fields to skim milk inoculated with Staphylococcus aureus. *Biosystems Engineering*, *87*(2), 137–144. doi:10.1016/j.biosystemseng.2003.11.005

Fang, Z., Zhao, Y., Warner, R. D., & Johnson, S. K. (2017). Active and intelligent packaging in meat industry. *Trends in Food Science & Technology*, *61*, 60–71. doi:10.1016/j.tifs.2017.01.002

Farber, J. M. (1991). Microbiological aspects of modified-atmosphere packaging technology - A review. *Journal of Food Protection*, *54*(1), 58–70. doi:10.4315/0362-028X-54.1.58 PMID:31051584

Fechner, G. T. (1860). Elemente der Psychophysik (Breitkopf und Hèartel, Leipzig, Germany). German.

Fernandes, F. A., Gallão, M. I., & Rodrigues, S. (2008). Effect of osmotic dehydration and ultrasound pre-treatment on cell structure: Melon dehydration. *Lebensmittel-Wissenschaft + Technologie*, *41*(4), 604–610. doi:10.1016/j.lwt.2007.05.007

Fernandes, F. A., Gallão, M. I., & Rodrigues, S. (2009). Effect of osmosis and ultrasound on pineapple cell tissue structure during dehydration. *Journal of Food Engineering*, *90*(2), 186–190. doi:10.1016/j.jfoodeng.2008.06.021

Fernandes, F. A., Oliveira, F. I., & Rodrigues, S. (2008). Use of ultrasound for dehydration of papayas. *Food and Bioprocess Technology*, *1*(4), 339–345. doi:10.100711947-007-0019-9

Fernandes, F. A., & Rodrigues, S. (2007). Ultrasound as pre-treatment for drying of fruits: Dehydration of banana. *Journal of Food Engineering*, *82*(2), 261–267. doi:10.1016/j.jfoodeng.2007.02.032

Fernández, M., Ganan, M., Guerra, C., & Hierro, E. (2014). Protein oxidation in processed cheese slices treated with pulsed light technology. *Food Chemistry*, *159*, 388–390. doi:10.1016/j.foodchem.2014.02.165 PMID:24767071

Fernández-Molina, J. J., Fernández-gutiérrez, S. A., Altunakar, B., Bermúdez-Aguirre, D., Swanson, B. G., & Barbosa-Cánovas, G. V. (2005). The combined effect of pulsed electric fields and conventional heating on the microbial quality and shelf life of skim milk. *Journal of Food Processing and Preservation*, *29*(5–6), 390–406. doi:10.1111/j.1745-4549.2005.00036.x

Fernando, S. S. N., Gunasekara, T. D. C. P., & Holton, J. (2018). Antimicrobial nanoparticles: Applications and mechanisms of action. *Sri Lankan Journal of Infectious Diseases*, *8*(1), 2–11. doi:10.4038ljid.v8i1.8167

Fierheller, M. G. (1991). Modified atmosphere packaging of miscellaneous products. In B. Ooraikul, & M. E. Stiles (Eds.), *Modified atmosphere packaging of food* (pp. 246–260). Boston, MA: Springer. doi:10.1007/978-1-4615-2117-4_8

Fiorentini, C., Demarchi, S. M., Ruiz, N. A. Q., Irigoyen, R. M. T., & Giner, S. A. (2015). Arrhenius activation energy for water diffusion during drying of tomato leathers: The concept of characteristic product temperature. *Biosystems Engineering*, *132*, 39–46. doi:10.1016/j.biosystemseng.2015.02.004

Fishman, M. L., Chau, H. K., Hotchkiss, A. T. Jr, White, A., Garcia, R. A., & Cooke, P. H. (2019). Effect of long-term cold storage and microwave extraction time on the physical and chemical properties of citrus pectin. *Food Hydrocolloids*, *92*, 104–116. doi:10.1016/j.foodhyd.2018.12.047

Fortunati, E., Puglia, D., Armentano, I., Valde, A., Ramos, M., Juarez, N., . . . Kenny, J. M. (2017). Multifunctional antimicrobial nanocomposites for food packaging applications. In A. M. Grumezescu (Ed.), Food preservation- Nanotechnology in the agri food industry (pp. 265-293). Academic Press, Elsevier. doi:10.1016/B978-0-12-804303-5.00008-0

Fuke, S., & Shimizu, T. (1993). Sensory and preference aspects of umami. *Trends in Food Science & Technology*, *4*(8), 246–251. doi:10.1016/0924-2244(93)90139-2

Galanakis, C. M. (2013). Emerging technologies for the production of nutraceuticals from agricultural by-products: A viewpoint of opportunities and challenges. *Food and Bioproducts Processing*, *91*(4), 575–579. doi:10.1016/j.fbp.2013.01.004

Gamboa-Santos, J., Soria, A. C., Pérez-Mateos, M., Carrasco, J. A., Montilla, A., & Villamiel, M. (2013). Vitamin C content and sensorial properties of dehydrated carrots blanched conventionally or by ultrasound. *Food Chemistry*, *136*(2), 782–788. doi:10.1016/j.foodchem.2012.07.122 PMID:23122127

Garcia-Amezquita, L. E., Primo-Mora, A. R., Barbosa-Cánovas, G. V., & Sepulveda, D. R. (2009). Effect of nonthermal technologies on the native size distribution of fat globules in bovine cheese-making milk. *Innovative Food Science & Emerging Technologies*, *10*(4), 491–494. doi:10.1016/j.ifset.2009.03.002

Garcia-Noguera, J., Oliveira, F. I., Gallão, M. I., Weller, C. L., Rodrigues, S., & Fernandes, F. A. (2010). Ultrasound-assisted osmotic dehydration of strawberries: Effect of pretreatment time and ultrasonic frequency. *Drying Technology*, *28*(2), 294–303. doi:10.1080/07373930903530402

García-Risco, M. R., Cortés, E., Carrascosa, A. V., & López-Fandiño, R. (1998). Microbiological and chemical changes in high-pressure-treated milk during refrigerated storage. *Journal of Food Protection*, *61*(6), 735–737. doi:10.4315/0362-028X-61.6.735 PMID:9709260

García-Risco, M. R., Olano, A., Ramos, M., & López-Fandiño, R. (2000). Micelar changes induced by high pressure. influence in the proteolytic activity and organoleptic properties of milk. *Journal of Dairy Science*, *83*(10), 2184–2189. doi:10.3168/jds.S0022-0302(00)75101-0 PMID:11049057

Garvey, M., & Rowan, N. J. (2019). Pulsed UV as a potential surface sanitizer in food production processes to ensure consumer safety. *Current Opinion in Food Science*, *26*, 65–70. doi:10.1016/j.cofs.2019.03.003

Gaucheron, F., Famelart, M. H., Mariette, F., Raulot, K., Michel, F., & Le Graetf, Y. (1997). Combined effects of temperature and high-pressure treatments on physicochemical characteristics of skim milk. *Food Chemistry*, *59*(3), 439–447. doi:10.1016/S0308-8146(96)00301-9

Gawith, J. A., & Robertson, T. R. (2000). Wrapping up packaging technology. *Journal of the Home Economics Institute of Australia*, *7*(2), 6–14.

Gayathri, D., & Rashmi, B. S. (2016). Anti-cancer properties of probiotics: A natural strategy for cancer prevention. *EC Nutrition*, *5*(4), 1191–1202.

Genschow, O., Reutner, L., & Wänke, M. (2012). The color red reduces snack food and soft drink intake. *Appetite*, *58*(2), 699–702. doi:10.1016/j.appet.2011.12.023 PMID:22245725

Gescheider, G. (1997). *The classical psychophysical methods. Psychophysics: The Fundamentals* (3rd ed.). Mahwah, NJ: Lawrence Erlbaum Associates.

Gescheider, G. (2013). *Psychophysics: the fundamentals*. Psychology Press. doi:10.4324/9780203774458

Gill, H. S., & Rutherfurd, K. J. (2001). Viability and dose–response studies on the effects of the immunoenhancing lactic acid bacterium Lactobacillus rhamnosus in mice. *British Journal of Nutrition*, *86*(2), 285–289. doi:10.1079/BJN2001402 PMID:11502243

Gilliland, S. E. (1990). Health and nutritional benefits from lactic acid bacteria. *FEMS Microbiology Letters*, *87*(1-2), 175–188. doi:10.1111/j.1574-6968.1990.tb04887.x PMID:2271223

Godoy, M. G., Gutarra, M. L. E., Castro, A. M., Machado, O. L. T., & Freire, D. M. G. (2011). Adding value to a toxic residue from the biodiesel industry: Production of two distinct pool of lipases from *Penicillium simplicissimum* in castor bean waste. *Journal of Industrial Microbiology & Biotechnology*, *38*(8), 945–953. doi:10.100710295-010-0865-8 PMID:20844923

Gök, V. (2007). Gıda paketleme sanayinde akıllı paketleme teknolojisi. *Gıda Teknolojileri Elektronik Dergisi, 2007*(1), 45-58.

Gold, K., Slay, B., Knackstedt, M., & Gaharwar, A. K. (2018). Antimicrobial activity of metal and metal-oxide based nanoparticles. *Advances in Therapy*, *1*(1700033), 1–15. doi:10.1002/adtp.201700033

Gomes, A. M., & Malcata, F. (1999). Bifidobacterium spp. and Lactobacillus acidophilus: Biological, biochemical, technological, and therapeutical properties relevant for use as probiotics. *Trends in Food Science & Technology*, *10*(4-5), 139–157. doi:10.1016/S0924-2244(99)00033-3

Gomes, C., Elena Castell-Perez, M., Chimbombi, E., Barros, F., Sun, S., Liu, J. D., ... Wright, A. O. (2009). Effect of oxygen-absorbing packaging on the shelf life of a liquid-based component of military operational rations. *Journal of Food Science*, *74*(4), 167–176. doi:10.1111/j.1750-3841.2009.01120.x PMID:19490321

Gomez-Estaca, J., Lopez-de-Dicastillo, C., Hernandez-Munoz, P., Catala, R., & Gavara, R. (2014). Advances in antioxidant active food packaging. *Trends in Food Science & Technology*, *35*(1), 42–51. doi:10.1016/j.tifs.2013.10.008

Göncü, A., & Özkal, S. G. (2017). Ekmeklerde aktif paketleme uygulamaları. *Türk Tarım - Gıda Bilim ve Teknoloji Dergisi, 5*(11), 1264-1273.

Granda-Restrepo, D. M., Soto-Valdez, H., Peralta, E., Troncoso-Rojas, R., Vallejo-Córdoba, B., Gámez-Meza, N., & Graciano-Verdugo, A. Z. (2009). Migration of α-tocopherol from an active multilayer film into whole milk powder. *Food Research International*, *42*(10), 1396–1402. doi:10.1016/j.foodres.2009.07.007

Green, E. R., & Mecsas, J. (2016). Bacterial secretion systems–an overview. *Microbiology Spectrum*, *4*(1). doi:10.1128/microbiolspec.VMBF-0012-2015 PMID:26999395

Grover, S., Kathuria, R. S., & Kaur, M. (2012). Energy values and technologies for non woody biomass: As a clean source of energy. *Indian Journal of Science and Technology*, *1*, 10–14.

Gruere, G. P., Narrod, C. A., & Abbott, L. (2011). Agriculture, food, and water nanotechnologies for the poor: Opportunities and constraints, Policy briefs 19, International Food Policy Research Institute (IFPRI). Retrieved from http://cdm15738.contentdm.oclc.org/utils/getfile/collection/p15738coll2/id/124891/filename/124892.pdf

Gueimonde, M., Flórez, A. B., van Hoek, A. H., Stuer-Lauridsen, B., Strøman, P., Clara, G., & Margolles, A. (2010). Genetic basis of tetracycline resistance in Bifidobacterium animalis subsp. lactis. *Applied and Environmental Microbiology*, *76*(10), 3364–3369. doi:10.1128/AEM.03096-09 PMID:20348299

Guerrero-Beltrán, J. Á., Sepulveda, D. R., Góngora-Nieto, M. M., Swanson, B., & Barbosa-Cánovas, G. V. (2010). Milk thermization by pulsed electric fields (PEF) and electrically induced heat. *Journal of Food Engineering*, *100*(1), 56–60. doi:10.1016/j.jfoodeng.2010.03.027

Guinard, J.-X., Pangborn, R. M., & Lewis, M. J. (1986). The time-course of astringency in wine upon repeated ingestion. *American Journal of Enology and Viticulture*, *37*(3), 184–189.

Gunther, N. W. IV, Abdul-Wakeel, A., Scullen, O. J., & Sommers, C. (2019). The evaluation of gamma irradiation and cold storage for the reduction of Campylobacter jejuni in chicken livers. *Food Microbiology*, *82*, 249–253. doi:10.1016/j.fm.2019.02.014 PMID:31027780

Guo, Z., Liu, X. M., Zhang, Q. X., Shen, Z., Tian, F. W., Zhang, H., ... Chen, W. (2011). Influence of consumption of probiotics on the plasma lipid profile: A meta-analysis of randomised controlled trials. *Nutrition, Metabolism, and Cardiovascular Diseases*, *21*(11), 844–850. doi:10.1016/j.numecd.2011.04.008 PMID:21930366

Gupta, C., Prakash, D., & Gupta, S. (2014). Natural useful therapeutic products from microbes. *Journal of Microbiology & Experimentation*, *1*(1), 30-37.

Gurol, C., Ekinci, F. Y., Aslan, N., & Korachi, M. (2012). Low temperature plasma for decontamination of E. coli in milk. *International Journal of Food Microbiology*, *157*(1), 1–5. doi:10.1016/j.ijfoodmicro.2012.02.016 PMID:22622128

Gutiérrez, L., Batlle, R., Andújar, S., Sánchez, C., & Nerín, C. (2011). Evaluation of antimicrobial active packaging to ıncrease shelf life of gluten-free sliced bread. *Packaging Technology & Science*, *24*(8), 485–494. doi:10.1002/pts.956

Hamad, A. F., Han, J. H., Kim, B. C., & Rather, I. A. (2017). The intertwine of nanotechnology with the food industry. *Saudi Journal of Biological Sciences*, *25*(1), 27–30. doi:10.1016/j.sjbs.2017.09.004 PMID:29379352

Han, Z., Cai, M., Cheng, J., & Sun, D. (2018). Effects of electric fields and electromagnetic wave on food protein structure and functionality: A review. *Trends in Food Science & Technology*, *75*(March), 1–9. doi:10.1016/j.tifs.2018.02.017

Happel, A. U., Barnabas, S. L., Froissart, R., & Passmore, J. S. (2018). Weighing in on the risks and benefits of probiotic use in HIV-infected and immunocompromised populations. *Beneficial Microbes*, *9*(2), 239–246. doi:10.3920/BM2017.0106 PMID:29345159

Harrison, S. K., & Bernhard, R. A. (1984). Time-intensity sensory characteristics of saccharin, xylitol and galactose and their effect on the sweetness of lactose. *Journal of Food Science*, *49*(3), 780–786. doi:10.1111/j.1365-2621.1984.tb13210.x

Harte, F., Luedecke, L., Swanson, B., & Barbosa-Cánovas, G. V. (2010). Low-fat set yogurt made from milk subjected to combinations of high hydrostatic pressure and thermal processing. *Journal of Dairy Science, 86*(4), 1074–1082. doi:10.3168/jds.S0022-0302(03)73690-X PMID:12741531

Härtel, C., Pagel, J., Rupp, J., Bendiks, M., Guthmann, F., Rieger-Fackeldey, E., ... Göpel, W. (2014). Prophylactic use of Lactobacillus acidophilus/Bifidobacterium infantis probiotics and outcome in very low birth weight infants. *The Journal of Pediatrics, 165*(2), 285–289.e1. doi:10.1016/j.jpeds.2014.04.029 PMID:24880888

Hashim, I. B., Resurreccion, A. V. A., & McWalters, K. H. (1995). Descriptive sensory analysis of irradiated frozen or refrigerated chicken. *Journal of Food Science, 60*(4), 664–666. doi:10.1111/j.1365-2621.1995.tb06202.x

Hecht, S. (1924). The visual discrimination of intensity and the Weber-Fechner law. *The Journal of General Physiology, 7*(2), 235–267. doi:10.1085/jgp.7.2.235 PMID:19872133

Heising, J. K., Dekker, M., Bartels, P. V., & Van Boekel, M. A. J. S. (2014). Monitoring the quality of perishable foods: Opportunities for ıntelligent packaging. *Critical Reviews in Food Science and Nutrition, 54*(5), 645–654. doi:10.1080/10408398.2011.600477 PMID:24261537

Heller, K. J. (2001). Probiotic bacteria in fermented foods: Product characteristics and starter organisms. *The American Journal of Clinical Nutrition, 73*(2), 374s–379s. doi:10.1093/ajcn/73.2.374s PMID:11157344

Hemeg, H. A. (2017). Nanomaterials for alternative antibacterial therapy. *International Journal of Nanomedicine, 12*, 8211–8225. doi:10.2147/IJN.S132163 PMID:29184409

Hempel, A. W., O'Sullivan, M. G., Papkovsky, D. B., & Kerry, J. P. (2013). Use of smart packaging technologies for monitoring and extending the shelf-life quality of modified atmosphere packaged (MAP) bread: Application of intelligent oxygen sensors and active ethanol emitters. *European Food Research and Technology, 237*(2), 117–124. doi:10.100700217-013-1968-z

Hesseltine, C. W., & Wang, H. L. (1967). Traditional fermented foods. *Biotechnology and Bioengineering, 9*(3), 275–288. doi:10.1002/bit.260090302

Heymann, H. (2019). A personal history of sensory science. *Food, Culture, & Society*, 1–21.

Hite, B. H. (1899). The effects of pressure in the preservation of milk. *Bulletin of the West Virginia University Agricultural Experimental Station Morgantown*, (58), 15–35.

Hoa, N. T., Baccigalupi, L., Huxham, A., Smertenko, A., Van, P. H., Ammendola, S., ... Cutting, S. M. (2000). Characterization of Bacillus species used for oral bacteriotherapy and bacterioprophylaxis of gastrointestinal disorders. *Applied and Environmental Microbiology, 66*(12), 5241–5247. doi:10.1128/AEM.66.12.5241-5247.2000 PMID:11097897

Hoel, S., Vadstein, O., & Jakobsen, A. N. (2018). Growth of mesophilic Aeromonas salmonicida in an experimental model of nigiri sushi during cold storage. *International Journal of Food Microbiology, 285*, 1–6. doi:10.1016/j.ijfoodmicro.2018.07.008 PMID:30005315

Hogan, S. A., & Kerry, J. P. (2008). Smart packaging of meat and poultry products. In J. Kerry & P. Butler (Eds.), *Smart packaging technologies for fast moving consumer goods* (pp. 33–59). West Sussex: John Wiley & Sons. doi:10.1002/9780470753699.ch3

Holker, U., Hofer, M., & Lenz, J. (2004). Biotechnological advantages of laboratory-scale solid-state fermentation with fungi. *Applied Microbiology and Biotechnology*, *64*, 175–186. doi:10.100700253-003-1504-3 PMID:14963614

Holst, J. J. (2007). The physiology of glucagon-like peptide 1. *Physiological Reviews*, *87*(4), 1409–1439. doi:10.1152/physrev.00034.2006 PMID:17928588

Holzapfel, W. (1997). Use of starter cultures in fermentation on a household scale. *Food Control*, *8*(5-6), 241–258. doi:10.1016/S0956-7135(97)00017-0

Horuz, E., Bozkurt, H., Karataş, H., & Maskan, M. (2018). Simultaneous application of microwave energy and hot air to whole drying process of apple slices: Drying kinetics, modeling, temperature profile and energy aspect. *Heat and Mass Transfer*, *54*(2), 425–436. doi:10.100700231-017-2152-y

Horuz, E., Jaafar, H. J., & Maskan, M. (2017). Ultrasonication as pretreatment for drying of tomato slices in a hot air–microwave hybrid oven. *Drying Technology*, *35*(7), 849–859. doi:10.1080/07373937.2016.1222538

Hoseinnejad, M., Jafari, S. M., & Katouzian, I. (2017). Inorganic and metal nanoparticles and their antimicrobial activity in food packaging applications. *Critical Reviews in Microbiology*, *44*(2), 161–181. doi:10.1080/1040841X.2017.1332001 PMID:28578640

Hoseinzadeh, E., Makhdoumi, P., Taha, P., Hossini, H., Stelling, J., Kamal, M. A., & Ashraf, G. M. (2017). A review on nano-antimicrobials: Metal nanoparticles, methods, and mechanisms. *Current Drug Metabolism*, *18*(2), 120–128. doi:10.2174/1389200217666161201111146 PMID:27908256

Hossain, A. B., & Fazliny, A. R. (2010). Creation of alternative energy by bio-ethanol production from pineapple waste and the usage of its properties for engine. *African Journal of Microbiological Research*, *4*, 813–819.

Hosseini, H., Shojaee-Aliabadi, S., Hosseini, S. M., & Mirmoghtadaie, L. (2017). Nanoantimicrobials in Food Industry. In A. E. Oprea, & A. M. Grumezescu (Eds.), *Nanotechnology applications in food-flavour, stability, nutrition, and safety* (pp. 223–238). London, UK: Academic Press; doi:10.1016/B978-0-12-811942-6.00011-X

Hotchkiss, J. H., & Chen, J. H. (1996). Microbiological effects of the direct addition of CO_2 to pasteurized milk. *Journal of Dairy Science*, *79*(Supplement 1), 87. PMID:8675787

Howkins, S. D. (1969). Diffusion rates and the effect of ultrasound. *Ultrasonics*, *7*(2), 129–130. doi:10.1016/0041-624X(69)90201-7

http://agritech.tnau.ac.in/agricultural_marketing/agrimark_cold%20storage.html(browsed on 20-6-2019)

https://www.winnesota.com/news/coldstorage

https://www.winnesota.com/news/coldstorage(browsed on 20-6-2019)

Huang, H. W., Wu, S. J., Lu, J. K., Shyu, Y. T., & Wang, C. Y. (2017). Current status and future trends of high-pressure processing in food industry. *Food Control*, *72*(12), 1–8. doi:10.1016/j.foodcont.2016.07.019

Huff, K. (2008). *Active and ıntelligent packaging: Innovations for the future*. Retrieved from http://www.iopp.org/files/public/VirginiaTechKarleighHuff.pdf

Humphrey, S. P., & Williamson, R. T. (2001). A review of saliva: Normal composition, flow, and function. *The Journal of Prosthetic Dentistry*, *85*(2), 162–169. doi:10.1067/mpr.2001.113778 PMID:11208206

Huppertz, T., Fox, P. F., & Kelly, A. L. (2003). High pressure-induced changes in the creaming properties of bovine milk, *4*(July), 349–359. doi:10.1016/S1466-8564

Hurme, E., Sipilainen-Malm, T., & Ahvenainen, R. V. T. T. (2002). Active and intelligent packaging. In T. Ohlsson, & N. Bengtsson (Eds.), Minimal processing technologies in the food industry (pp. 87-123). Cambridge: CRC Press.

Hust, R. M., Betts, G. D., & Earnshaw, R. G (1995). The antimicrobial effect of power ultrasound. R&D Report, 4.

Imperial, I. C., & Ibana, J. A. (2016). Addressing the antibiotic resistance problem with probiotics: Reducing the risk of its double-edged sword effect. *Frontiers in Microbiology*, *7*, 1983. doi:10.3389/fmicb.2016.01983 PMID:28018315

Innocente, N., Segat, A., Manzocco, L., Marino, M., Maifreni, M., Bortolomeoli, I., ... Nicoli, M. C. (2014). Effect of pulsed light on total microbial count and alkaline phosphatase activity of raw milk. *International Dairy Journal*, *39*(1), 108–112. doi:10.1016/j.idairyj.2014.05.009

Irkin, R., & Esmer, O. K. (2015). Novel food packaging systems with natural antimicrobial agents. *Journal of Food Science and Technology*, *52*(10), 6095–6111. doi:10.100713197-015-1780-9 PMID:26396358

Izli, N., Yıldız, G., Ünal, H., Işık, E., & Uylaşer, V. (2014). Effect of different drying methods on drying characteristics, colour, total phenolic content and antioxidant capacity of Goldenberry (P hysalis peruviana L.). *International Journal of Food Science & Technology*, *49*(1), 9–17. doi:10.1111/ijfs.12266

Jaeger, H., Meneses, N., & Knorr, D. (2014). Pulsed electric field technology. In Y. Motarjemi, G. Moy, & E. Todd (Eds.), *Encyclopedia of food safety* (Vol. 3, pp. 239–244). Elsevier. doi:10.1016/B978-0-12-378612-8.00260-2

Jager, G., Schlich, P., Tijssen, I., Yao, J., Visalli, M., De Graaf, C., & Stieger, M. (2014). Temporal dominance of emotions: Measuring dynamics of food-related emotions during consumption. *Food Quality and Preference*, *37*, 87–99. doi:10.1016/j.foodqual.2014.04.010

Jamal, P., Akbar, I., Yumi, Z., & Irwandi, J. (2016). Process development for maximum lycopene production from selected fruit waste and its antioxidant and antiradical activity. *Journal of Food Processing & Technology*, *7*(4), 576–581. doi:10.4172/2157-7110.1000576

James, S., & James, C. (2010). The food cold-chain and climate change. *Food Research International*, *43*(7), 1944–1956. doi:10.1016/j.foodres.2010.02.001

Jeong, D. W., Heo, S., Lee, B., Lee, H., Jeong, K., Her, J. Y., ... Lee, J. (2017). Effects of the predominant bacteria from meju and doenjang on the production of volatile compounds during soybean fermentation. *International Journal of Food Microbiology*, *19*(262), 8–13. doi:10.1016/j.ijfoodmicro.2017.09.011 PMID:28950164

Jiang, J.-F., Li, S.-F., & Liu, Z.-H. (2018). Study on heat transfer and cold storage characteristics of a falling film type of cold energy regenerator with PCM. *Applied Thermal Engineering*, *143*, 676–687. doi:10.1016/j.applthermaleng.2018.07.127

Ji, Y., Kim, H., Park, H., Lee, J., Yeo, S., Yang, J., ... Bomba, A. (2012). Modulation of the murine microbiome with a concomitant anti-obesity effect by Lactobacillus rhamnosus GG and Lactobacillus sakei NR28. *Beneficial Microbes*, *3*(1), 13–22. doi:10.3920/BM2011.0046 PMID:22348905

Jones, M. L., Martoni, C. J., Parent, M., & Prakash, S. (2012). Cholesterol-lowering efficacy of a microencapsulated bile salt hydrolase-active Lactobacillus reuteri NCIMB 30242 yoghurt formulation in hypercholesterolaemic adults. *British Journal of Nutrition*, *107*(10), 1505–1513. doi:10.1017/S0007114511004703 PMID:22067612

Ju, H. Y., Law, C. L., Fang, X. M., Xiao, H. W., Liu, Y. H., & Gao, Z. J. (2016). Drying kinetics and evolution of the sample's core temperature and moisture distribution of yam slices (Dioscorea alata L.) during convective hot-air drying. *Drying Technology*, *34*(11), 1297–1306. doi:10.1080/07373937.2015.1105814

Kähkönen, P., Tuorila, H., & Hyvönen, L. (1995). Dairy fat content and serving temperature as determinants of sensory and hedonic characteristics in cheese soup. *Food Quality and Preference*, *6*(2), 127–133. doi:10.1016/0950-3293(95)98555-W

Kalra, E. K. (2003). Nutraceutical--definition and introduction. *AAPS PharmSci*, *5*(3), E25. doi:10.1208/ps050325 PMID:14621960

Kanda, H., Tateya, S., Tamori, Y., Kotani, K., Hiasa, K. I., Kitazawa, R., ... Kasuga, M. (2006). MCP-1 contributes to macrophage infiltration into adipose tissue, insulin resistance, and hepatic steatosis in obesity. *The Journal of Clinical Investigation*, *116*(6), 1494–1505. doi:10.1172/JCI26498 PMID:16691291

Kang, S. J., Seo, J. Y., Cho, K. M., Lee, C. K., Kim, J. H., & Kim, J. S. (2016). Antioxidant and neuroprotective effects of Doenjang Prepared with Rhizopus, Pichia, and Bacillus. *Preventive Nutrition and Food Science*, *21*(3), 221–226. doi:10.3746/pnf.2016.21.3.221 PMID:27752498

Kapilan, R. (2015). Solid state fermentation for microbial products: A review. *Archives of Applied Science Research*, *7*(8), 21–25.

Karagöz, Ş., & Demirdöven, A. (2017). Gıda ambalajlamada güncel uygulamalar: Modifiye atmosfer, aktif, akıllı ve nanoteknolojik ambalajlama uygulamaları. *Gaziosmanpaşa Bilimsel Araştırma Dergisi*, *6*(1), 9–21.

Kartal, S. (2010). *Çileğin raf ömrünün mikroperfore filmler ve oksijen tutucular kullanılarak denge modifiye atmosfer ile arttırılması.* (Unpublished master's thesis). Çanakkale Onsekiz Mart University, Çanakkale, Turkey.

Kasahara, I., Carrasco, V., & Aguilar, L. (2015). Inactivation of Escherichia coli in goat milk using pulsed ultraviolet light. *Journal of Food Engineering*, *152*, 43–49. doi:10.1016/j.jfoodeng.2014.11.012

Kastelein, J. J., Akdim, F., Stroes, E. S., Zwinderman, A. H., Bots, M. L., Stalenhoef, A. F., ... Groot, E. D. (2008). Simvastatin with or without Ezetimibe in Familial Hypercholesterolemia. *The New England Journal of Medicine*, *358*(14), 1431–1443. doi:10.1056/NEJMoa0800742 PMID:18376000

Keast, R., Dalton, P., & Breslin, P. A. S. (2004). *Flavor interactions at the sensory level. Flavor Perception*. Wiley Online Library.

Kek, S. P., Chin, N. L., & Yusof, Y. A. (2013). Direct and indirect power ultrasound assisted pre-osmotic treatments in convective drying of guava slices. *Food and Bioproducts Processing*, *91*(4), 495–506. doi:10.1016/j.fbp.2013.05.003

Kelen, A., Ress, S., Nagy, T., Pallai, E., & Pintye-Hodi, K. (2006). Mapping of temperature distribution in pharmaceutical microwave vacuum drying. *Powder Technology*, *162*(2), 133–137. doi:10.1016/j.powtec.2005.12.001

Kerry, R. G., Patra, J. K., Gouda, S., Park, Y., Shin, H., & Das, G. (2018). Benefaction of probiotics for human health: A review. *Journal of Food and Drug Analysis*, *26*(3), 927–939. doi:10.1016/j.jfda.2018.01.002 PMID:29976412

Khan, M. K., Ahmad, K., Hassan, S., Imran, M., Ahmad, N., & Xu, C. (2018). Effect of novel technologies on polyphenols during food processing. *Innovative Food Science and Emerging Technologies*, *45*(December 2017), 361–381. doi:10.1016/j.ifset.2017.12.006

Khanal, S. K., Grewell, D., Sung, S., & Van Leeuwen, J. (2007). Ultrasound applications in wastewater sludge pretreatment: A review. *Critical Reviews in Environmental Science and Technology*, *37*(4), 277–313. doi:10.1080/10643380600860249

Khani, S., Motamedifar, M., Golmoghaddam, H., Hosseini, H. M., & Hashemizadeh, Z. (2012). In vitro study of the effect of a probiotic bacterium Lactobacillus rhamnosus against herpes simplex virus type 1. *The Brazilian Journal of Infectious Diseases*, *16*(2), 129–135. doi:10.1016/S1413-8670(12)70293-3 PMID:22552453

Kilcast, D. (1999). Sensory techniques to study food texture. In A. J. Rosenthal (Ed.), *Food texture: Measurement and perception* (pp. 30–60). Gaithersburg, MD: Aspen Publishers.

Kılınç, B., & Çaklı, Ş. (2004). Su ürünlerinin modifiye atmosferde paketlenmesi. *E.U. Su Ürünleri Dergisi*, *21*(3-4), 349–353.

Kim, H. J., Yong, H. I., Park, S., Kim, K., Choe, W., & Jo, C. (2015). Microbial safety and quality attributes of milk following treatment with atmospheric pressure encapsulated dielectric barrier discharge plasma. *Food Control*, *47*, 451–456. doi:10.1016/j.foodcont.2014.07.053

Kim, S. H., Huh, C. S., Choi, I. D., Jeong, J. W., Ku, H. K., Ra, J. H., ... Ahn, Y. T. (2014). The anti-diabetic activity of Bifidobacterium lactis HY 8101 in vitro and in vivo. *Journal of Applied Microbiology*, *117*(3), 834–845. doi:10.1111/jam.12573 PMID:24925305

Kim, T., Lee, J., Kim, S., Park, M., Chang, H. C., & Kim, H. (2009). Analysis of microbial communities in doenjang, a Korean fermented soybean paste, using nested PCR-denaturing gradient gel electrophoresis. *International Journal of Food Microbiology*, *131*(2-3), 265–271. doi:10.1016/j.ijfoodmicro.2009.03.001 PMID:19324443

Klaic, R., Sallet, D., Foletto, E. L., Jacques, R. J. S., Guedes, J. V. C., Kuhn, R. C., & Mazutti, M. A. (2017). Optimization of solid-state fermentation for bioherbicide production by *Phoma sp. Brazilian Journal of Chemical Engineering*, *34*(2), 377–384. doi:10.1590/0104-6632.20170342s20150613

Knorr, D., Zenker, M., Heinz, V., & Lee, D.-U. (2004). Applications and potential of ultrasonics in food processing. *Trends in Food Science & Technology*, *15*(5), 261–266. doi:10.1016/j.tifs.2003.12.001

Knudsen, J. C., & Skibsted, L. H. (2010). High pressure effects on the structure of casein micelles in milk as studied by cryo-transmission electron microscopy. *Food Chemistry*, *119*(1), 202–208. doi:10.1016/j.foodchem.2009.06.017

Korachi, M., Ozen, F., Aslan, N., Vannini, L., Guerzoni, M. E., Gottardi, D., & Ekinci, F. Y. (2015). Biochemical changes to milk following treatment by a novel, cold atmospheric plasma system. *International Dairy Journal*, *42*, 64–69. doi:10.1016/j.idairyj.2014.10.006

Koutchma, T. (2014). *Adapting high hydrostatic pressure (HPP) for food processing operations. Adapting high hydrostatic pressure (HPP) for food processing operations*. doi:10.1016/C2013-0-12997-2

Koyani, R. D., & Rajput, K. S. (2015). Solid state fermentation: comprehensive tool for utilization of lignocellulosic through biotechnology. *Journal of Bioprocessing & Biotechniques*, *5*(10), 258. doi:10.4172/2155-9821.1000258

Kozak, Y., Farid, M., & Ziskind, G. (2017). Experimental and comprehensive theoretical study of cold storage packages containing PCM. *Applied Thermal Engineering*, *115*, 899–912. doi:10.1016/j.applthermaleng.2016.12.127

Kramer, B., Wunderlich, J., & Muranyi, P. (2015). Pulsed light decontamination of endive salad and mung bean sprouts and impact on color and respiration activity. *Journal of Food Protection*, *78*(2), 340–348. doi:10.4315/0362-028X.JFP-14-262 PMID:25710149

Krishnamurthy, K., Demirci, A., & Irudayaraj, J. M. (2007). Inactivation of Staphylococcus aureus in milk using flow-through pulsed UV-light treatment system. *Journal of Food Science*, *72*(7), M233–M239. doi:10.1111/j.1750-3841.2007.00438.x PMID:17995646

Krop, E. M., Hetherington, M. M., Nekitsing, C., Miquel, S., Postelnicu, L., & Sarkar, A. (2018). Influence of oral processing on appetite and food intake–A systematic review and meta-analysis. *Appetite*, *125*, 253–269. doi:10.1016/j.appet.2018.01.018 PMID:29408331

Kuddus, M., Singh, P., Thomas, G., & Al-Hazimi, A. (2013). Recent developments in production and biotechnological applications of C-Phycocyanin. *BioMed Research International*, *2013*, 1–9. doi:10.1155/2013/742859 PMID:24063013

Kumar, K., Kumar, S., Kumar, S., Kumar, V., & Vyas, P. (2016, October). *Role of nanotechnology in food processing and preservation.* Paper presented at National Conference on Technologies in Sustainable Food Systems, Longowal, Sangrur, Punjab, India.

Kumar, R. S., Varman, D. R., Kanmani, P., Yuvaraj, N., Paari, K. A., Pattukumar, V., & Arul, V. (2010). Isolation, characterization, and identification of a potential probiont from South Indian fermented foods (Kallappam, Koozh, and Mor Kuzhambu) and its use as biopreservative. *Probiotics and Antimicrobial Proteins*, *2*(3), 145–151. doi:10.100712602-010-9052-5 PMID:26781237

Kümmel, J., Stessl, B., Gonano, M., Walcher, G., Bereuter, O., Fricker, M., ... Ehling-Schulz, M. (2016). Staphylococcus aureus entrance into the dairy chain: Tracking S. aureus from dairy cow to cheese. *Frontiers in Microbiology*, *7*(OCT), 1–11. doi:10.3389/fmicb.2016.01603 PMID:27790200

Ku, S., Park, M., Ji, G., & You, H. (2016). Review on bifidobacterium bifidum bgn4: Functionality and nutraceutical applications as a probiotic microorganism. *International Journal of Molecular Sciences*, *17*(9), 1544. doi:10.3390/ijms17091544 PMID:27649150

Labuza, T. P., & Breene, W. M. (1989). Applications of "active packaging" for improvement of shelf-life and nutritional quality of fresh and extended shelf-life foods. *Journal of Food Processing and Preservation*, *13*(1), 1–69. doi:10.1111/j.1745-4549.1989.tb00090.x

Lam, V., Su, J., Koprowski, S., Hsu, A., Tweddell, J. S., Rafiee, P., ... Baker, J. E. (2012). Intestinal microbiota determine severity of myocardial infarction in rats. *The FASEB Journal*, *26*(4), 1727–1735. doi:10.1096/fj.11-197921 PMID:22247331

Laroussi, M. (2005). Low temperature plasma-based sterilization: Overview and state-of-the-art. *Plasma Processes and Polymers*, *2*(5), 391–400. doi:10.1002/ppap.200400078

Larsen, N., Vogensen, F. K., van den Berg, F. W., Nielsen, D. S., Andreasen, A. S., Pedersen, B. K., ... Jakobsen, M. (2010). Gut microbiota in human adults with type 2 diabetes differs from non-diabetic adults. *PLoS One*, *5*(2), e9085. doi:10.1371/journal.pone.0009085 PMID:20140211

Larson-Powers, N., & Pangborn, R. M. (1978). Paired comparison and time-intensity measurements of the sensory properties of beverages and gelatins containing sucrose or synthetic sweeteners. *Journal of Food Science, 43*(1), 41–46. doi:10.1111/j.1365-2621.1978.tb09732.x

Lassen, J., & Yazdankhah, S. (2015). Assessment of probiotics in infant formula and cereal based baby foods containing Bifidobacterium lactis Bb12– Update 2014. *European Journal of Nutrition & Food Safety, 5*(2), 101–103. doi:10.9734/EJNFS/2015/14818

Lawless, H. T., & Heymann, H. (2013). *Sensory evaluation of food: principles and practices.* Springer Science & Business Media.

Le Révérend, B., & Hartmann, C. (2014). Numerical modeling of human mastication, a simplistic view to design foods adapted to mastication abilities. *Physiology & Behavior, 124*, 61–64. doi:10.1016/j.physbeh.2013.10.012 PMID:24471180

Lee, B., Kim, J., Kang, Y. M., Lim, J., Kim, Y., Lee, M., ... Je, J. (2010). Antioxidant activity and γ-aminobutyric acid (GABA) content in sea tangle fermented by Lactobacillus brevis BJ20 isolated from traditional fermented foods. *Food Chemistry, 122*(1), 271–276. doi:10.1016/j.foodchem.2010.02.071

Lee, L., Arul, J., Lencki, R., & Castaigne, F. (1996). A review on modified atmosphere packaging and preservation of fresh fruits and vegetables: Physiological basis and practical aspects-Part II. *Packaging Technology & Science, 9*(1), 1–17. doi:10.1002/(SICI)1099-1522(199601)9:1<1::AID-PTS349>3.0.CO;2-W

Lee, S. J., Bose, S., Seo, J. G., Chung, W. S., Lim, C. Y., & Kim, H. (2014). The effects of co-administration of probiotics with herbal medicine on obesity, metabolic endotoxemia and dysbiosis: A randomized double-blind controlled clinical trial. *Clinical Nutrition (Edinburgh, Lothian), 33*(6), 973–981. doi:10.1016/j.clnu.2013.12.006 PMID:24411490

Lee, S. J., & Rahman, A. T. M. M. (2014). Intelligent packaging for food products. In J. H. Han (Ed.), *Innovations in food packaging* (pp. 171–209). Academic Press. doi:10.1016/B978-0-12-394601-0.00008-4

Lee, W., Pathanibul, P., Quarterman, J., Jo, J., Han, N., Miller, M. J., ... Seo, J. (2012). Whole cell biosynthesis of a functional oligosaccharide, 2′-fucosyllactose, using engineered Escherichia coli. *Microbial Cell Factories, 11*(1), 48. doi:10.1186/1475-2859-11-48 PMID:22545760

Leighton, T. G. (1998). The principles of cavitation. *Ultrasound in food processing, 12.*

Le, T. K., Hosaka, T., Nguyen, T. T., Kassu, A., Dang, T. O., Tran, H. B., ... Pham, X. D. (2015). Bifidobacterium species lower serum glucose, increase expressions of insulin signaling proteins, and improve adipokine profile in diabetic mice. *Biomedical Research, 36*(1), 63–70. doi:10.2220/biomedres.36.63 PMID:25749152

Li, C., Ding, Q., Nie, S.-P., Zhang, Y.-S., Xiong, T., & Xie, M.-Y. (2014). Carrot juice fermented with *Lactobacillus plantarum* NCU116 ameliorates type 2 diabetes in rats. *Journal of Agricultural and Food Chemistry, 62*(49), 11884–11891. doi:10.1021/jf503681r PMID:25341087

Li, G., Hwang, Y., Radermacher, R., & Chun, H.-H. (2013). Review of cold storage materials for subzero applications. *Energy, 51*, 1–17. doi:10.1016/j.energy.2012.12.002

Liu, R. H. (2004). Potential synergy of phytochemicals in cancer prevention: Mechanism of action. *The Journal of Nutrition, 134*(12), 3479S–3485S. doi:10.1093/jn/134.12.3479S PMID:15570057

Liu, Z., Hemar, Y., Tan, S., Sanguansri, P., Niere, J., Buckow, R., & Augustin, M. A. (2015). Pulsed electric field treatment of reconstituted skim milks at alkaline pH or with added EDTA. *Journal of Food Engineering, 144*, 112–118. doi:10.1016/j.jfoodeng.2014.06.033

Liu, Z., Juliano, P., Williams, R. P., Niere, J., & Augustin, M. A. (2014). Ultrasound effects on the assembly of casein micelles in reconstituted skim milk. *The Journal of Dairy Research, 81*(2), 146–155. doi:10.1017/S0022029913000721 PMID:24351847

Li, X., & Farid, M. (2016). A review on recent development in non-conventional food sterilization technologies. *Journal of Food Engineering, 182*, 33–45. doi:10.1016/j.jfoodeng.2016.02.026

Ludwig, D. S. (1999). Dietary fiber, weight gain, and cardiovascular disease risk factors in young adults. *Journal of the American Medical Association, 282*(16), 1539. doi:10.1001/jama.282.16.1539 PMID:10546693

Lu, W., Liu, G., Xing, X., & Wang, H. (2019). Investigation on ternary salt-water solutions as phase change materials for cold storage. *Energy Procedia, 158*, 5020–5025. doi:10.1016/j.egypro.2019.01.662

Lv, Y., Tahir, I. I., & Olsson, M. E. (2019). Effect of ozone application on bioactive compounds of apple fruit during short-term cold storage. *Scientia Horticulturae, 253*, 49–60. doi:10.1016/j.scienta.2019.04.021

Magiorakos, A., Srinivasan, A., Carey, R., Carmeli, Y., Falagas, M., Giske, C., ... Monnet, D. (2012). Multidrug-resistant, extensively drug-resistant and pan drug-resistant bacteria: An international expert proposal for interim standard definitions for acquired resistance. *Clinical Microbiology and Infection, 18*(3), 268–281. doi:10.1111/j.1469-0691.2011.03570.x PMID:21793988

Mahendran, R., Ramanan, K. R., Barba, F. J., Lorenzo, J. M., López-Fernández, O., Munekata, P. E. S., ... Tiwari, B. K. (2018, December). Recent advances in the application of pulsed light processing for improving food safety and increasing shelf life. *Trends in Food Science & Technology, 88*, 67–79. doi:10.1016/j.tifs.2019.03.010

Majid, I., Nayik, G. A., Dar, S. M., & Nanda, V. (2018). Novel food packaging technologies: Innovations and future prospective. *Journal of the Saudi Society of Agricultural Sciences, 17*(4), 454–462. doi:10.1016/j.jssas.2016.11.003

Maldonado, M. C., & Strasser de Saad, A. M. (1998). Production of pectin estrase and polygalactouronase by *Aspergillus niger* in submerged and solid state systems. *Journal of Industrial Microbiology & Biotechnology, 20*(1), 34–38. doi:10.1038j.jim.2900470 PMID:9523455

Malpure, P. P., Shah, A. S., & Juvekar, A. R. (2006). Antioxidant and anti-inflammatory activity of extract obtained from *Aspergillus candidus* MTCC 2202 broth filtrate. *Indian Journal of Experimental Biology*, *44*, 468–473. PMID:16784117

Manas, P., Pagan, R., Raso, J., Sala, F. J., & Condon, S. (2000). Inactivation of Salmonella enteritidis, Salmonella Typhimurium and Salmonella seftenberg by ultrasonic waves under pressure, *Journal of Food Protection*, 63, 451-456.

Manas, P., & Pagan, R. (2003). A review on microbial inactivation by new technologies of food preservation. *Journal of Applied Microbiology*, *98*(6), 1387–1399. doi:10.1111/j.1365-2672.2005.02561.x PMID:15916651

Manpreet, S., Sawraj, S., Sachin, D., Pankaj, S., & Banerjee, U. C. (2005). Influence of process parameters on the production of metabolites in solid-state fermentation. *Malaysian Journal of Microbiology*, *1*(2), 1–9.

Marchesini, G., Fasolato, L., Novelli, E., Balzan, S., Contiero, B., Montemurro, F., ... Segato, S. (2015). Ultrasonic inactivation of microorganisms: A compromise between lethal capacity and sensory quality of milk. *Innovative Food Science & Emerging Technologies*, *29*, 215–221. doi:10.1016/j.ifset.2015.03.015

Martens, J. H., Barg, H., Warren, M., & Jahn, D. (2002). Microbial production of vitamin B 12. *Applied Microbiology and Biotechnology*, *58*(3), 275–285. doi:10.100700253-001-0902-7 PMID:11935176

Mason, T. J., Paniwnyk, L., & Lorimer, J. P. (1996). The uses of ultrasound in food technology. *Ultrasonics Sonochemistry*, *3*(3), S253–S260. doi:10.1016/S1350-4177(96)00034-X

Massaro, M., Scoditti, E., & Carluccio, M. A., & DeCaterina, R. (2010). Nutraceuticals and prevention of Atherosclerosis: Focus on ω-3 polyunsaturated fatty acids and mediterranean diet polyphenols. *Cardiovascular Therapeutics*, *28*(4), e13–e19. doi:10.1111/j.1755-5922.2010.00211.x PMID:20633019

Mattila-Sandholm, T., Ahvenainen, R., Hurme, E., & Jarvi-Kaariainen, T. (1995). *Finnish Patent No. FI-94802*. Leakage Indicator, VTT Biotechnology and Food Research.

Mattila-Sandholm, T., Ahvenainen, R., Hurme, E., & Jarvi-Kaariainen, T. (1995). *Finnish Patent No. FI-94802*. Leakage Indicator, VTT Biotechnology, and Food Research.

McClements, J. (1995). Advances in the application of ultrasound in food analysis and processing. *Trends in Food Science & Technology*, *6*(9), 293–299. doi:10.1016/S0924-2244(00)89139-6

McKellar, R. C., Mohareb, E., & Piyasena, P. (2003). Inactivation of microbes using ultrasound: A review. *International Journal of Food Microbiology*, *87*(3), 207–216. doi:10.1016/S0168-1605(03)00075-8 PMID:14527793

Meilgaard, M. C., Civille, G. V., & Carr, B. T. (2007). *Sensory evaluation techniques* (3rd ed.). Boca Raton, FL: Taylor & Francis.

Mela, D. J., Langley, K. R., & Martin, A. (1994). No effect of oral or sample temperature on sensory assessment of fat content. *Physiology & Behavior*, *56*(4), 655–658. doi:10.1016/0031-9384(94)90222-4 PMID:7800728

Midilli, A., Kucuk, H., & Yapar, Z. (2002). A new model for single-layer drying. *Drying Technology*, *20*(7), 1503–1513. doi:10.1081/DRT-120005864

Miraghajani, M., Dehsoukhteh, S. S., Rafie, N., Hamedani, S. G., Sabihi, S., & Ghiasvand, R. (2017). Potential mechanisms linking probiotics to diabetes: A narrative review of the literature. *Sao Paulo Medical Journal*, *135*(2), 169–178. doi:10.1590/1516-3180.2016.0311271216 PMID:28538869

Misra, N. N., Koubaa, M., Roohinejad, S., Juliano, P., Alpas, H., Inácio, R. S., ... Barba, F. J. (2017). Landmarks in the historical development of twenty first century food processing technologies. *Food Research International*, *97*(May), 318–339. doi:10.1016/j.foodres.2017.05.001 PMID:28578057

Misra, N. N., Oliver, K., & Schlüter, P. J. C. (2016). *Cold plasma in food and agriculture*. Academic Press; doi:10.1016/b978-0-12-801365-6.09991-1

Mitchell, D. A., & Berovic, N. K. M. (Eds.). (2006). Solid-state fermentation bioreactors fundamentals of design and operation. Berlin, Germany: Springer-Verlag. doi:10.1007/3-540-31286-2

Momin, J. K., Jayakumar, C., & Prajapati, J. B. (2013). Potential of nanotechnology in functional foods. *Emirates Journal of Food and Agriculture*, *25*(1), 10–19. doi:10.9755/ejfa.v25i1.9368

Moraes, A. R. F., Gouveia, L. E. R., Soares, N. F. F., Santos, M. M. S., & Gonçalves, M. P. J. C. (2007). Development and evaluation of antimicrobial film on butter conservation. *Food Science and Technology (Campinas)*, *27*, 33–36. doi:10.1590/S0101-20612007000500006

Morales-de la Peña, M., Welti-Chanes, J., & Martín-Belloso, O. (2019). Novel technologies to improve food safety and quality. *Current Opinion in Food Science*, *30*, 1–7. doi:10.1016/j.cofs.2018.10.009

Moreau, M., Orange, N., & Feuilloley, M. G. J. (2008). Non-thermal plasma technologies: New tools for bio-decontamination. *Biotechnology Advances*, *26*(6), 610–617. doi:10.1016/j.biotechadv.2008.08.001 PMID:18775485

Moreno, S. E. H. (2017). Distribución de pesticidas en sistemas de abejas, polen, cera y miel. Análisis por LC-MS/MS. Retrieved from http://repositorio.ual.es/bitstream/handle/10835/6453/16628_Memoria_TFG FINAL2.pdf?sequence=1&isAllowed=y

Mota, C. L., Luciano, C., Dias, A., Barroca, M. J., & Guiné, R. P. F. (2010). Convective drying of onion: Kinetics and nutritional evaluation. *Food and Bioproducts Processing*, *88*(2-3), 115–123. doi:10.1016/j.fbp.2009.09.004

Motevali, A., Minaei, S., Banakar, A., Ghobadian, B., & Darvishi, H. (2016). Energy analyses and drying kinetics of chamomile leaves in microwave-convective dryer. *Journal of the Saudi Society of Agricultural Sciences*, *15*(2), 179–187. doi:10.1016/j.jssas.2014.11.003

Mothibe, K. J., Zhang, M., Nsor-atindana, J., & Wang, Y. C. (2011). Use of ultrasound pretreatment in drying of fruits: Drying rates, quality attributes, and shelf life extension. *Drying Technology*, *29*(14), 1611–1621. doi:10.1080/07373937.2011.602576

Mullan, W. M. A. (2002). *Science and technology of modified atmosphere packaging*. Retrieved from https://www.dairyscience.info/index.php/packaging/117-modified-atmosphere-packaging.html

Murooka, Y., Piao, Y., Kiatpapan, P., & Yamashita, M. (2005). Production of tetrapyrrole compounds and vitamin B12 using genetically engineering of Propionibacterium freudenreichii. An overview. *Le Lait*, *85*(1-2), 9–22. doi:10.1051/lait:2004035

Murray, J. M., Delahunty, C. M., & Baxter, I. A. (2001). Descriptive sensory analysis: Past, present and future. *Food Research International*, *34*(6), 461–471. doi:10.1016/S0963-9969(01)00070-9

Murthy, T. P. K., & Manohar, B. (2014). Hot air-drying characteristics of mango ginger: Prediction of drying kinetics by mathematical modeling and artificial neural network. *Journal of Food Science and Technology*, *51*(12), 3712–3721. doi:10.100713197-013-0941-y PMID:25477637

Musoni, M., Destain, J., Thonart, P., Bahama, J. B., & Delvigne, F. (2015). Bioreactor design and implementation strategies for the cultivation of filamentous fungi and the production of fungal metabolites: From traditional methods to engineered systems. *Biotechnologie, Agronomie, Société et Environnement*, *19*(4), 430–442.

Mussa, D. M., & Ramaswamy, H. S. (1997). Ultra high pressure pasteurization of milk: Kinetics of microbial destruction and changes in Physico-chemical characteristics. *Lebensmittel-Wissenschaft + Technologie*, *30*(6), 551–557. doi:10.1006/fstl.1996.0223

Mussatto, S. I., Ballesteros, L. F., Martins, S., & Teixeira, J. A. (2012). Use of agro-industrial wastes in solid-state fermentation processes. Industrial Waste (pp. 121–141). In Tech: Rijeka, Croatia.

Nagpal, R., Kumar, A., Kumar, M., Behare, P. V., Jain, S., & Yadav, H. (2012). Probiotics, their health benefits and applications for developing healthier foods: A review. *FEMS Microbiology Letters*, *334*(1), 1–15. doi:10.1111/j.1574-6968.2012.02593.x PMID:22568660

Nair, H. B., Sung, B., Yadav, V. R., Kannappan, R., Chaturvedi, M. M., & Aggarwal, B. B. (2010). Delivery of antiinflammatory nutraceuticals by nanoparticles for the prevention and treatment of cancer. *Biochemical Pharmacology*, *80*(12), 1833–1843. doi:10.1016/j.bcp.2010.07.021 PMID:20654584

Nakai, S., & Li-chan, E. (1988). *Hydrophobic interactions in food systems authors*. Boca Raton, FL: CRC Press.

Nalwa, H. S. (2005). *Encyclopedia of Nanoscience & Nanotechnology, 1*. Valencia, CA: American Scientific Publishers.

Nasri, H., Baradaran, A., Shirzad, H., & Rafieian-Kopaei, M. (2014). New concepts in nutraceuticals as alternative for pharmaceuticals. *International Journal of Preventive Medicine*, *5*(12), 1487. PMID:25709784

Newman, D. J., & Cragg, G. M. (2007). Natural products as sources of new drugs over the last 25 years. *Journal of Natural Products*, *70*(3), 461–477. doi:10.1021/np068054v PMID:17309302

Neyraud, E., Palicki, O., Schwartz, C., Nicklaus, S., & Feron, G. (2012). Variability of human saliva composition: Possible relationships with fat perception and liking. *Archives of Oral Biology*, *57*(5), 556–566. doi:10.1016/j.archoralbio.2011.09.016 PMID:22024405

Nicholson, M. D. (1998). The role of natural antimicrobials in food/packaging biopreservation. *Journal of Plastic Film & Sheeting*, *14*(3), 234–241. doi:10.1177/875608799801400306

Niemira, B. A. (2012). Cold plasma decontamination of foods. *Annual Review of Food Science and Technology*, *3*(1), 125–142. doi:10.1146/annurev-food-022811-101132 PMID:22149075

Nigam, P. S., & Pandey, A. (2009). Solid-state fermentation technology for bioconversion of biomass and agricultural residues. In P. Singh nee' Nigam, & A. Pandey (Eds.), Biotechnology for agro-industrial residues utilisation (pp. 197-223). Springer Science+Business Media B.V.

Nishimura, M., Ohkawara, T., Tetsuka, K., Kawasaki, Y., Nakagawa, R., Satoh, H., ... Nishihira, J. (2016). Effects of yogurt containing Lactobacillus plantarum HOKKAIDO on immune function and stress markers. *Journal of Traditional and Complementary Medicine*, *6*(3), 275–280. doi:10.1016/j.jtcme.2015.07.003 PMID:27419093

Noble, A. C. (1994). Bitterness in wine. *Physiology & Behavior*, *56*(6), 1251–1255. doi:10.1016/0031-9384(94)90373-5 PMID:7878098

Nowacka, M., Wiktor, A., Śledź, M., Jurek, N., & Witrowa-Rajchert, D. (2012). Drying of ultrasound pretreated apple and its selected physical properties. *Journal of Food Engineering*, *113*(3), 427–433. doi:10.1016/j.jfoodeng.2012.06.013

Odriozola-Serrano, I., Bendicho-Porta, S., & Martín-Belloso, O. (2010). Comparative study on shelf life of whole milk processed by high-intensity pulsed electric field or heat treatment. *Journal of Dairy Science*, *89*(3), 905–911. doi:10.3168/jds.S0022-0302(06)72155-5 PMID:16507684

Ohshima, T., Tanino, T., Kameda, T., & Harashima, H. (2016). Engineering of operation condition in milk pasteurization with PEF treatment. *Food Control*, *68*, 297–302. doi:10.1016/j.foodcont.2016.03.047

Ojha, K. S., Mason, T. J., O'Donnell, C. P., Kerry, J. P., & Tiwari, B. K. (2017). Ultrasound technology for food fermentation applications. *Ultrasonics Sonochemistry*, *34*, 410–417. doi:10.1016/j.ultsonch.2016.06.001 PMID:27773263

Ojima, T., Rahman, M. M., Kumagai, Y., Nishiyama, R., Narsico, J., & Inoue, A. (2018). Polysaccharide-degrading enzymes from marine gastropods. *Marine Enzymes and Specialized Metabolism - Part B Methods in Enzymology,* 457-497.

Okubo, T., Takemura, N., Yoshida, A., & Sonoyama, K. (2013). KK/Ta mice administered Lactobacillus plantarum strain no. 14 have lower adiposity and higher insulin sensitivity. *Bioscience of Microbiota, Food and Health*, *32*(3), 93–100. doi:10.12938/bmfh.32.93 PMID:24936367

Oliveira, F. I., Gallão, M. I., Rodrigues, S., & Fernandes, F. A. N. (2011). Dehydration of Malay apple (Syzygium malaccense L.) using ultrasound as pre-treatment. *Food and Bioprocess Technology*, *4*(4), 610–615. doi:10.100711947-010-0351-3

Oliveira, M., Abadias, M., Usall, J., Torres, R., Teixido, N., & Vinas, I. (2015). Application of modified atmosphere packaging as a safety approach to fresh-cut fruits and vegetables - A review. *Trends in Food Science & Technology*, *46*(1), 13–26. doi:10.1016/j.tifs.2015.07.017

Onwude, D. I., Hashim, N., Janius, R. B., Nawi, N. M., & Abdan, K. (2016). Modeling the thin-layer drying of fruits and vegetables: A review. *Comprehensive Reviews in Food Science and Food Safety*, *15*(3), 599–618. doi:10.1111/1541-4337.12196

Opalić, M., Domitran, Z., Komes, D., Belščak, A., Horžić, D., & Karlović, D. (2009). The effect of ultrasound pre-treatment and air-drying on the quality of dried apples. *Czech Journal of Food Sciences*, *27*(4), 297. doi:10.17221/606-CJFS

Orrhage, K., Sillerström, E., Gustafsson, J. A., Nord, C. E., & Rafter, J. (1994). Binding of mutagenic heterocyclic amines by intestinal and lactic acid bacteria. *Mutation Research. Fundamental and Molecular Mechanisms of Mutagenesis*, *311*(2), 239–248. doi:10.1016/0027-5107(94)90182-1 PMID:7526189

Osborn, J. W., & Mao, J. (1993). A thin bite-force transducer with three-dimensional capabilities reveals a consistent change in bite-force direction during human jaw-muscle endurance tests. *Archives of Oral Biology*, *38*(2), 139–144. doi:10.1016/0003-9969(93)90198-U PMID:8476343

Otles, S., & Yalcin, B. (2008). Intelligent food packaging. *LogForum*, *4*(3), 1–9.

Özdemir, M., & Floros, J. D. (2004). Active food packaging technologies. *Critical Reviews in Food Science and Nutrition*, *44*(3), 185–193. doi:10.1080/10408690490441578 PMID:15239372

Ozturk, B., Uzun, S., & Karakaya, O. (2019). Combined effects of aminoethoxyvinylglycine and MAP on the fruit quality of kiwifruit during cold storage and shelf life. *Scientia Horticulturae*, *251*, 209–214. doi:10.1016/j.scienta.2019.03.034

Paéz, R., Lavari, L., Vinderola, G., Audero, G., Cuatrin, A., Zaritzky, N., & Reinheimer, J. (2012). Effect of heat treatment and spray drying on lactobacilli viability and resistance to simulated gastrointestinal digestion. *Food Research International*, *48*(2), 748–754. doi:10.1016/j.foodres.2012.06.018

Pandey, A., Soccol, C. R., Leo, J. A. R., & Nigam, P. (2001). *Solid-state fermentation in Biotechnology* (pp. 221–230). New Delhi, India: Asiatech Publishers.

Pandey, P. K., Ramaswamy, H. S., & Idziak, E. (2003). High pressure destruction kinetics of in raw milk at two temperatures. *Journal of Food Process Engineering*, *26*(3), 265–283. doi:10.1111/j.1745-4530.2003.tb00601.x

Parker, B. A., Sturm, K., MacIntosh, C. G., Feinle, C., Horowitz, M., & Chapman, I. M. (2004). Relation between food intake and visual analogue scale ratings of appetite and other sensations in healthy older and young subjects. *European Journal of Clinical Nutrition, 58*(2), 212–218. doi:10.1038j.ejcn.1601768 PMID:14749739

Park, K. Y., Jung, K. O., Rhee, S. H., & Choi, Y. H. (2003). Antimutagenic effects of doenjang (Korean fermented soypaste) and its active compounds. *Mutation Research. Fundamental and Molecular Mechanisms of Mutagenesis, 523*, 43–53. doi:10.1016/S0027-5107(02)00320-2 PMID:12628502

Patel, R., Prajapati, J. P., & Balakrishnan, S. (2015). *Recent trends in packaging of dairy and food products.* Paper presented at the meeting National seminar on Indian Dairy Industry - Opportunities and Challenges. Gujarat, India.

Patel, R. M., & Denning, P. W. (2013). Therapeutic use of prebiotics, probiotics, and postbiotics to prevent necrotizing enterocolitis: What is the current evidence? *Clinics in Perinatology, 40*(1), 11–25. doi:10.1016/j.clp.2012.12.002 PMID:23415261

Pathakoti, K., Manubolu, M., & Hwang, H. M. (2017). Nanostructures: Current uses and future applications in food science. *Journal of Food and Drug Analysis, 25*(2), 245–253. doi:10.1016/j.jfda.2017.02.004 PMID:28911665

Patra, J. K., Das, G., Paramithiotis, S., & Shin, H. S. (2016). Kimchi and other widely consumed traditional fermented foods of Korea: A review. *Frontiers in Microbiology, 7*, 1493. doi:10.3389/fmicb.2016.01493 PMID:27733844

Patterson, M. F., & Kilpatrick, D. J. (1998). The combined effect of high hydrostatic pressure and mild heat on inactivation of pathogens in milk and poultry. *Journal of Food Protection, 61*(4), 432–436. doi:10.4315/0362-028X-61.4.432 PMID:9709206

Pereira de Abreu, D. A., Cruz, J. M., & Paseiro Losada, P. (2012). Active and intelligent packaging for the food industry. *Food Reviews International, 28*(2), 146–187. doi:10.1080/87559129.2011.595022

Pereira, D. I., & Gibson, G. R. (2002). Effects of consumption of probiotics and prebiotics on serum lipid levels in humans. *Critical Reviews in Biochemistry and Molecular Biology, 37*(4), 259–281. doi:10.1080/10409230290771519 PMID:12236466

Peryam, D. R. (1954). CHAIRMAN THURSTONE We will next hear the paper by Mr. David R. Peryam. Field Testing of Armed Forces Rations. In *Food Acceptance Testing Methodology: A Symposium Sponsored by the Quartermaster Food and Container Institute for the Armed Forces, Quartermaster Research and Development Command, US Army Quartermaster Corps [at The] Palmer House, Chicago, Oct. 8-9, 1953* (Vol. 1, p. 75). National Academies.

Pili, J., Danielli, A., Nyari, N. L., Zeni, J., Cansian, R. L., Backes, G. T., & Valduga, E. (2017). Biotechnological potential of agro-industrial waste in the synthesis of pectin lyase from*Aspergillus brasiliensis. Food Science & Technology International*. doi:10.1177/1082013217733574

Pineau, N., Schlich, P., Cordelle, S., Mathonnière, C., Issanchou, S., Imbert, A., ... Köster, E. (2009). Temporal dominance of sensations: Construction of the TDS curves and comparison with time–intensity. *Food Quality and Preference*, *20*(6), 450–455. doi:10.1016/j.foodqual.2009.04.005

Piqueras-Fiszman, B., Alcaide, J., Roura, E., & Spence, C. (2012). Is it the plate or is it the food? Assessing the influence of the color (black or white) and shape of the plate on the perception of the food placed on it. *Food Quality and Preference*, *24*(1), 205–208. doi:10.1016/j.foodqual.2011.08.011

Piqueras-Fiszman, B., Giboreau, A., & Spence, C. (2013). Assessing the influence of the color of the plate on the perception of a complex food in a restaurant setting. *Flavour (London)*, *2*(1), 24. doi:10.1186/2044-7248-2-24

Piqueras-Fiszman, B., & Spence, C. (2012). The influence of the color of the cup on consumers' perception of a hot beverage. *Journal of Sensory Studies*, *27*(5), 324–331. doi:10.1111/j.1745-459X.2012.00397.x

Povey, J. W. (1989). Ultrasonic in food engineering II. Applications. *Journal of Food Engineering*, *9*(1), 1–20. doi:10.1016/0260-8774(89)90047-2

Prabhakaran, D., Jeemon, P., & Roy, A. (2016). Cardiovascular diseases in India. *Circulation*, *133*(16), 1605–1620. doi:10.1161/CIRCULATIONAHA.114.008729 PMID:27142605

Prabu, S., Suriyaprakash, T. K., Kumar, C., & Kumar, S. (2012). Nutraceuticals and their medicinal importance. *International Journal of Health & Allied Sciences*, *1*(2), 47. doi:10.4103/2278-344X.101661

Prescott, S. L., Wickens, K., Westcott, L., Jung, W., Currie, H., Black, P. N., ... Crane, J. (2008). Supplementation with Lactobacillus rhamnosus or Bifidobacterium lactis probiotics in pregnancy increases cord blood interferon-γ and breast milk transforming growth factor-β and immunoglobin A detection. *Clinical and Experimental Allergy*, *38*(10), 1606–1614. doi:10.1111/j.1365-2222.2008.03061.x PMID:18631345

Proulx, J., Agustin, M., Sullivan, G., VanWees, S., Jian, J., Hilton, S. T., & Moraru, C. I. (2016). Short communication: Influence of pulsed light treatment on the quality and sensory characteristics of Cheddar cheese. *Journal of Dairy Science*, *100*(2), 1004–1008. doi:10.3168/jds.2016-11579 PMID:28012618

Proulx, J., Hsu, L. C., Miller, B. M., Sullivan, G., Paradis, K., & Moraru, C. I. (2015). Pulsed-light inactivation of pathogenic and spoilage bacteria on cheese surface. *Journal of Dairy Science*, *98*(9), 5890–5898. doi:10.3168/jds.2015-9410 PMID:26162787

Proulx, J., Sullivan, G., Marostegan, L. F., VanWees, S., Hsu, L. C., & Moraru, C. I. (2017). Pulsed light and antimicrobial combination treatments for surface decontamination of cheese: Favorable and antagonistic effects. *Journal of Dairy Science*, *100*(3), 1664–1673. doi:10.3168/jds.2016-11582 PMID:28109595

Putnik, P., Kresoja, Ž., Bosiljkov, T., Režek Jambrak, A., Barba, F. J., Lorenzo, J. M., ... Bursać Kovačević, D. (2018, July). Comparing the effects of thermal and non-thermal technologies on pomegranate juice quality: A review. *Food Chemistry*, *279*, 150–161. doi:10.1016/j.foodchem.2018.11.131 PMID:30611474

Qidwai, A., Pandey, A., Kumar, R., Shukla, S. K., & Dikshit, A. (2018). Advances in biogenic nanoparticles and the mechanisms of antimicrobial effects. *Indian Journal of Pharmaceutical Sciences*, *80*(4), 592–603. doi:10.4172/pharmaceutical-sciences.1000398

Rademacher, B., & Hinrichs, J. (2006). Effects of high-pressure treatment on indigenous enzymes in bovine milk: Reaction kinetics, inactivation and potential application. *International Dairy Journal*, *16*(6), 655–661. doi:10.1016/j.idairyj.2005.10.021

Rademacher, B., Pfeiffer, B., & Kessler, H. G. (1998). *Inactivation of microorganisms and enzymes in pressure-treated raw milk. High pressure food science, bioscience, and chemistry*. The Royal Society of Chemistry; doi:10.1533/9781845698379.3.145

Rane, A. N., Baikar, V. V., Ravi Kumar, V., & Deopurkar, R. L. (2017). Agro-industrial wastes for production of biosurfactant by *Bacillus subtilis ANR 88* and its application in synthesis of silver and gold nanoparticles. *Frontiers in Microbiology*, *8*, 492. doi:10.3389/fmicb.2017.00492 PMID:28392783

Razak, D. L. A., Rashid, N. Y. A., Jamaluddin, A., Sharifudin, S. A., Kahar, A. A., & Long, K. (2017). Cosmeceutical potentials and bioactive compounds of rice bran fermented with single and mix culture of *Aspergillus oryzae* and *Rhizopus oryzae. Journal of the Saudi Society of Agricultural Sciences*, *16*(2), 127–134. doi:10.1016/j.jssas.2015.04.001

Ricce, C., Rojas, M. L., Miano, A. C., Siche, R., & Augusto, P. E. D. (2016). Ultrasound pre-treatment enhances the carrot drying and rehydration. *Food Research International*, *89*, 701–708. doi:10.1016/j.foodres.2016.09.030 PMID:28460968

Riva, M., Piergiovanni, L., & Schiraldi, A. (2001). Performances of time–temperature indicators in the study of temperature exposure of packaged fresh food. *Packaging Technology & Science*, *14*(1), 1–9. doi:10.1002/pts.521

Robertson, G. L. (2012). *Food packaging, principles and practice*. London, UK: CRC Press.

Rodríguez, Ó., Eim, V., Rosselló, C., Femenia, A., Cárcel, J. A., & Simal, S. (2018). Application of power ultrasound on the convective drying of fruits and vegetables: Effects on quality. *Journal of the Science of Food and Agriculture*, *98*(5), 1660–1673. doi:10.1002/jsfa.8673 PMID:28906555

Roh, S. W., Kim, K., Nam, Y., Chang, H., Park, E., & Bae, J. (2009). Investigation of archaeal and bacterial diversity in fermented seafood using barcoded pyrosequencing. *The ISME Journal*, *4*(1), 1–16. doi:10.1038/ismej.2009.83 PMID:19587773

Rowan, N. J. (2019). Pulsed light as an emerging technology to cause disruption for food and adjacent industries – Quo vadis? *Trends in Food Science & Technology*, *88*(March), 316–332. doi:10.1016/j.tifs.2019.03.027

Ruiz-Garcia, L., & Lunadai, L. (2011). The role of RFID in agriculture: Applications, limitations, and challenges. *Computers and Electronics in Agriculture*, *79*(1), 42–50. doi:10.1016/j.compag.2011.08.010

Saba, M. K., & Amini, R. (2017). Nano-ZnO/carboxymethyl cellulose-based active coating impact on ready-to-use pomegranate during cold storage. *Food Chemistry*, *232*, 721–726. doi:10.1016/j.foodchem.2017.04.076 PMID:28490133

Sabu, A., Augur, C., Swati, C., & Pandey, A. (2006). Tannase production by *Lactobacillus* sp. ASR-S1 under solid-state fermentation, Process. *Bioche*, *41*, 575–580.

Sadh, P. K., Kumar, S., Chawla, P., & Duhan, J. S. (2018). Fermentation: A boon for production of bioactive compounds by processing of food industries wastes (by-products). *Molecules (Basel, Switzerland)*, *23*(2560), 1–33. PMID:30297619

Sadh, P. K., Chawla, P., & Duhan, J. S. (2018). Fermentation approach on phenolic, antioxidants and functional properties of peanut press cake. *Food Bioscience*, *22*, 113–120. doi:10.1016/j.fbio.2018.01.011

Sagar, V. R., & Kumar, P. S. (2010). Recent advances in drying and dehydration of fruits and vegetables: A review. *Journal of Food Science and Technology*, *47*(1), 15–26. doi:10.100713197-010-0010-8 PMID:23572596

Sait, H. H. (2018). Design and analysis of a flooded tube for cold energy storage. *International Journal of Refrigeration*, *94*, 151–160. doi:10.1016/j.ijrefrig.2018.07.018

Sajas, J. F., & Gorbatow, W. M. (1978). *The use of ultrasonics in meat technology*. Germany: Fleischwirtschaft.

Sala, F. J., Burgos, J., Condon, S., Lopez, P., & Raso, J. (1995). *Effect of heat and ultrasound on microorganism and enzymes. In G. W. Gound (Ed.), New methods of food preservation* (pp. 176–204). London, UK: Blackie Academic & Professional.

Sánchez, A., Artola, A., Font, X., Gea, T., Barrena, R., Gabriel, D., ... Mondini, C. (2015). Greenhouse gas emissions from organic waste composting. *Environmental Chemistry Letters*, *13*(3), 223–238. doi:10.100710311-015-0507-5

Sanchez-Garcia, M. D., & Lagaron, J. M. (2012). Nanocomposites for food and beverage packaging materials. In B. Bhandari, & Y. H. Roos (Eds.), *Food Materials, Science, and Engineering*. Chichester, UK: Blackwell Publishing. doi:10.1002/9781118373903.ch11

Sanders, M. E., Akkermans, L. M., Haller, D., Hammerman, C., Heimbach, J. T., Hörmannsperger, G., & Huys, G. (2010). Safety assessment of probiotics for human use. *Gut Microbes*, *1*(3), 164–185. doi:10.4161/gmic.1.3.12127 PMID:21327023

Sandhya. (2010). Modified atmosphere packaging of fresh produce: Current status and future needs. *Food Science and Technology, 43*, 381-392.

Santhirasegaram, V., Razali, Z., & Somasundram, C. (2016). *Safety improvement of fruit juices by novel thermal and nonthermal processing. Food hygiene and toxicology in ready-to-eat foods*. Elsevier Inc.; doi:10.1016/B978-0-12-801916-0.00012-1

Santivarangkna, C., Kulozik, U., & Foerst, P. (2006). Effect of carbohydrates on the survival of Lactobacillus helveticus during vacuum drying. *Letters in Applied Microbiology*, *42*(3), 271–276. doi:10.1111/j.1472-765X.2005.01835.x PMID:16478516

Santos, A., San Mauro, M., Sanchez, A., Torres, J. M., & Marquina, D. (2003). The antimicrobial properties of different strains of Lactobacillus spp. isolated from kefir. *Systematic and Applied Microbiology*, *26*(3), 434–437. doi:10.1078/072320203322497464 PMID:14529186

Santos, C. L., Albuquerque, A. J. R., Sampaio, F. C., & Keyson, D. (2013). Nanomaterials with antimicrobial properties: Applications in health sciences. In A. Mendez-Vilas (Ed.), *Microbial pathogens and strategies for combating them: Science, Technology, and Education* (pp. 143–154). Formatex.

Sarkar, S. (2007). Potential of kefir as a dietetic beverage – a review. *British Food Journal*, *109*(4), 280–290. doi:10.1108/00070700710736534

Saxena, J., & Dash, K. K. (2015). Drying kinetics and moisture diffusivity study of ripe Jackfruit. *International Food Research Journal*, *22*(1), 414.

Schläfer, O., Onyeche, T., Bormann, H., Schröder, C., & Sievers, M. (2002). Ultrasound stimulation of micro-organisms for enhanced biodegradation. *Ultrasonics, 40*(1-8), 25-29.

Schnettler, B., Crisostomo, G., Mora, M., Lobos, G., Miranda, H., & Grunert, K. G. (2013). Acceptance of nanotechnology applications and satisfaction with food-related life in southern Chile. *Food Science and Technology (Campinas)*, *34*(1), 157–163. doi:10.1590/S0101-20612014005000001

Schottenfeld, D., & Beebe-Dimmer, J. (2015). The cancer burden attributable to biologic agents. *Annals of Epidemiology*, *25*(3), 183–187. doi:10.1016/j.annepidem.2014.11.016 PMID:25523895

Schrader, K., Buchheim, W., & Morr, C. V. (1997). High pressure effects on the colloidal calcium phosphate and the structural integrity of micellar casein in milk. Part 1 phosphate in heated milk systems. *Food / Nahrung, 41*(3), 133–138.

Scott, K. P., Martin, J. C., Duncan, S. H., & Flint, H. J. (2013). Prebiotic stimulation of human colonic butyrate-producing bacteria and bifidobacteria, in vitro. *FEMS Microbiology Ecology*, *87*(1), 30–40. doi:10.1111/1574-6941.12186 PMID:23909466

Segat, A., Misra, N. N., Cullen, P. J., & Innocente, N. (2016). Effect of atmospheric pressure cold plasma (ACP) on activity and structure of alkaline phosphatase. *Food and Bioproducts Processing*, *98*, 181–188. doi:10.1016/j.fbp.2016.01.010

Şengül, M., Erkaya, T., Başlar, M., & Ertugay, M. F. (2011). Effect of photosonication treatment on inactivation of total and coliform bacteria in milk. *Food Control*, *22*(11), 1803–1806. doi:10.1016/j.foodcont.2011.04.015

Shahbaz, H. M., Kim, J. U., Kim, S.-H., & Park, J. (2018). *Advances in nonthermal processing technologies for enhanced microbiological safety and quality of fresh fruit and juice products. food processing for increased quality and consumption.* Elsevier; doi:10.1016/b978-0-12-811447-6.00007-2

Shaibani, A. R., Keshtkar, M. M., & Sardari, P. T. (2019). Thermo-economic analysis of a cold storage system in full and partial modes with two different scenarios: A case study. *Journal of Energy Storage*, *24*, 100783. doi:10.1016/j.est.2019.100783

Sharma, A., Gupta, V., Khan, M., Balda, S., Gupta, N., Capalash, N., & Sharma, P. (2017). Flavonoid-rich agro-industrial residues for enhanced bacterial laccase production by submerged and solid-state fermentation. *Biotech*, *7*, 200–212. PMID:28667639

Sharma, P., Bremer, P., Oey, I., & Everett, D. W. (2014). Bacterial inactivation in whole milk using pulsed electric field processing. *International Dairy Journal*, *35*(1), 49–56. doi:10.1016/j.idairyj.2013.10.005

Sheng, L., Tsai, H.-C., Zhu, H., & Zhu, M.-J. (2019). Survival of Listeria monocytogenes on blueberries post-sanitizer treatments and subsequent cold storages. *Food Control*, *100*, 138–143. doi:10.1016/j.foodcont.2019.01.019

Shewale, R. N., Sawale, P. D., Khedkar, C. D., & Singh, A. (2014). Selection criteria for probiotics: A review. *International Journal of Probiotics & Prebiotics*, *9*(1).

Shi, F., Zhou, X., Yao, M.-M., Tan, Z., Zhou, Q., Zhang, L., & Ji, S.-J. (2018). miRNAs play important roles in aroma weakening during the shelf life of 'Nanguo' pear after cold storage. *Food Research International, 116,* 942-952. doi:10.1101/247932

Shobharani, P., & Agrawal, R. (2011). Enhancement of cell stability and viability of probiotic Leuconostoc mesenteroides MTCC 5209 on freeze drying. *International Journal of Dairy Technology*, *64*(2), 276–287. doi:10.1111/j.1471-0307.2010.00640.x

Siegrist, M., Cousin, M. E., Kastenholz, H., & Wiek, A. (2007). Public acceptance of nanotechnology foods and food packaging: The influence of affect and trust. *Appetite*, *49*(2), 459–466. doi:10.1016/j.appet.2007.03.002 PMID:17442455

Siegrist, M., Stampfli, N., Kastenholz, H., & Keller, C. (2008). Perceived risks and perceived benefits of different nanotechnology foods and nanotechnology food packaging. *Appetite*, *51*(2), 283–290. doi:10.1016/j.appet.2008.02.020 PMID:18406006

Silva, S. B., Cantarelli, V. V., & Ayub, M. A. (2013). Production and optimization of poly-γ-glutamic acid by Bacillus subtilis BL53 isolated from the Amazonian environment. *Bioprocess and Biosystems Engineering*, *37*(3), 469–479. doi:10.100700449-013-1016-1 PMID:23872848

Sim, W., Barnard, R. T., Blaskovich, M. A. T., & Ziora, Z. M. (2018). Antimicrobial silver in medicinal and consumer applications: A patent review of the past decade (2007–2017). *Antibiotics (Basel, Switzerland)*, *7*(93), 1–15. doi:10.3390/antibiotics7040093 PMID:30373130

Singhania, R. R., Patel, A. K., Soccol, C. R., & Pandey, A. (2009). Recent advances in solid-state fermentation. *Biochemical Engineering Journal*, *44*(1), 13–18. doi:10.1016/j.bej.2008.10.019

Singh, J., Dutta, T., Kim, K. H., Rawat, M., Samddar, P., & Kumar, P. (2018). Green' synthesis of metals and their oxide nanoparticles: Applications for environmental remediation. *Journal of Nanobiotechnology*, *16*(84), 2–24. doi:10.118612951-018-0408-4 PMID:30373622

Singh, J., & Sinha, S. (2012, January-March). Classification, regulatory acts, and applications of nutraceuticals for health. *International Journal of Pharma and Bio Sciences*, *2*(1), 177–187.

Singh, R. K., Mishra, S. K., & Kumar, N. (2010). Optimization of culture conditions for amylase production by thermophilic *Bacillus sp.* in submerged fermentation. *Research Journal of Pharmaceutical, Biological and Chemical Sciences*, *1*(4), 867–876.

Sivertsvik, M., Rosnes, J. T., & Bergslien, H. (2002). Modified atmosphere packaging. In T. Ohlsson & N. Bengtsson (Eds.), *Minimal processing technologies in the food industry* (pp. 61–86). Cambridge: CRC Press. doi:10.1533/9781855736795.61

Sivertsvik, M., Rosnes, J. T., & Jeksrud, W. K. (2004). Solubility and absorption rate of carbon dioxide into non-respiring foods. Part 2: Raw fish fillets. *Journal of Food Engineering*, *63*(4), 451–458. doi:10.1016/j.jfoodeng.2003.09.004

Skinner, E. Z. (1989). Commentary, sensory evaluation. In celebration of our beginnings, sensory evaluation of materials and products. Philadelphia, PA: American Society for Testing and Materials.

Slattery, C., Cotter, P. D., & O'Toole, P. W. (2019). Analysis of health benefits conferred by *Lactobacillus* Species from Kefir. *Nutrients*, *11*(6), E1252. doi:10.3390/nu11061252 PMID:31159409

Smith, J. L., & Alford, J. A. (1970). Presence of antioxidant materials in bacteria. *Lipids*, *5*(10), 795–799. doi:10.1007/BF02531970 PMID:4992304

Smolander, M. (2003). The use of freshness ındicators in packaging. In R. Ahvenainen (Ed.), *Novel food packaging techniques* (pp. 127–143). Cambridge: Woodhead Publishing. doi:10.1533/9781855737020.1.127

Smolander, M. (2008). Freshness indicators and food packaging. In J. Kerry, & P. Butler (Eds.), *Smart packaging technologies for fast moving consumer goods* (pp. 111–127). West Sussex: John Wiley & Sons. doi:10.1002/9780470753699.ch7

Sokolowsky, M., & Fischer, U. (2012). Evaluation of bitterness in white wine applying descriptive analysis, time-intensity analysis, and temporal dominance of sensations analysis. *Analytica Chimica Acta*, *732*, 46–52. doi:10.1016/j.aca.2011.12.024 PMID:22688033

Song, A.-X., Mao, Y.-H., Siu, K.-C., Tai, W. C. S., & Wu, J.-Y. (2019). Protective effects of exopolysaccharide of a medicinal fungus on probiotic bacteria during cold storage and simulated gastrointestinal conditions. *International Journal of Biological Macromolecules*, *133*, 957–963. doi:10.1016/j.ijbiomac.2019.04.108 PMID:31028812

Song, H., Yuan, W., Jin, P., Wang, W., Wang, X., Yang, L., & Zhang, Y. (2016). Effects of chitosan/nano-silica on postharvest quality and antioxidant capacity of loquat fruit during cold storage. *Postharvest Biology and Technology*, *119*, 41–48. doi:10.1016/j.postharvbio.2016.04.015

Sousa, B. A., & Correia, R. T. P. (2012). Phenolic content, antioxidant activity and antiamylolytic activity of extracts obtained from bioprocessed pineapple and guava wastes. *Brazilian Journal of Chemical Engineering*, *29*(1), 25–30. doi:10.1590/S0104-66322012000100003

Sportelli, M. C., Picca, R. A., & Cioffi, N. (2015). Nano-antimicrobials based on metals. In D. A. Phoenix, F. Harris, & S. R. Dennison, (Eds.), Novel antimicrobial agents and strategies (pp. 181-218). Wiley-VCH Verlag GmbH & Co. KGaA: Weinheim, Germany.

Srinivasan, K. (2005). Role of spices beyond food flavoring: Nutraceuticals with multiple health effects. *Food Reviews International*, *21*(2), 167–188. doi:10.1081/FRI-200051872

Srividya, N., Ghoora, M. D., & Padmanabh, P. R. (2016). Antimicrobial nanotechnology-research implications and prospects in food safety. In A. Grumezescu (Ed.), *Food preservation-nanotechnology in agri-food industry* (pp. 125–157). Academic Press, Elsevier; doi:10.1016/B978-0-12-804303-5.00014-6

Starr, M. P. (1958). The blue pigment of Corynebacterium insidiosum. *Archiv für Mikrobiologie*, *30*(4), 325–334. doi:10.1007/BF00411227 PMID:13595813

Stenman, L. K., Waget, A., Garret, C., Briand, F., Burcelin, R., Sulpice, T., & Lahtinen, S. (2015). Probiotic B420 and prebiotic polydextrose improve efficacy of antidiabetic drugs in mice. *Diabetology & Metabolic Syndrome*, *7*(1), 75. doi:10.118613098-015-0075-7 PMID:26366205

Stone, H., & Sidel, J. (1992). Descriptive Analysis. In *Sensory evaluation practices* (2nd ed., pp. 202–242). Academic Press.

Stone, H., Sidel, J., Oliver, S., Woolsey, A., & Singleton, R. C. (2008). Sensory evaluation by quantitative descriptive analysis. *Descriptive Sensory Analysis in Practice*, *28*, 23–34.

Stritih, U., & Butala, V. (2007). Energy saving in building with PCM cold storage. *International Journal of Energy Research*. doi:10.1002/er.1318

Styles, M. F., Hoover, D. G., & Farkas, D. F. (1991). Response of Listeria monocytogenes and Vibrio parahaemolyficus to high hydrostatic pressure. *Journal of Food Science*, *56*(5), 1404–1407. doi:10.1111/j.1365-2621.1991.tb04784.x

Subramaniyam, R., & Vimala, R. (2012). Solid state and submerged fermentation for the production of bioactive substances: A comparative study. *International Journal of Science and Nature*, *3*(3), 480–486.

Sui, Q., Roginski, H., Williams, R. P. W., Versteeg, C., & Wan, J. (2011). Effect of pulsed electric field and thermal treatment on the physicochemical and functional properties of whey protein isolate. *International Dairy Journal*, *21*(4), 206–213. doi:10.1016/j.idairyj.2010.11.001

Sund-Levander, M., Forsberg, C., & Wahren, L. K. (2002). Normal oral, rectal, tympanic and axillary body temperature in adult men and women: A systematic literature review. *Scandinavian Journal of Caring Sciences*, *16*(2), 122–128. doi:10.1046/j.1471-6712.2002.00069.x PMID:12000664

Sun, Y., Chen, J., Zhang, S., Li, H., Lu, J., Liu, L., ... Jiaping, L. (2014). Effect of power ultrasound pre-treatment on the physical and functional properties of reconstituted milk protein concentrate. *Journal of Food Engineering*, *124*, 11–18. doi:10.1016/j.jfoodeng.2013.09.013

Sun, Y., Singh, Z., Tokala, V. Y., & Heather, B. (2019). Harvest maturity stage and cold storage period influence lemon fruit quality. *Scientia Horticulturae*, *249*, 322–328. doi:10.1016/j.scienta.2019.01.056

Suppakul, P., Miltz, J., Sonneveld, K., & Bigger, S. W. (2003). Active packaging technologies with an emphasis on antimicrobial packaging and its applications. *Journal of Food Science*, *68*(2), 408–420. doi:10.1111/j.1365-2621.2003.tb05687.x

Sürengil, G., & Kılınç, B. (2011). Gıda - Ambalaj sektöründe nanoteknolojik uygulamalar ve su ürünleri açısından önemi. *Journal of Fisheries Sciences Com*, *5*(4), 317–325.

Sürengil, G., & Kılınç, B. (2011). Gıda - Ambalaj sektöründe nanoteknolojik uygulamalar ve su ürünleri açısından önemi. *Journal of Fisheries Sciences Com.*, *5*(4), 317–325.

Suslick, K. S. (1988). *Ultrasound: Its chemical, physical and biological effects*. New York: VCH Publishers.

Swain, M. R., & Ray, R. C. (2008). Optimization of cultural conditions and their statistical interpretation for production of indole-3-acetic acid by *Bacillus subtilis CM5* using cassava fibrous residue. *Journal of Scientific and Industrial Research (New Delhi, India)*, *67*, 622–628.

Swartz, M. (1980). Sensory screening of synthetic sweeteners using time-intensity evaluations. *Journal of Food Science*, *45*(3), 577–581. doi:10.1111/j.1365-2621.1980.tb04105.x

Szczesniak, A. (1963). Classification of textural characteristics. *Journal of Food Science*, *28*(4), 385–389. doi:10.1111/j.1365-2621.1963.tb00215.x

Tabuchi, M., Ozaki, M., Tamura, A., Yamada, N., Ishida, T., Hosoda, M., & Hosono, A. (2003). Antidiabetic effect of Lactobacillus GG in streptozotocin-induced diabetic rats. *Bioscience, Biotechnology, and Biochemistry*, *67*(6), 1421–1424. doi:10.1271/bbb.67.1421 PMID:12843677

Tajima, M. (2007). Strategic value of RFID in supply chain management. *Journal of Purchasing and Supply Management*, *13*(4), 261–273. doi:10.1016/j.pursup.2007.11.001

Talukdar, S., Afroz, H. M. M., Hossain, M. A., Aziz, M., & Hossain, M. M. (2019). Heat transfer enhancement of charging and discharging of phase change materials and size optimization of a latent thermal energy storage system for solar cold storage application. *Journal of Energy Storage*, *24*, 100797. doi:10.1016/j.est.2019.100797

Tamang, J. (1998). Role of microorganisms in traditional fermented foods. *Indian Food Industry*, *17*, 162–166.

Taniwaki, M. H., Hocking, A. D., Pitt, J. I., & Fleet, G. H. (2001). Growth of fungi and mycotoksin production on cheese under modified atmospheres. *International Journal of Food Microbiology*, *68*(1-2), 125–133. doi:10.1016/S0168-1605(01)00487-1 PMID:11545212

Tan, Q., Xu, H., Aguilar, Z. P., Peng, S., Dong, S., Wang, B., ... Wei, H. (2013). Safety assessment and probiotic evaluation of enterococcus faeciumyf5 isolated from sourdough. *Journal of Food Science*, *78*(4), M587–M593. doi:10.1111/1750-3841.12079 PMID:23488799

Taoukis, P. S., & Labuza, T. P. (2003). Time-temperature indicators. In R. Ahvenainen (Ed.), *Novel food packaging techniques* (pp. 103–126). Cambridge: Woodhead Publishing. doi:10.1533/9781855737020.1.103

Taoukis, P., & Tsironi, T. (2016). Smart packaging for monitoring and managing food and beverage shelf life. In P. Subramaniam, & P. Wareing (Eds.), *The stability and shelf life of food* (pp. 141–168). Woodhead Publishing. doi:10.1016/B978-0-08-100435-7.00005-8

Taskin, O., Izli, G., & Izli, N. Convective drying kinetics and quality parameters of European Cranberrybush. *Journal of Agricultural Sciences*, *24*(3), 349–358.

Taylor, A. J., & Roozen, J. P. (1996). Volatile flavor release from foods during eating. *Critical Reviews in Food Science and Nutrition*, *36*(8), 765–784. doi:10.1080/10408399609527749 PMID:8989509

Taylor, M. J., & Richardson, T. (1980). Antioxidant activity of skim milk: Effect of sonication.. *Journal of Dairy Science*, *63*(11), 1938–1942. doi:10.3168/jds.S0022-0302(80)83161-4 PMID:6893712

Thompson, M. P., & Kurzrock, R. (2004). Epstein-Barr virus and cancer. *Clinical Cancer Research*, *10*(3), 803–821. doi:10.1158/1078-0432.CCR-0670-3 PMID:14871955

Thorngate, J. H., & Noble, A. C. (1995). Sensory evaluation of bitterness and astringency of 3R (−)-epicatechin and 3S (+)-catechin. *Journal of the Science of Food and Agriculture*, *67*(4), 531–535. doi:10.1002/jsfa.2740670416

Tian, F., Decker, E. A., & Goddard, J. M. (2012). Development of an iron chelating polyethylene film for active packaging applications. *Journal of Agricultural and Food Chemistry*, *60*(8), 2046–2052. doi:10.1021/jf204585f PMID:22288894

Tian, S., Shao, S., & Liu, B. (2019). Investigation on transient energy consumption of cold storages: Modeling and a case study. *Energy*, *180*, 1–9. doi:10.1016/j.energy.2019.04.217

Tomlins, K. I. (2000). *Methods for the sensory evaluation of food and drink products*. Chatham Maritime, UK: Natural Resources Institute.

Trapecar, M., Leouffre, T., Faure, M., Jensen, H. E., Granum, P. E., Cencic, A., & Hardy, S. P. (2011). The use of a porcine intestinal cell model system for evaluating the food safety risk of Bacillus cereus probiotics and the implications for assessing enterotoxigenicity. *APMIS*, *119*(12), 877–884. doi:10.1111/j.1600-0463.2011.02797.x PMID:22085364

Traprcaar Yu, L., & Osullivan, D. (2014). Production of galactooligosaccharides using a hyperthermophilic β-galactosidase in permeabilized whole cells of Lactococcus lactis. *Journal of Dairy Science*, *97*(2), 694–703. doi:10.3168/jds.2013-7492 PMID:24359820

Trucco, M. (2009). Gene-environment interaction in type 1 diabetes mellitus. *Endocrinología y Nutrición*, *56*, 56–59. doi:10.1016/S1575-0922(09)73521-1 PMID:20629235

Trujillo, A. J., Ferragut, B., Juan, B., Roig-Sagués, A. X., & Guamis, B. (2016). Processing of dairy products utilizing high pressure. In V. M. Balasubramaniam, G. V. Barbosa-Cánovas, & H. L. M. Lelieveld (Eds.), *High pressure processing of food* (pp. 553–590). New York: Springer; doi:10.1007/978-1-4939-3234-4_25

Tseng, W., Fang, T., Cho, C., Chen, P., & Tsai, C. (2012). Assessments of growth conditions on the production of cyanophycin by recombinant Escherichia coli strains expressing cyanophycin synthetase gene. *Biotechnology Progress*, *28*(2), 358–363. doi:10.1002/btpr.1513 PMID:22252992

Tufarelli, V., & Laudadio, V. (2016). An overview on the functional food concept: Prospectives and applied researches in probiotics, prebiotics and synbiotics. *The Journal of Experimental Biology*, *4*, 3S.

Turner, J. R. (2017). Metal-based antimicrobial strategies. *Microbial Biotechnology*, *10*(5), 1062–1065. doi:10.1111/1751-7915.12785 PMID:28745454

Tyagi, P. K. (2016). Production of metal nanoparticles from biological resources. *International Journal of Current Microbiology and Applied Sciences*, *5*(3), 548–558. doi:10.20546/ijcmas.2016.503.064

Üçüncü, M. (2011). *Gıda ambalajlanma teknolojisi*. İstanbul, Turkey: Ambalaj Sanayiciler Derneği.

Valero, M., Recrosio, N., Saura, D., Mun˜oz, N., Martı´, N., & Lizama, V. (2007). Effects of ultrasonic treatments in orange juice processing. *Journal of Food Engineering*, *80*(2), 509–516. doi:10.1016/j.jfoodeng.2006.06.009

Van der Steen, C., Jacxsens, L., Devlieghere, F., & Debevere, J. (2002). Combining high oxygen atmospheres with low oxygen modified atmosphere packaging to improve the keeping quality of strawberries and raspberries. *Postharvest Biology and Technology*, *26*(1), 49–58. doi:10.1016/S0925-5214(02)00005-4

Vanderroost, M., Ragaert, P., Devlieghere, F., & Meulenaer, B. D. (2014). Intelligent food packaging: The next generation. *Trends in Food Science & Technology*, *39*(1), 47–62. doi:10.1016/j.tifs.2014.06.009

Vastrad, B. M., & Neelagund, S. E. (2011). Optimization and production of neomycin from different agro industrial wastes in solid state fermentation. *International Journal of Pharmaceutical Sciences and Drug Research*, *3*, 104–111.

Vastrad, B. M., & Neelagund, S. E. (2012). Optimization of process parameters for rifamycin b production under solid state fermentation from *Amycolatopsis mediterranean MTCC14. Int. J. Curr. Pharm. Res*, *4*, 101–108.

Velu, S., Abu Bakar, F., Mahyudin, N. A., Saari, N., & Zaman, M. Z. (2013). Effect of modified atmosphere packaging on microbial flora changes in fishery products. *International Food Research Journal*, *20*(1), 17–26.

Vermeiren, L., Devlieghere, F., Van Beest, M., De Kruijf, N., & Debevere, J. (1999). Developments in the active packaging of foods. *Trends in Food Science & Technology*, *10*(3), 77–86. doi:10.1016/S0924-2244(99)00032-1

Verwaal, R., Wang, J., Meijnen, J., Visser, H., Sandmann, G., Berg, J. A., & Ooyen, A. J. (2007). High-level production of Beta-Carotene in Saccharomyces cerevisiae by successive transformation with Carotenogenic Genes from Xanthophyllomyces dendrorhous. *Applied and Environmental Microbiology*, *73*(13), 4342–4350. doi:10.1128/AEM.02759-06 PMID:17496128

Vickers, Z. M. (1982). Relationships of chewing sounds to judgments of crispness, crunchiness, and hardness. *Journal of Food Science*, *47*(1), 121–124. doi:10.1111/j.1365-2621.1982.tb11041.x

Vidhyalakshmi, R., Vallinachiyar, C., & Radhika, R. (2012). Production of xanthan from agro-industrial waste. *J. Adv. Sci. Res*, *3*, 56–59.

Vijayakumar, S., Grewell, D., Annandarajah, C., Benner, L., & Clark, S. (2015). Quality characteristics and plasmin activity of thermosonicated skim milk and cream. *Journal of Dairy Science*, *98*(10), 6678–6691. doi:10.3168/jds.2015-9429 PMID:26233461

Vijayaraghavan, P., Vincent, S. G. P., Arasu, M. V., & Al-Dhabi, N. A. (2016). Bioconversion of agro-industrial wastes for the production of fibrinolytic enzyme from *Bacillus halodurans IND18*: Purification and biochemical characterization. *Electronic Journal of Biotechnology*, *20*, 1–8. doi:10.1016/j.ejbt.2016.01.002

Vintila, T., Dragomirescu, M., Jurcoane, S., Vintila, D., Caprita, R., & Maniu, M. (2009). Production of cellulase by submerged and solid-state cultures and yeasts selection for conversion of lignocellulose to ethanol. *Romanian Biotechnological Letters*, *14*, 4275–4281.

Wang, C., He, Z., Li, H., Wennerstern, R., & Sun, Q. (2017). Evaluation on performance of a phase change material based cold storage house. *Energy Procedia*, *105*, 3947–3952. doi:10.1016/j.egypro.2017.03.820

Wang, D., Jiang, J., Tao, L., Kou, Z., & Yao, L. (2017). Experimental investigation on a novel cold storage defrosting device based on electric heater and reverse cycle. *Applied Thermal Engineering*, *127*, 1267–1273. doi:10.1016/j.applthermaleng.2017.08.122

Wang, J., Zhao, X., Tian, Z., Yang, Y., & Yang, Z. (2015). Characterization of an exopolysaccharide produced by *Lactobacillus plantarum* YW11 isolated from Tibet Kefir. *Carbohydrate Polymers*, *125*, 16–25. doi:10.1016/j.carbpol.2015.03.003 PMID:25857955

Wang, L., Hu, C., & Shao, L. (2017). The antimicrobial activity of nanoparticles: Present situation and prospects for the future. *International Journal of Nanomedicine*, *12*, 1227–1249. doi:10.2147/IJN.S121956 PMID:28243086

Wang, X., Dennis, M., Jiang, J., Zhou, L., Zhai, X., & Lipiński, W. (2019). Performance of a novel cold thermal storage material in an emulated air conditioning system using different storage strategies. *International Journal of Refrigeration*, *104*, 259–269. doi:10.1016/j.ijrefrig.2019.05.038

Wang, Y., Yu, R., & Chou, C. (2004). Viability of lactic acid bacteria and bifidobacteria in fermented soymilk after drying, subsequent rehydration and storage. *International Journal of Food Microbiology*, *93*(2), 209–217. doi:10.1016/j.ijfoodmicro.2003.12.001 PMID:15135959

Wei, Q., Wang, X., Sun, D.-W., & Pu, H. (2019). Rapid detection and control of psychrotrophic microorganisms in cold storage foods: A review. *Trends in Food Science & Technology*, *86*, 453–464. doi:10.1016/j.tifs.2019.02.009

Wei, S. H., Chen, Y. P., & Chen, M. J. (2015). Selecting probiotics with the abilities of enhancing GLP-1 to mitigate the progression of type 1 diabetes in vitro and in vivo. *Journal of Functional Foods*, *18*, 473–486. doi:10.1016/j.jff.2015.08.016

Wendin, K., Janestad, H., & Hall, G. (2003). Modelling and analysis of dynamic sensory data. *Food Quality and Preference*, *14*(8), 663–671. doi:10.1016/S0950-3293(02)00208-2

Wessling, C., Nielsen, T., Leufvén, A., & Jägerstad, M. (1998). Mobility of α-tocopherol and BHT in LDPE in contact with fatty food simulants. *Food Additives and Contaminants*, *15*(6), 709–715. doi:10.1080/02652039809374701 PMID:10209582

Weymouth, F. W. (1958). Visual sensory units and the minimal angle of resolution. *American Journal of Ophthalmology*, *46*(1), 102–113. doi:10.1016/0002-9394(58)90042-4 PMID:13545337

Whillock, G. O. H., & Harvey, B. F. (1997). Ultrasonically enhanced corrosion of 304L stainless steel: The effect of temperature and hydrostatic pressure. *Ultrasonics Sonochemistry*, *4*(1), 23–31. doi:10.1016/S1350-4177(96)00014-4 PMID:11233921

WHO publishes list of bacteria for which new antibiotics are urgently needed (2017, Feb. 27). Retrieved from https://www.paho.org/hq/index.php?lang=en

Wind, R. D., Tolboom, H., Klare, I., Huys, G., & Knol, J. (2010). Tolerance and safety of the potentially probiotic strain Lactobacillus rhamnosus PRSF-L477: A randomised, double-blind placebo-controlled trial in healthy volunteers. *British Journal of Nutrition*, *104*(12), 1806–1816. doi:10.1017/S0007114510002746 PMID:20691131

Woods, A. T., Poliakoff, E., Lloyd, D. M., Dijksterhuis, G. B., & Thomas, A. (2010). Flavor expectation: The effect of assuming homogeneity on drink perception. *Chemosensory Perception*, *3*(3–4), 174–181. doi:10.100712078-010-9080-2

Wrigley, D. M., & Llorca, N. G. (1992). Decrease of Salmonella typhimurium in skim milk and egg by heat and ultrasonic wave treatment. *Journal of Food Protection*, *55*(9), 678–680. doi:10.4315/0362-028X-55.9.678 PMID:31084132

Yadav, H., Lee, J. H., Lloyd, J., Walter, P., & Rane, S. G. (2013). Beneficial metabolic effects of a probiotic via butyrate-induced GLP-1 hormone secretion. *The Journal of Biological Chemistry*, *288*(35), 25088–25097. doi:10.1074/jbc.M113.452516 PMID:23836895

Yalemtesfa, B., Alemu, T., & Santhanam, A. (2010). Solid substrate fermentation and conversion of orange waste into fungal biomass using *Aspergillus niger KA-06* and *Chaetomium Spp KC-06. African Journal of Microbiological Research, 4*, 1275–1281.

Yam, K. L., Takhistov, P. T., & Miltz, J. (2006). Intelligent packaging: concepts and applications. *Journal of Food Science, 70*(1), 1–10. doi:10.1111/j.1365-2621.2005.tb09052.x

Yan, F., Cao, H., Cover, T. L., Whitehead, R., Washington, M. K., & Polk, D. B. (2007). Soluble proteins produced by probiotic bacteria regulate intestinal epithelial cell survival and growth. *Gastroenterology, 132*(2), 562–575. doi:10.1053/j.gastro.2006.11.022 PMID:17258729

Yang, F., Zhang, M., Mujumdar, A. S., Zhong, Q., & Wang, Z. (2018). Enhancing drying efficiency and product quality using advanced pretreatments and analytical tools—An overview. *Drying Technology, 36*(15), 1824–1838. doi:10.1080/07373937.2018.1431658

Yang, H. J., Lee, J. H., Won, M., & Song, K. B. (2016). Antioxidant activities of distiller dried grains with solubles as protein films containing tea extracts and their application in the packaging of pork meat. *Food Chemistry, 196*, 174–179. doi:10.1016/j.foodchem.2015.09.020 PMID:26593480

Yao, M., Li, B., Ye, H., Huang, W., Luo, Q., Xiao, H., ... Li, L. (2018). Enhanced viability of probiotics (Pediococcus pentosaceus Li05) by encapsulation in microgels doped with inorganic nanoparticles. *Food Hydrocolloids, 83*, 246–252. doi:10.1016/j.foodhyd.2018.05.024

Yong, H. I., Kim, H. J., Park, S., Kim, K., Choe, W., Yoo, S. J., & Jo, C. (2015). Pathogen inactivation and quality changes in sliced cheddar cheese treated using flexible thin-layer dielectric barrier discharge plasma. *Food Research International, 69*, 57–63. doi:10.1016/j.foodres.2014.12.008

Yüksel, M. E., & Zaim, A. H. (2009). *Yeni nesil teknoloji olarak RFID, RFID sistem yapıları ve bir RFID sistem tasarımı yaklaşımı.* Paper presented at the meeting 5th International Advanced Technologies Symposium. Karabük, Turkey.

Yu, L. J., Ngadi, M., & Raghavan, G. S. V. (2009). Effect of temperature and pulsed electric field treatment on rennet coagulation properties of milk. *Journal of Food Engineering, 95*(1), 115–118. doi:10.1016/j.jfoodeng.2009.04.013

Zaika, L. L., & Smith, J. L. (1975). Antioxidants and pigments of Aspergillus niger. *Journal of the Science of Food and Agriculture, 26*(9), 1357–1369. doi:10.1002/jsfa.2740260915

Zambrano-Zaragoza, M. L., Gonzalez-Reza, R., Mendoza-Munoz, N., Miranda-Linares, V., Bernal-Couoh, T. F., Mendoza-Elvira, S., & Quintanar-Guerrero, D. (2018). Nanosystems in edible coatings: A novel strategy for food preservation. *International Journal of Molecular Sciences, 19*(3), 705. doi:10.3390/ijms19030705 PMID:29494548

Zhang, Q., Rao, Y., Jiao, Y., Li, L., Li, Y., & Jin, L. (2017). preparation and performance of composite building materials with phase change material for thermal storage. *Energy Procedia, 143*, 125–130. doi:10.1016/j.egypro.2017.12.659

Zhang, Y., Wei, J., Yuan, Y., & Yue, T. (2019). Diversity and characterization of spoilage-associated psychrotrophs in food in cold chain. *International Journal of Food Microbiology*, *290*, 86–95. doi:10.1016/j.ijfoodmicro.2018.09.026 PMID:30317110

Zhao, L., Zhang, S., Uluko, H., Liu, L., Lu, J., Xue, H., ... Lv, J. (2014). Effect of ultrasound pretreatment on rennet-induced coagulation properties of goat's milk. *Food Chemistry*, *165*, 167–174. doi:10.1016/j.foodchem.2014.05.081 PMID:25038663

Zhao, W., Yang, R., Shen, X., Zhang, S., & Chen, X. (2013). Lethal and sublethal injury and kinetics of Escherichia coli, Listeria monocytogenes and Staphylococcus aureus in milk by pulsed electric fields. *Food Control*, *32*(1), 6–12. doi:10.1016/j.foodcont.2012.11.029

Zheng, J., Li, S., Xu, Y., & Zheng, X. (2019). Effect of oxalic acid on edible quality of bamboo shoots (Phyllostachys prominens) without sheaths during cold storage. *Lwt*, *109*, 194–200. doi:10.1016/j.lwt.2019.04.014

Zheng, L., Zhang, W., & Liang, F. (2017). Experiment study on thermal conductivity of microcapsule phase change suspension applied to solar powered air conditioning cold storage system. *Procedia Engineering*, *205*, 1237–1244. doi:10.1016/j.proeng.2017.10.364

Zheng, Y., Lu, Y., Wang, J., Yang, L., Pan, C., & Huang, Y. (2013). Probiotic properties of Lactobacillus strains isolated from Tibetan kefir grains. *PLoS One*, *8*(7), e69868. doi:10.1371/journal.pone.0069868 PMID:23894554

Zhou, X., Tan, J., Gou, Y., Liao, Y., Xu, F., Li, G., ... Chen, Z. (2019). The biocontrol of postharvest decay of table grape by the application of kombucha during cold storage. *Scientia Horticulturae*, *253*, 134–139. doi:10.1016/j.scienta.2019.04.025

Zhou, X., Zhou, D.-Y., Liu, Z.-Y., Yin, F.-W., Liu, Z.-Q., Li, D.-Y., & Shahidi, F. (2019). Hydrolysis and oxidation of lipids in mussel Mytilus edulis during cold storage. *Food Chemistry*, *272*, 109–116. doi:10.1016/j.foodchem.2018.08.019 PMID:30309519

Ziegler, M., Kent, D., Stephan, R., & Guldimann, C. (2019). Growth potential of Listeria monocytogenes in twelve different types of RTE salads: Impact of food matrix, storage temperature and storage time. *International Journal of Food Microbiology*, *296*, 83–92. doi:10.1016/j.ijfoodmicro.2019.01.016 PMID:30851644

About the Contributors

Seydi Yıkmış was born in 1982 in Adıyaman. In 2005, he graduated from Inönü University, Faculty of Engineering, Department of Food Engineering. In 2013, he graduated from Süleyman Demirel University, Institute of Science, Department of Food Engineering. In 2018, he finished his PhD education at Istanbul University, Institute of Health Sciences, Department of Food Hygiene and Technology. He is still working as a lecturer in Nutrition and Dietetics Department of Namık Kemal University.

* * *

Eda Adal is from the Gaziantep University. She has completed her MSc and Ph.D on encapsulation, functional foods and compositional analysis. Besides that, she has worked on ultrasound applications in food systems at Helsinki University in Finland. She has a background about ultrasound, encapsulation vacuum, sprays and freeze-drying and statistic calculation in food systems. She is now working on several projects about functional foods.

Tugba Aktar is an Assistant Professor of Food Engineering at Alanya Alaaddin Keykubat University. She majored in Food Engineering at the Erciyes University and earned a degree of MSc and Ph.D. in Food Science and Nutrition from the University of Leeds in 2012 and 2015 respectively. Following her Ph.D. she joined Alanya Alaaddin Keykubat University. Her research interests include sensory studies, texture analysis, rheology analysis, modeling, consumer behavior, as well as human sensation, perception mechanisms, and mechanoreceptors. Tugba is currently serving as department chair since 2017 and is involved in several infrastructural and research projects. She currently works on experimental research in the above-mentioned areas.

Hami Alpas is a Food Engineer and also holds an MBA degree from Dept. of Business Administration on Total Quality Management. His main research areas are: Unit Operations in Food Engineering, Non-thermal Food Processing Technologies, Food Security under Climate change effects, Threats to the global food supply chain, Food Safety and Food Security through Total Food Protection and Food Defense, Food Loss and Waste. He has served as a "visiting scholar" in 1996 and 1998 at University of Wyoming, USA; as a "visiting scientist" in 2001 and 2002 at Ohio State University, USA and as a "visiting professor" in 2006, 2007 and 2008 at University of Bordeaux I, France. He has participated to the Food Defense Collaborative Exchange Programme as an expert at the Food Protection and Defense Institute (USA) in 2013 and in 2017. He has supervised 5 Ph.D and 16 M.Sc. thesis. He has over 90 international journal articles (SCI) and over 1700 citations (ISI-Web of Science; h-factor 25) as well as close to 80 academic presentations in 55 different international meetings. He has Co-Chaired international conferences and gave keynote and invited speeches. He has completed 15 national, 4 international projects including EU/JRC, CNRS-EGIDE and NATO ARW/ATC projects. He has authored 15 chapters in internationally edited books and has edited 3 international books by Springer on Food Chain Security, Eco-terrorism and Food Defense. He has also organized and co-directed 3 NATO-workshops (ARW-ATC). He was the Co-director of NATO-SPS project on "Food Chain Security" (2004-2009) and EU-FP7 project on "Plant Food Security". He has worked as a consultant to World Bank and FAO projects. Currently, he is acting as the Academic Consultant to COMCEC Agriculture Working Group (AWG) and is the Coordinator of project on Reduction of Food Loss and Waste in the OIC Countries (COMCEC-2018-TURAGRIC-569). http://fde.metu.edu.tr/prof-hami-alpas.

Şebnem Öztürkoğlu Budak received her BSc diploma in Food Engineering from Ankara University (Turkey) in 2001. She obtained her Master of Science Degree in Food Engineering from Middle East Technical University (Turkey) in 2004. In September 2007, she started her PhD in Food Engineering Department of Ankara University (Turkey), and graduated in March 2014. In 2011, she began her second PhD in Utrecht University, and performed her laboratory studies in CBS-KNAW Fungal Biodiversity Center (recently named Westerdijk Biodiversity Center, The Netherlands). She obtained her second PhD degree in September 2016. In September 2010, Dr. Öztürkoğlu Budak joined the Dairy Technology in Faculty of Agriculture in Ankara University, where she contributes the teaching and research program on the dairy technology program. She also continued her research on the molecular biology, molecular fungal physiology and enzyme technology in Westerdijk Biodiversity Center (The Netherlands). Her areas of expertise include fermented dairy products, enzyme technology and dairy microbiology including starter and adjunct cultures.

Basak Gokce graduated from Istanbul Veterinary Faculty, and is working on Food hygiene and technology.

N. Kapilan is presently working as a professor and Head, Department of Mechanical Engineering, Nagarjuna College of Engineering and Technology, Devanahalli, Bengaluru, India. Having 22 years of teaching and 8 years of research experience. Published several research papers in journals and conferences. Completed 4 sponsored research projects and carrying out 2 research projects. Filed one patent and authored 2 book chapters.

Alperen Koker received his B.Sc. degree in Food Engineering from Middle East Technical University in 2017. He is still working as a Research Assistant in Food Engineering Department at the Middle East Technical University. His research interests include non-thermal technologies, particularly high hydrostatic pressure processing, and non-bovine milk and milk products.

İlhami Okur is a research assistant at METU. He received a Master Degree in food engineering from METU. His recent publications include " Effect of high hydrostatic pressure in physicochemical properties and in vitro digestibility of cornstarch by nuclear magnetic resonance relaxometry". His research interest includes non-thermal technologies such as high hydrostatic pressure and ultrasound, nuclear magnetic resonance, FTIR relaxometry, and modeling.

K. A. Paari has got his doctoral degree from Pondichhery Central University,India during the year 2013. He has good research experience in the field of marine biotechnology, food preservation and probiotics. He has published many research articles in various peer reviewed journals which reflects his sincerity and enthusiasm for the subject. He has achieved his postdoctoral experience from Gangneung-Wonju national University, South Korea.Currently, he is working as Assistant professor in Christ College (Deemed to be University), Bangalore and guiding students for achieving their masters and doctoral degrees.

Zoomi Singh is a Research Scholar, Centre of Food Technology, University of Allahabad.

Nazli Turkmen has been working as a Research Assistant at the Department of Dairy Technology, Faculty of Agriculture, Ankara University since 2012. Her M.Sc. thesis was about using odourless fish oil in fruit yoghurt production. She studied on a traditional product "Maras ice cream" within the scope of doctorate thesis topic.

Her areas of expertise include milk of small ruminants, ice cream and fermented dairy products technology.

Ilavarasan Vickraman graduated from Christ (Deemed to be University) in May 2019 with a bachelor's of science in Chemistry, Botany, and Zoology. He had done his dissertation on "Isolation, characterization of methanogens and its potential in biogas production" as a part of his bachelor's degree. He holds keen interests in biofuels, stress relief by probiotics and evolutionary molecular biology.

Index

U

V

Z

www.ingramcontent.com/pod-product-compliance
Ingram Content Group UK Ltd.
Pitfield, Milton Keynes, MK11 3LW, UK
UKHW051452290926
15471UKWH00008B/1155

9 781799 819257